T0211741

Lecture Notes in Artificial Intelligence   9286

Subseries of Lecture Notes in Computer Science

LNAI Series Editors

Randy Goebel
  University of Alberta, Edmonton, Canada
Yuzuru Tanaka
  Hokkaido University, Sapporo, Japan
Wolfgang Wahlster
  DFKI and Saarland University, Saarbrücken, Germany

LNAI Founding Series Editor

Joerg Siekmann
  DFKI and Saarland University, Saarbrücken, Germany

More information about this series at http://www.springer.com/series/1244

Albert Bifet · Michael May
Bianca Zadrozny · Ricard Gavalda
Dino Pedreschi · Francesco Bonchi
Jaime Cardoso · Myra Spiliopoulou (Eds.)

Machine Learning and Knowledge Discovery in Databases

European Conference, ECML PKDD 2015
Porto, Portugal, September 7–11, 2015
Proceedings, Part III

 Springer

Editors

Albert Bifet
Huawei Noah's Ark Lab
Shatin
Hong Kong SAR

Dino Pedreschi
Università di Pisa
Pisa
Italy

Michael May
Siemens AG Corporate Technology
München
Germany

Francesco Bonchi
Eurecat / Yahoo Labs
Barcelona
Spain

Bianca Zadrozny
IBM Research Brazil
Rio de Janeiro
Brazil

Jaime Cardoso
University of Porto - INESC TEC
Porto
Portugal

Ricard Gavalda
Universitat Politècnica de Catalunya
Barcelona
Spain

Myra Spiliopoulou
Otto-von-Guericke University
Magdeburg
Germany

ISSN 0302-9743   ISSN 1611-3349 (electronic)
Lecture Notes in Artificial Intelligence
ISBN 978-3-319-23460-1   ISBN 978-3-319-23461-8 (eBook)
DOI 10.1007/978-3-319-23461-8

Library of Congress Control Number: 2015947118

LNCS Sublibrary: SL7 – Artificial Intelligence

Foreword to the ECML PKDD 2015 Industry, Governmental, and NGO Track

For the first time, the Industry, Governmental, and NGO Track of ECML PKDD 2015 had a separate Program Committee and its papers are published in its own proceedings inside this volume. The track brought together participants from academia, industry, government, and NGOs (non-governmental organizations) in a venue that highlighted practical and real-world studies in machine learning, knowledge discovery, and data mining. The industry, governmental, and NGO track was distinct from the research track in that the works presented solved real-world problems and focused on engineering systems, applications, and challenges, following three main areas: "Engineering Systems," "Data Science," or "Challenges." The main technical program included three plenary talks by three invited speakers: Hang Li (Huawei), Andreas Antrup (Zalando), and Wei Fan (Baidu Big Data Lab) and 12 technical talks. The Industry, Government, and NGO Track of ECML PKDD 2015 was a highly selective venue. This year, only 12 of the 42 papers were accepted, which corresponds to an acceptance rate of 28.6 %. Every paper received at least three reviews and most papers were thoroughly discussed before a PC senior member made a recommendation. The final decision was made by the Program Chairs based on those recommendations. The topics of the papers were extremely exciting, as they were applications of online analysis, clustering, kernel regression, active learning, network security, recommendation systems, fraud detection, semi-supervised learning, and reinforcement learning. These proceedings of the 2015 European Conference on Machine Learning and Principles and Practice of Knowledge Discovery in Databases contain the papers of the works presented at the industry, government, and NGO track. We thank the Conference Chairs, João Gama and Alípio Jorge, for their support, and Proceedings Chairs, Michelangelo Ceci and Paulo Cortez, for their help with putting these proceedings together. Most importantly, of course, we thank the authors for their contributions, and the PC senior members and PC members for their substantial efforts to guarantee and sometimes even improve the quality of these proceedings. We wish the reader an enjoyable experience exploring the many exciting research results presented in these proceedings of the 2015 edition of the Industry, Government, and NGO Track of the European Conference on Machine Learning and Principles and Practice of Knowledge Discovery in Databases.

July 2015

Albert Bifet
Bianca Zadrozny
Michael May

Foreword to the ECML PKDD 2015 Nectar Track

The goal of the ECML PKDD Nectar Track, started in 2012, is to offer conference attendees a compact overview of recent scientific advances at the frontier of machine learning and data mining with other disciplines, as already published in related conferences and journals. Authors were invited to submit 4-page summaries of their published work. Particularly welcome was work illustrating the pervasiveness of data-driven exploration and modeling in science and technology, as well as innovative applications.

We received 29 submissions and each was reviewed by at least three PC members. Finally, 14 were selected for inclusion in the proceedings and presentation at the conference. Topics covered, among others, music recommendation, analysis of urban mobility data, geospatial text streams analysis, privacy issues in data mining, use of formal concept analysis in data mining, epidemics prediction from social data, and applications to particle physics and astronomy, as well as more traditional machine learning and data mining research issues.

We thank all authors of submitted papers and all PC members for their excellent work.

July 2015

Ricard Gavaldà
Dino Pedreschi

Foreword to the ECML PKDD 2015 Demo Track

It is our great pleasure to introduce the Demo Track of ECML PKDD 2015. This year's track continues its tradition of providing a forum for researchers and practitioners to demonstrate novel systems and research prototypes, using data mining and machine learning techniques in a variety of application domains. Besides the live demonstrations during the conference period, each selected demo is allocated a 4-page paper in the proceedings.

The Demo Track of ECML PKDD 2015 solicited working systems based on state-of-the-art machine learning and data mining technology. Both innovative prototype implementations and mature systems were welcome, provided that they used machine learning techniques and knowledge discovery processes in a real setting. The evaluation criteria encompassed innovation, interestingness for the target users, and whether it would be of interest mainly for researchers, mainly for practitioners, or both. Each submission was evaluated by at least two reviewers.

This year we received 32 submissions. Seventeen demos will be presented at the Demo Session during the conference in Porto. The demo descriptions also contain links to videos and some of them to live software, so that interested readers can inspect and even try them. The accepted demos cover a wide range of machine learning and data mining techniques, as well as a very diverse set of real-world application domains.

The success of the Demo Track of ECML PKDD 2015 is due to the effort of several people. Foremostly we thank the authors for their submissions and their engagement in turning data mining and machine learning methods to software that can be presented and tried by others. We want to thank all members of our Program Committee for helping us in the difficult task of selecting the most interesting submissions. Finally, we would like like to thank the ECML PKDD General Chairs and the Program Chairs for entrusting us with this track, and the whole Organizing Committee for the practical support and the logistics for the Demo Track.

We hope that the readers will enjoy this set of short papers and the demonstrated systems, and that the Demo Session will inspire further ECML PKDD participants in turning their research ideas in working prototypes that can be used by other researchers and practitioners in machine learning and data mining.

July 2015

Francesco Bonchi
Jaime Cardoso
Myra Spiliopoulou

Preface

We are delighted to introduce the proceedings of the 2015 edition of the European Conference on Machine Learning and Principles and Practice of Knowledge Discovery in Databases, or ECML PKDD for short. This conference stems from the former ECML and PKDD conferences, the two premier European conferences on, respectively, Machine Learning and Knowledge Discovery in Databases. Originally independent events, the two conferences were organized jointly for the first time in 2001. The sinergy between the two led to increasing integration, and eventually the two merged in 2008. Today, ECML PKDD is a world-wide leading scientific event that aims at exploiting the synergies between Machine Learning and Data Mining, focusing on the development and application of methods and tools capable of solving real-life problems.

ECML PKDD 2015 was held in Porto, Portugal, during September 7–11. This was the third time Porto hosted the major European Machine Learning event. In 1991, Porto was host to the fifth EWSL, the precursor of ECML. More recently, in 2005, Porto was host to a very successful ECML PKDD. We were honored that the community chose to again have ECML PKDD 2015 in Porto, just ten years later. The 2015 ECML PKDD was co-located with "Intelligent System Applications to Power Systems", ISAP 2015, a well-established forum for scientific and technical discussion, aiming at fostering the widespread application of intelligent tools and techniques to the power system network and business. Moreover, it was collocated, for the first time, with the Summer School on "Data Sciences for Big Data."

ECML PKDD traditionally combines the research-oriented extensive program of the scientific and journal tracks, which aim at being a forum for high quality, novel research in Machine Learning and Data Mining, with the more focused programs of the demo track, dedicated to presenting real systems to the community, the PhD track, which supports young researchers, and the nectar track, dedicated to bringing relevant work to the community. The program further includes an industrial track, which brings together participants from academia, industry, government, and non-governmental organizations in a venue that highlights practical and real-world studies of machine learning, knowledge discovery, and data mining. The industrial track of ECML PKDD 2015 has a separate Program Committee and separate proceedings volume. Moreover, the conference program included a doctoral consortium, three discovery challenges, and various workshops and tutorials.

The research program included five plenary talks by invited speakers, namely, Hendrik Blockeel (University of Leuven and Leiden University), Pedro Domingos (University of Washington), Jure Leskovec (Stanford University), Nataša Milić-Frayling (Microsoft Research), and Dino Pedreschi (Università di Pisa), as well as one ISAP +ECML PKDD joint plenary talk by Chen-Ching Liu (Washington State University). Three invited speakers contributed to the industrial track: Andreas Antrup (Zalando and

University of Edinburgh), Wei Fan (Baidu Big Data Lab), and Hang Li (Noah's Ark Lab, Huawei Technologies).

Three discovery challenges were announced this year. They focused on "MoRe-BikeS: Model Reuse with Bike rental Station data," "On Learning from Taxi GPS Traces", and "Activity Detection Based on Non-GPS Mobility Data," respectively.

Twelve workshops were held, providing an opportunity to discuss current topics in a small and interactive atmosphere: "MetaSel - Meta-learning and Algorithm Selection," "Parallel and Distributed Computing for Knowledge Discovery in Databases," "Interactions between Data Mining and Natural Language Processing," "New Frontiers in Mining Complex Patterns," "Mining Ubiquitous and Social Environments," "Advanced Analytics and Learning on Temporal Data," "Learning Models over Multiple Contexts," "Linked Data for Knowledge Discovery," "Sports Analytics," "BigTargets: Big Multi-target Prediction", "DARE: Data Analytics for Renewable Energy Integration," and "Machine Learning in Life Sciences."

Ten tutorials were included in the conference program, providing a comprehensive introduction to core techniques and areas of interest for the scientific community: "Similarity and Distance Metric Learning with Applications to Computer Vision," "Scalable Learning of Graphical Models," "Meta-learning and Algorithm Selection," "Machine Reading the Web - Beyond Named Entity Recognition and Relation Extraction," "VC-Dimension and Rademacher Averages: From Statistical Learning Theory to Sampling Algorithms," "Making Sense of (Multi-)Relational Data," "Collaborative Filtering with Binary, Positive-Only Data," "Predictive Maintenance," "Eureka! - How to Build Accurate Predictors for Real-Valued Outputs from Simple Methods," and "The Space of Online Learning Problems."

The main track received 380 paper submissions, of which 89 were accepted. Such a high volume of scientific work required a tremendous effort by the Area Chairs, Program Committee members, and many additional reviewers. We managed to collect three highly qualified independent reviews per paper and one additional overall input from one of the Area Chairs. Papers were evaluated on the basis of significance of contribution, novelty, technical quality, scientific and technological impact, clarity, repeatability, and scholarship. The industrial, demo, and nectar tracks were equally successful, attracting 42, 32, and 29 paper submissions, respectively.

For the third time, the conference used a double submission model: next to the regular conference tracks, papers submitted to the Springer journals Machine Learning (MACH) and Data Mining and Knowledge Discovery (DAMI) were considered for presentation at the conference. These papers were submitted to the ECML PKDD 2015 special issue of the respective journals, and underwent the normal editorial process of these journals. Those papers accepted for one of these journals were assigned a presentation slot at the ECML PKDD 2015 conference. A total of 191 original manuscripts were submitted to the journal track during this year. Some of these papers are still being refereed. Of the fully refereed papers, 10 were accepted in DAMI and 15 in MACH, together with 4+4 papers from last year's call, which were also scheduled for presentation at this conference. Overall, this resulted in a number of 613 submissions (to the scientific track, industrial track and journal track), of which 126 were selected for presentation at the conference, making an overall acceptance rate of about 21 %.

Part I and Part II of the proceedings of the ECML PKDD 2015 conference contain the full papers of the contributions presented in the scientific track, the abstracts of the scientific plenary talks, and the abstract of the ISAP+ECML PKDD joint plenary talk. Part III of the proceedings of the ECML PKDD 2015 conference contains the full papers of the contributions presented in the industrial track, short papers describing the demonstrations, the nectar papers, and the abstracts of the industrial plenary talks.

The scientific track program results from continuous collaboration between the scientific tracks and the general chairs. Throughout we had the unfaltering support of the Local Chairs, Carlos Ferreira, Rita Ribeiro, and João Moreira, who managed this event in a thoroughly competent and professional way. We thank the Social Media Chairs, Dunja Mladenić and Márcia Oliveira, for tweeting the new face of ECML PKDD, and the Publicity Chairs, Ricardo Campos and Carlos Ferreira, for their excellent work in spreading the news. The beautiful design and quick response time of the web site is due to the work of our Web Chairs, Sylwia Bugla, Rita Ribeiro, and João Rodrigues. The beautiful image on all the conference materials is based on the logo designed by Joana Amaral e João Cravo, inspired by Porto landmarks. It has been a pleasure to collaborate with the Journal, Industrial, Demo, Nectar, and PhD Track Chairs. ECML PKDD would not be complete if not for the efforts of the Tutorial Chairs, Fazel Famili, Mykola Pechenizkiy, and Nikolaj Tatti, the Workshop Chairs, Stan Matwin, Bernhard Pfahringer, and Luís Torgo, and the Discovery Challenge Chairs, Michel Ferreira, Hillol Kargupta, Luís Moreira-Matias, and João Moreira. We thank the Awards Committee Chairs, Pavel Brazdil, Sašo Džerosky, Hiroshi Motoda, and Michèle Sebag, for their hard work in selecting papers for awards. A special meta thanks to Pavel: ECML PKDD at Porto is only possible thanks to you. We gratefully acknowledge the work of Sponsorship Chairs, Albert Bifet and André Carvalho, for their key work. Special thanks go to the Proceedings Chairs, Michelangelo Ceci and Paulo Cortez, for the difficult task of putting these proceedings together. We appreciate the support of Artur Aiguzhinov, Catarina Félix Oliveira, and Mohammad Nozari (U. Porto) for helping to check this front matter. We thank the ECML PKDD Steering Committee for kindly sharing their experience, and particularly the General Steering Committee Chair, Fosca Giannotti. The quality of ECML PKDD is only possible due to the tremendous efforts of the Program Committee; our sincere thanks for all the great work in improving the quality of these proceedings. Throughout, we relied on the exceptional quality of the Area Chairs. Our most sincere thanks for their support, with a special thanks to the members who contributed in difficult personal situations, and to Paulo Azevedo for stepping in when the need was there. Last but not least, we would like to sincerely thank all the authors who submitted their work to the conference.

July 2015

Annalisa Appice
Pedro Pereira Rodrigues
Vítor Santos Costa
Carlos Soares
João Gama
Alípio Jorge

Organization

ECML/PKDD 2015 Organization

Conference Co-chairs

João Gama   University of Porto, INESC TEC, Portugal
Alipío Jorge   University of Porto, INESC TEC, Portugal

Program Co-chairs

Annalisa Appice   University of Bari Aldo Moro, Italy
Pedro Pereira Rodrigues   University of Porto, CINTESIS, INESC TEC, Portugal
Vítor Santos Costa   University of Porto, INESC TEC, Portugal
Carlos Soares   University of Porto, INESC TEC, Portugal

Journal Track Chairs

Concha Bielza   Technical University of Madrid, Spain
João Gama   University of Porto, INESC TEC, Portugal
Alipío Jorge   University of Porto, INESC TEC, Portugal
Indré Žliobaité   Aalto University and University of Helsinki, Finland

Industrial Track Chairs

Albert Bifet   Huawei Noah's Ark Lab, China
Michael May   Siemens, Germany
Bianca Zadrozny   IBM Research, Brazil

Local Organization Chairs

Carlos Ferreira   Oporto Polytechnic Institute, INESC TEC, Portugal
João Moreira   University of Porto, INESC TEC, Portugal
Rita Ribeiro   University of Porto, INESC TEC, Portugal

Tutorial Chairs

Fazel Famili   CNRC, France
Mykola Pechenizkiy   TU Eindhoven, The Netherland
Nikolaj Tatti   Aalto University, Finland

Workshop Chairs

Stan Matwin   Dalhousie University, NS, Canada
Bernhard Pfahringer   University of Waikato, New Zealand
Luís Torgo   University of Porto, INESC TEC, Portugal

Awards Committee Chairs

Pavel Brazdil   INESC TEC, Portugal
Sašo Džeroski   Jožef Stefan Institute, Slovenia
Hiroshi Motoda   Osaka University, Japan
Michèle Sebag   Université Paris Sud, France

Nectar Track Chairs

Ricard Gavaldà   UPC, Spain
Dino Pedreschi   Università di Pisa, Italy

Demo Track Chairs

Francesco Bonchi   Yahoo! Labs, Spain
Jaime Cardoso   University of Porto, INESC TEC, Portugal
Myra Spiliopoulou   Otto-von-Guericke University Magdeburg, Germany

PhD Chairs

Jaakko Hollmén   Aalto University, Finland
Panagiotis Papapetrou   Stockholm University, Sweden

Proceedings Chairs

Michelangelo Ceci   University of Bari, Italy
Paulo Cortez   University of Minho, Portugal

Discovery Challenge Chairs

Michel Ferreira   University of Porto, INESC TEC, Geolink, Portugal
Hillol Kargupta   Agnik, MD, USA
Luís Moreira-Matias   NEC Research Labs, Germany
João Moreira   University of Porto, INESC TEC, Portugal

Sponsorship Chairs

Albert Bifet   Huawei Noah's Ark Lab, China
André Carvalho   University of São Paulo, Brazil
Pedro Pereira Rodrigues   University of Porto, Portugal

Publicity Chairs

Ricardo Campos	Polytechnic Institute of Tomar, INESC TEC, Portugal
Carlos Ferreira	Oporto Polytechnic Institute, INESC TEC, Portugal

Social Media Chairs

Dunja Mladenić	JSI, Slovenia
Márcia Oliveira	University of Porto, INESC TEC, Portugal

Web Chairs

Sylwia Bugla	INESC TEC, Portugal
Rita Ribeiro	University of Porto, INESC TEC, Portugal
João Rodrigues	INESC TEC, Portugal

ECML PKDD Steering Committee

Fosca Giannotti	ISTI-CNR Pisa, Italy
Michèle Sebag	Université Paris Sud, France
Francesco Bonchi	Yahoo! Research, Spain
Hendrik Blockeel	KU Leuven, Belgium and Leiden University, The Netherlands
Katharina Morik	University of Dortmund, Germany
Tobias Scheffer	University of Potsdam, Germany
Arno Siebes	Utrecht University, The Netherlands
Peter Flach	University of Bristol, UK
Tijl De Bie	University of Bristol, UK
Nello Cristianini	University of Bristol, UK
Filip Železný	Czech Technical University in Prague, Czech Republic
Siegfried Nijssen	LIACS, Leiden University, The Netherlands
Kristian Kersting	Technical University of Dortmund, Germany
Rosa Meo	Università di Torino, Italy
Toon Calders	Eindhoven University of Technology, The Netherlands
Chedy Raïssi	INRIA Nancy Grand-Est, France

Industrial Track Program Committee

Senior members

Michael Berthold	Francesco Bonchi	Hillol Kargupta
Ralf Klinkenberg	Nuria Oliver	Michal Rosen-Zvi

Members

Kareem Aggour	Arthur Von Eschen	Cheng Weiwei
Björn Bringmann	Ding Wei	Kun Zhang
Kamal Ali	Oksana Yakhnenko	Yu Zheng
Rui Cai	Zheng Zhao	Duen Horng Chau
Nicola Barberi	Arijit Chatterjee	Jiefeng Cheng
Berkant Barla	Wei Chen	Kamalika Das
Cambazoglu	Brian Dalessandro	Seymour Douglas
Liangliang Cao	Gianmarco De Francisci	Simon Fischer
Haifeng Chen	Morales	Cesare Furlanello
Michelangelo D'Agostino	Wang Fei	Joao Gomes
Jesse Davis	Enrique Frias	Georges Hébrail
Pinelli Fabio	Aris Gkoulalas-Divanis	Zhexue Huang
Avrilia Floratou	Francesco Gullo	Anne Kao
Dinesh Garg	Bing Hu	Nicolas Kourtellis
Amit Goyal	Jianying Hu	Mounia Lalmas
Shawndra Hill	Hongxia Jin	Bangyong Liang
Alex Jaimes	Deguang Kong	Arun Maiya
Alexandros Karatzoglou	Mohit Kumar	Luis Matias
Shonali Krishnaswamy	Kuang-chih Lee	Veena B Mendiratta
Ni Lao	Ping Luo	Vivek Narasayya
Jiebo Luo	Manish Marwah	Stelios Paparizos
Silviu Maniu	Sameep Mehta	Daniele Quercia
Dimitrios Mavroeidis	Iris Miliaraki	Jose San Pedro
Berlingerio Michele	Feng Pan	Alkis Simitsis
Xia Ning	Fabio Pinelli	Papadimitriou Spiros
Fernando Perez-Cruz	Barna Saha	Jilei Tian
Matthew Rattigan	Amy Shi-Nash	Michail Vlachos
Krishna Sankar	Scott Smith	Xiang Wang
Tomas Singliar	Shu Tao	Donghui Wu
L Subramaniam	Volker Tresp	YanChang Zhao
Guangjian Tian	Pinghui Wang	

Industrial Track External Reviewers

Le An	Abon Chaudhuri	Derek Doran
Vijay K. Gurbani	Sen Wu	Bing Xu

Nectar Track Program Committee

Luiza Antonie	Rosa Meo	Luc De Raedt
Raissi Chedy	Katharina Morik	Ricard Gavaldà
Ernestina Menasalvas	Pierre Geurts	Frank Hutter
Yuhong Guo	David Taniar	Andreas Maletti
Yiannis Koutis	Tony Bagnall	Michele Berlingerio

Weike Pan   Joao Gomes   Dino Pedreschi
Balaraman Ravindran   Kristian Kersting   Lars Schmidt-Thieme
Sourav Bhowmick   Wagner Meira
Dino Ienco   Nanni Mirco

Demo Track Program Committee

Bettina Berendt   Thomas Buelow   Gustavo Carneiro
Dirk Elias   Ricard Gavalda   Markus Harz
John Hipwell   Mark Last   Vincent Lemaire
Ernestina Menasalvas   Themis Palpanas   Joao Papa
Mykola Pechenizkiy   Bernhard Pfahringer   Jerzy Stefanowski
Luis Teixeira   Grigorios Tsoumakas

Sponsors

Platinum Sponsor

BNP PARIBAS   http://www.bnpparibas.com/
ONR Global   www.onr.navy.mil/science-technology/onr-global.aspx

Gold Sponsors

Zalando   https://www.zalando.co.uk/
HUAWEI   http://www.huawei.com/en/

Silver Sponsors

Deloitte   http://www2.deloitte.com/
Amazon   http://www.amazon.com/

Bronze Sponsors

Xarevision   http://xarevision.pt/
Farfetch   http://www.farfetch.com/pt/
NOS   http://www.nos.pt/particulares/Pages/home.aspx

Award Sponsor

Machine Learning   http://link.springer.com/journal/10994
Data Mining and   http://link.springer.com/journal/10618
   Knowledge Discovery
Deloitte   http://www2.deloitte.com/

Lanyard Sponsor

KNIME   http://www.knime.org/

Invited Talk Sponsor

ECCAI   http://www.eccai.org/
Cliqz   https://cliqz.com/

Technicolor   http://www.technicolor.com/
University of Bari Aldo   http://www.uniba.it/english-version
   Moro

Additional Supporters

INESCTEC   https://www.inesctec.pt/
University of Porto,   http://sigarra.up.pt/fep/pt/web_page.inicial
   Faculdade de Economia
Springer   http://www.springer.com/
University of Porto   http://www.up.pt/

Official Carrier

TAP   http://www.flytap.com/

Invited Talks Abstracts
(Industrial Track)

AI Research at Huawei Technologies

Hang Li

Noah's Ark Lab, Huawei Technologies, Shenzhen, China

Abstract. To enable computers to listen, speak, see, and learn like humans, and even more to build computers that analyze, infer, predict, and make decisions better than humans is the ultimate dream of artificial intelligence. In the near future, we envision that with advanced AI technologies each person and each enterprise will have multiple intelligent computer assistants to help accomplish various tasks with results that are as good or even better than if they were performed by humans. Huawei Technologies, one of the major telecommunication equipment and service companies in the world, is also conducting research toward this noble goal of AI. In this talk, I will introduce some progresses which Huawei, and particularly its Noah' Ark Lab, have made recently, including a platform for deep learning research, natural language processing using deep learning, construction and utilization of large-scale knowledge base in telecommunication domain.

Bio. Hang Li is director of the Noah's Ark Lab of Huawei Technologies. His research areas include natural language processing, information retrieval, statistical machine learning, and data mining. He graduated from Kyoto University in 1988 and earned his PhD from the University of Tokyo in 1998. He worked at the NEC lab in Japan during 1991 and 2001, and Microsoft Research Asia during 2001 and 2012.

Algorithmic Fashion

Andreas Antrup

Zalando, The University of Edinburgh, Edinburgh, UK

Abstract. Fashion and the fashion industry are commonly considered the realm of intuition and experts rather than of machine learning and data. But that is not so; in fact, I am going to argue that the decision-making both on the consumer side and the industry side is being made increasingly transparent and open to change by the Internet. Instead of focussing on a particular methodology, the talk is going to be centred on the clever orchestration of data and algorithms. It is a peek into our journey at Zalando.

Bio. Andreas Antrup heads Data Science at Zalando - Europe's leading online shop for shoes and fashion. After brief stints in entrepreneurship and banking he joined Zalando in 2011 to build analytics and data-driven automation. Andreas studied ecnonomics at the University of Edinburgh to graduate with an MSc in 2008 and a PhD in 2011. Together with his team he now drives predictive analytics across the value chain of Zalando.

Deep Medical Learning

Wei Fan

Baidu Big Data Lab, Beijing, China

Abstract. Recent market intelligences shows that when people encounter medical or healthcare related problems, 90 % of them first go to the internet for help. The percentage of search engine queries on medical and healthcare are in double digits. Currently, search engine is still not the right place to look for medical and healthcare related information, unless the user is very clear on names of medical, health conditions or treatments. However, a majority of medical queries are on new conditions where the users may only have some vague ideas on their discomforts or symptoms. Their intention is also not as clear as obtaining knowledge on "food, entertainment, electronics, etc" where both the users are trained to do those search and search engines are also tuned to match these searches well. In this talk, we will discuss deep learning techniques that are used to (1) understand users' intention (2) build a complete symptom to disease prediction model and (3) a Q-A system that speaks to the users to either provide useful or authoritative information or asks for additional information when their queries or intentions are unclear. The deep learning techniques leverages DNN, CNN and bi-directional RNN, which are used for feature constructions from raw queries, modeling from training Q&A pairs as well as modeling.

Bio. Wei Fan is currently the director and deputy head of Baidu Big Data Lab in Sunnyvale, California. He received his PhD in Computer Science from Columbia University in 2001. His main research interests and experiences are in various areas of data mining and database systems, such as, deep learning, stream computing, high performance computing, extremely skewed distribution, cost-sensitive learning, risk analysis, ensemble methods, easy-to-use nonparametric methods, graph mining, predictive feature discovery, feature selection, sample selection bias, transfer learning, time series analysis, bioinformatics, social network analysis, novel applications and commercial data mining systems. His co-authored paper received ICDM'2006 Best Application Paper Award, he led the team that used his Random Decision Tree (www.dice.com) method to win 2008 ICDM Data Mining Cup Championship. He received 2010 IBM Outstanding Technical Achievement Award for his contribution to IBM Infosphere Streams. He is the associate editor of ACM Transaction on Knowledge Discovery and Data Mining (TKDD). During his times as the Associate Director in Huawei Noah's Ark Lab in Hong Kong from August 2012 to December 2014, he has led his colleagues to develop Huawei Stream-SMART - a streaming platform for online and real-time processing, query and mining of very fast streaming data. StreamSMART is 3 to 5 times faster than STORM and 10 times faster than SparkStreaming, and was used in Beijing Telecom, Saudi Arabia STC, Norway Telenor and a few other mobile carriers in Asia.

Since joining Baidu Big Data Lab, Wei has been working on O2O, particularly, on medical and healthcare applications that converges from online queries to offline visits.

Contents – Part III

Demo Track

Industrial Track

Autonomous HVAC Control,
A Reinforcement Learning Approach

Enda Barrett[1,2] and Stephen Linder[1,2(✉)]

[1] Schneider Electric, Cityeast Business Park, Galway, Ireland
[2] Schneider Electric, 800 Federal Street, Andover, MA 01810-1067, USA
{Enda.Barrett,Stephen.Linder}@schneider-electric.com
http://www.schneider-electric.com/

Abstract. Recent high profile developments of autonomous learning thermostats by companies such as Nest Labs and Honeywell have brought to the fore the possibility of ever greater numbers of intelligent devices permeating our homes and working environments into the future. However, the specific learning approaches and methodologies utilised by these devices have never been made public. In fact little information is known as to the specifics of how these devices operate and learn about their environments or the users who use them. This paper proposes a suitable learning architecture for such an intelligent thermostat in the hope that it will benefit further investigation by the research community. Our architecture comprises a number of different learning methods each of which contributes to create a complete autonomous thermostat capable of controlling a HVAC system. A novel state action space formalism is proposed to enable a Reinforcement Learning agent to successfully control the HVAC system by optimising both occupant comfort and energy costs. Our results show that the learning thermostat can achieve cost savings of 10% over a programmable thermostat, whilst maintaining high occupant comfort standards.

Keywords: HVAC control · Reinforcement learning · Bayesian learning

1 Introduction

Thermostats for controlling Heating, Ventilation and Air Conditioning (HVAC) systems in the home and office can largely be broken into two main categories: programmable and manual. Programmable thermostats allow the user to schedule heating and cooling to achieve patterns that work best for one's schedule. A thermal set-point is specified by a user and it governs the temperature and humidity levels that must be reached when the controller is active. Manual thermostats are non-programmable and require an external operator (human) to turn on and off the functions of heating and cooling as required. Manual thermostats are usually cheaper than their programmable counterparts.

Recently there has been a surge in the development of intelligent thermostats which boost the ability to autonomously control HVAC systems. These include

© Springer International Publishing Switzerland 2015
A. Bifet et al. (Eds.): ECML PKDD 2015, Part III, LNAI 9286, pp. 3–19, 2015.
DOI: 10.1007/978-3-319-23461-8_1

offers from companies such as Nest Labs[3] and Honeywell[2]. They often only require the user to enter temperature set-points, while the schedule is learned automatically, with the objective of minimizing energy consumption while still allowing for occupant comfort. The unit attempts to learn a user's preference over time based on their manual adjustments and produce a schedule which is deemed optimal for the observed patterns of occupancy. The principal characteristics of these units is that they promote some notion of self-learning and automation; however, the user can usually override the learned schedule with a pre-fixed one.

In order to learn and effectively make decisions these systems rely on observations from sensory inputs about their environment. Commonly they use: temperature/humidity and motion sensors; an internal clock/calendar to track date/time; and external data sources such as the local weather conditions. Over time the goal of the learning thermostat is to refine its knowledge, update its understanding of the environment and make optimal decisions in accordance with its defined objective functions of maintaining occupant comfort and minimising cost.

To date, the specific learning approaches employed by companies such as Nest and Honeywell have never been publicly released and are guarded as trade secrets. In addition, there has been little activity from the research community to devise suitable open architectures for solving such problems. Therefore, this paper proposes a suitable learning architecture which utilises a number of learning methods capable of learning an optimal or near optimal control policy over time. The solution comprises a *Bayesian Learning* approach to accurately predict room occupancy over time and a *Reinforcement Learning* (Q-learning) method to learn a control policy for the thermostat unit itself. The reinforcement learning agent samples the output from the room occupancy prediction module to enable a better control solution.

In summary the principal contributions of this work are

- A learning architecture which can support occupancy prediction and HVAC control, concurrently optimising both user comfort and energy costs
- A novel state action space formalism for the individual learning approach which enables a multi-criteria optimisation solution.

The rest of this paper is structured as follows: *Background Research* provides an overview of relevant and related work in this field, including work specific to the learning approaches used and other applied learning work. *Markov Decision Processes & Learning Methods* describes the concepts and learning approaches used in this work. *HVAC Control* details specifics relating to how to apply these methods to the real world problem. *Initial Results* details our preliminary findings, leading finally to *Conclusions & Future Work.*

2   Background Research

A standard HVAC system can be considered to comprise two principal components, a heating/cooling element and a fan for circulating the air. In order to operate the system, the user generally specifies a setpoint value on the thermostat interface denoting the room conditions they require. Using the output from a temperature sensor the thermostat monitors the changing room conditions as a result of turning on the HVAC. Once the room temperature has reached the target setpoint the system is switched off. In more optimised systems the fan speeds can be adjusted to enable further optimisations over this simple scenario.

To autonomously control HVAC systems, a number of methods are required to enable the features depicted on modern controllers such as the Nest. Firstly a learning method must learn when to turn on and off heating and cooling, we refer to this as the thermostat control policy for which we employ a reinforcement learning method known as Q-learning. The goal of the learning process is to control the HVAC system i.e. turn it on and off, to ensure that the user's setpoint is maintained at the lowest possible cost. In addition as an aid to this control policy, Bayesian inference is utilised to predict room occupancy allowing for greater cost savings where unoccupied rooms need no longer be heated. An accurate prediction method is necessary in order to preemptively heat and cool rooms prior to being occupied. Occupancy sensors can only tell you when they detect whether or not a room is occupied, not when it is going to be occupied, thus only heating and cooling based on occupancy detection will likely result in a low comfort rating from users who will have to wait until the room reaches the set point temperature. The combination of these techniques provides an overall architecture capable of providing a solution to the problem. One of the key value propositions is the ability of each component to build up knowledge whilst operating directly with the environment, without prior experience.

Whilst there has been substantial activity in the commercial space with numerous patents filed in this area for both automating the control of HVAC systems entirely or partially through varying components of these systems, there has been little activity in the research community. To the best of the authors' knowledge this paper is the first application of Reinforcement Learning to this problem domain.

In the 1990s manufacturers such as Mitsubishi [15] developed advanced fuzzy rule bases for controlling air conditioning systems in buildings which greatly outperformed the ubiquitous bang-bang controllers. The fuzzy rule systems allowed for intermediary control states where the air conditioning system could alternate between different fan speeds, humidity and temperature based on the environmental observations to reduce energy consumption and improve occupant comfort. Fuzzy systems rely on user defined rules which are collectively termed the fuzzy rule base. The output of the rules are combined to produce a smooth control response which creates smaller deviations around the temperature set points. Patents [9] [1] [4] describe a variety of fuzzy logic control methods ranging from the determination of thermal set-points in a HVAC system to methods for controlling HVAC to maximise occupant comfort in the automotive sector.

More recently, a patent filed by Nest [13] on 19 October 2012 describes a thermostat which uses machine learning and offers a taxonomy of learning approaches over which its claims are held. However the disclosure does not present a description of how these methods are applied or even which methods are used in their implementation. On the Web there have been a number of hardware teardowns of the Nest thermostat, but to date little is known about the software controling the device.

In a user trial on a number of homes in the US and UK by Scott et. al. [18] (entitled PreHeat) RFID tags were added to the house keys of each occupant. Their home heating solution used occupancy sensing and prediction to better estimate when to heat the homes. Their results demonstrated substantial savings over a pre-scheduled heating solution for heating and cooling rooms in a house.

In a more general context, approaches from learning theory have been successfully applied to automated control problems across a range of domains. Dutreilh et. al. [12] devised a Q-learning approach for allocating resources to applications in the cloud. Gerald Tesauro created TD-Gammon [21], a reinforcement learning artificially intelligent agent capable of playing backgammon to international level. Other notable successes include workflow scheduling [6], traffic light control [24] and application scaling [7] in computational clouds. The important novelty common to these works is not so much their extension to learning theory but more so their application of learning theory to solve a real world problem.

3 Markov Decision Processes and Learning Methods

3.1 Markov Decision Processes

Markov Decision Processes (MDPs) are a particular mathematical framework suited to modelling decision making under uncertainty. A MDP can typically be represented as a four tuple consisting of states, actions, transition probabilities and rewards.

- S, represents the environmental state space;
- A, represents the total action space;
- $p(.|s, a)$, defines a probability distribution governing state transitions $s_{t+1} \sim p(.|s_t, a_t)$;
- $q(.|s, a)$, defines a probability distribution governing the rewards received $R(s_t, a_t) \sim q(.|s_t, a_t)$;

S is the set of all possible states represents the agent's observable world. The agent learning experience can be broken up into discrete time periods. At the end of each time period t the agent occupies state $s_t \in S$. The agent chooses an action $a_t \in A(s_t)$, where $A(s_t)$ is the set of all possible actions within state s_t. The execution of the chosen action, results in a state transition to s_{t+1} and an immediate numerical reward $R(s_t, a_t)$. The state transition probability $p(s_{t+1}|s_t, a_t)$ governs the likelihood that the agent will transition to state s_{t+1} as a result of choosing a_t in s_t. The numerical reward received upon arrival at the next state is governed by $q(s_{t+1}|s_t, a_t)$ and is indicative as to the benefit of choosing a_t whilst in s_t.

The solution of a MDP results in the output of a policy π, denoting a mapping from states to actions, guiding the agent's decisions over the entire learning period.

In the specific case where a complete environmental model is known, i.e. (S, A, p, q) are fully observable, the problem reduces to a planning problem [16] and can be solved using traditional dynamic programming techniques such as value iteration. However if there is no complete model available, which is often common with real world problems, then one must either attempt to approximate the missing model (Model Based Reinforcement Learning) or directly estimate the value function or policy (Model Free Reinforcement Learning). Model based methods use statistical techniques in order to approximate the missing model [14], whereas model free learners attempt to directly approximate a control policy through environmental interactions.

In this work we choose to utilise a model free learning method known as Q-learning. The approach has been widely applied to real-world problems, which allows for stricter comparisons with previous work and is capable of finding an optimal or near optimal control policy in a reasonable time.

3.2   Reinforcement Learning

Modeling the HVAC control problem as a MDP enables us to design a solution which can effectively handle environmental uncertainty. However as with most real world learning problems, we have no prior knowledge of the complete environmental model, the distribution of rewards or transition probabilities. Therefore, solutions from Dynamic Programming such as Value Iteration or Policy Iteration cannot be used to generate an optimal policy π for these problems. As an alternative to Dynamic Programming, model free Reinforcement Learning methods such as Q-learning [23] can be used to generate optimal policies in the absence of a complete environmental model.

Q-learning belongs to a collection of algorithms called Temporal Difference (TD) methods. Not requiring a complete model of the environment, TD methods possess a significant advantage and have the capability of being able to make predictions incrementally and in an online fashion. We choose to use Q-learning for this research, not for its demonstrated efficacy within the domain but more for its wide applicability to applied domains published previously [10] [8] [22]. The update rule for Q-learning is defined as

$$Q(s,a) \leftarrow Q(s,a) + \alpha[r + \gamma Q(s',a') - Q(s,a)] \tag{1}$$

and calculated each time a state is reached which is nonterminal. Approximations of $Q^{\pi}(s,a)$ which are indicative as to the benefit of taking action a while in state s, are calculated after each time interval. Actions are chosen based on π, the policy being followed. A number of action selection policies can be used to decide what action to select whilst occupying a particular state, examples include ϵ-greedy, softmax and unbiased sampling [20]. The goal of these selection strategies is often to carefully balance exploration and exploitation to yield the best possible results in the shortest possible time frame. Over time the actions selected should converge to the optimal where the agents consistently choose actions which present it with the greatest amount of cumulative reward over the course of the interaction. In the case of ϵ-greedy, the goal is to choose the best action most of the time except for a certain amount of time governed by ϵ when the agent chooses an exploratory action. Let $A'(s) \subseteq A(s)$, be the set of all non-greedy actions. The probability of selection for each non-greedy action is reduced to $\frac{\epsilon}{|A'(s)|}$, resulting in a probability of $1 - \epsilon$ for the greedy strategy.

Estimated action values for each state action pair $Q^\pi(s, a)$ can be represented in tabular form or as part of a generalised function approximator. The goal of the learning agent is to maximize its returns in the long run, often forgoing short term gains in place of long term benefits. By introducing a discount factor γ, $(0 < \gamma < 1)$, an agent's degree of myopia can be controlled. A value close to 1 for γ assigns a greater weight to future rewards, while a value close to 0 considers only the most recent rewards. Reinforcement learning based approaches are capable of reasoning over multiple actions, choosing only those which yield the greatest cumulative reward over the entire duration of the episode. The steps involved in Q-learning are depicted by Algorithm 1.

Algorithm 1. Reinforcement Learning Algorithm (Q-learning)

Initialize $Q(s, a)$ arbitrarily
  Repeat (for each episode)
  Initialize s
  repeat
    Choose a from s using policy derived from Q (ϵ-greedy)
    Take action a and observe r, s'
    $Q(s, a) \leftarrow Q(s, a) + \alpha[r + \gamma \max_{a'} Q(s', a') - Q(s, a)]$
    $s \leftarrow s'$;
  until s is terminal

Q-learning can often require significant experience within a given environment in order to learn a good policy. This is largely determined by the size of the state and action space. In particular, tabular Q-learning methods require continuous updating of the value estimates through repeatedly revisiting the states and choosing actions in the environment. As the size of the state action space grows, this problem can become more pronounced (often referred to as the curse of dimensionality), where each additional state or action variable added, increases the problem size exponentially. For the purposes of this work we utilise tabular Q-learning methods but convergence times could be improved by utilising techniques such as parallel learning[5] or function approximation.

3.3   Bayesian Inference

The final part of the problem requires a solution for occupancy prediction in order to make better judgements as to when one is required to heat and cool the space under control.

For this work we employ a Bayesian inference technique in order to make predictions. Bayes theorem is a mathematical framework which allows for the integration of one's observations into one's beliefs. The posterior probability $P'(X = x|e)$, denoting the probability that a random variable X has a value equal to x given experience e can be computed via

$$P(Y|X) = \frac{P(X|Y)P(Y)}{P(X)} \qquad (2)$$

which requires one conditional probability $P(X|Y)$ and two unconditional probabilities $(P(Y), P(X))$ to compute a single conditional posterior probability $P(Y|X)$ [17].

Bayesian learning algorithms generally combine Bayesian inference (Bayes rule) and agent learning to build up probabilistic knowledge about a given domain. Statistical inference methods can prove particularly useful when attempting to approximate the likelihood of an event occurring given past experiences. By providing an estimated occupancy model our overall solution is capable of reducing the energy consumption by only heating and cooling when necessary.

4   HVAC Control

This section discusses the specifics of applying each technology to the domain. We present a novel state action space formalism for Q-learning which enables it to effectively control heating and cooling in an online manner. In addition we describe a method to predict occupancy using a modified Bayes rule and corresponding update function.

4.1   Occupancy Prediction

The specific inference rule applied for occupancy prediction was originally defined by David Spiegelhalter [19] and further extended by Prashant Doshi [11]. It employs a modified Bayes rule, where all that is required to compute the posterior probability is an initial prior probability and subsequent environmental experience. The approach involves maintaining an experience counter $Expc$ for each observation and updating the distribution according to equations (3) and (4). These equations define the update rules for approximating the likelihood of occupancy based on past experience [1]

$$P'(s = s'|a, s = s) = \frac{P(s = s'|a, s = s) \times Expc + 1}{Expc'} \qquad (3)$$

where equation 4 ensures that the probability distribution over the total number of possibilities sums up to 1. y represents the set of all possible next states achievable from s minus s' the actual next state resulting from action selection a.

---

[1] $Expc'$ is the incremented counter, $Expc' = Expc + 1$.

$$P'(s = y|a, s = s) = \frac{P(s = y|a, s = s) \times Expc}{Expc'} \tag{4}$$

From an implementation perspective the approach requires an occupancy sensor to provide it with the necessary evaluative feedback in order to update the model over time. Every minute the learning agent queries the sensor which returns a boolean result (true or false) depending on whether or not the room was occupied at that time. Based on the response a binomial distribution is updated accordingly using equations (3) and (4). This simple solution is surprisingly efficient at making predictions and doesn't require large amounts of environmental experience.

4.2 HVAC Control Using Q-learning

The HVAC system employs Q-learning by framing the environment as a MDP. In order to accurately solve the problem we must first define the set of states S and actions A i.e. the agent's observable world and the actions it can take in it:

- rt : is the room temperature (*source*: temperature sensor, *unit*: °C);
- tto : is the time to occupancy (*source*: occupancy predictor, *unit*: minutes);
- ot : is the outside temperature (*source*: weather station, *unit*: °C);

The second thing we define is the action space A which consists of the following four choices:

- $Heat_{on}$: turns on heating;
- $Heat_{off}$: turns off heating;
- $Cool_{on}$: turns on cooling;
- $Cool_{off}$: turns off cooling;

The idea is to try to keep the number of states and actions low so that the problem remains within the bounds of tractability. However even though there are only three state variables, each state variable can take on a wide range of values quickly creating a relatively large state space. For instance the indoor temperature could range from the low teens to the mid to high twenties (12°C to 27°C). The outside temperature could vary from region to region, but in places such as North America it would not be uncommon to experience highs of 40°C in the Summer and lows of −20°C in the depths of Winter. In addition the time to occupancy tto (minutes) at any particular moment may be a number of hours away, substantially increasing the size of the state space.

HVAC controller actions are executed at discrete time intervals known as epoches. For instance an epoch of 5 minutes assumes that a controlling action for the HVAC system may be executed at either 10:00 or 10:05, but not at 10:02. The granularity is a configurable parameter and can be adjusted to ensure an optimal configuration such as at minutely intervals. At the end of each epoch the learning agent observes the current state of the environment and chooses whether or not to execute an automated HVAC action (turn on or off).

The transition probabilities T i.e. the likelihood of transitioning between states after executing particular actions is not known apriori so this problem cannot be solved using Dynamic Programming methods such as Policy or Value iteration. In addition we do not attempt to estimate T, instead Q-learning observes the consequences of T and adjusts accordingly.

The rewards achievable by the learning agent are distributed in accordance with certain scenarios that arise and are scalar in value. A setpoint variable sp specifies the user defined objective temperature setting. Rewards are calculated as follows:

1. (Room.occupied $= false$) & (Action $= Heat_{on}$) & (rt $>$ sp $||$ rt $<$ sp); $R = -1$
2. (Room.occupied $= true$) & (Action $= Heat_{on}$) & (rt $>$ sp $||$ rt $<$ sp); $R = -3$
3. (Room.occupied $= false$) & (Action $= Heat_{on}$) & (rt $=$ sp); $R = -1$
4. (Room.occupied $= true$) & (Action $= Heat_{on}$) & (rt $=$ sp); $R = -1$
5. (Room.occupied $= false$) & (Action $= Heat_{off}$) & (rt $<$ sp $||$ sp $>$ rt); $R = 0$
6. (Action $= Heat_{off}$) & (rt $=$ sp); $R = 0$
7. (Room.occupied $= true$) & (Action $= Heat_{off}$) & (rt $<$ sp $||$ sp $>$ rt); $R = -3$

We assume a threshold around the setpoint of plus or minus 1°C. So if the user specifies a setpoint temperature of 23°C, the variable sp will range from $22°C - 24°C$. Each scenario listed above determines the rewards achievable as a result of choosing an action a within a particular state s. We haven't included cooling as part of the scenario, but the same rules will govern its action selection also. *Scenarios 1,3,4* result in a reward of -1. This cost is representative of the cost that would be incurred for operating the heating control of the HVAC unit per time step. This could easily be extended to include real time energy pricing costs if necessary. For *Scenarios 2 and 7* a fixed penalty is applied resulting in a reward of -3. The penalty chosen does not need to be specifically -3 but it must be greater than the unit cost of the HVAC operation i.e. (-1). For *Scenarios 2 and 7*, irregardless of the action chosen i.e. $Heat_{on}, Heat_{off}$ the penalty is applied because the setpoint temperature specified by the user has not been met and the room is presently occupied. *Scenarios 5 and 6* result in a reward of 0, i.e. no cost is incurred as the heating is turned off and either the room is not occupied *(Scenario 5)* or the setpoint has already been met *(Scenario 6)*.

5 Initial Results

This section describes our initial results with the autonomous thermostat controller. For the purposes of this research we conducted evaluations via simulation only. We present results for both occupancy prediction and thermostat control, demonstrating empirically the efficacy of the solutions as possible approaches for solving the problem.

5.1 Occupancy Prediction

We evaluated our occupancy prediction method by creating an occupancy model, which simulates when the room is occupied. We assume an occupancy sensor is always available, however we vary the accuracy of this sensor using a Gaussian distribution of mean zero and standard deviation one. For a Gaussian or normal distribution, 70% of the time a random variable X takes on a value x which will fall within one standard deviation either side of the mean. 95% of the time the value will fall within two standard deviations and 99% of the time it will fall within three standard deviations of the mean. Thus we can vary the accuracy of the sensor by introducing statistical noise and returning either a false positive or negative depending on the actual result. This is to replicate scenarios which may cause false positives such as a cat/dog moving or false negatives such as if a person is in the sensor blind spot. We argue that a prediction approach should be capable of handling such sensing errors which would arise under normal operating conditions.

The goal of the Bayesian learner is to approximate the user's patterns of occupancy as closely as possible based solely on the learners observations. To do this, the learner continuously updates its beliefs, represented probabilistically, over time. In order to evaluate the approach we employ the Kullback-Liebler (KL) divergence to determine the difference between the binomial distribution of the learner and the true values as learning progresses. The KL divergence, sometimes referred to as information gain or relative entropy gives a measure of the distance between two probability distributions. For two probability distributions P and Q, the KL divergence is

$$D_{KL}(P \parallel Q) = \sum_i ln \frac{P(i)}{Q(i)} P(i) \tag{5}$$

Note that the KL divergence is not a true metric and is not symmetrical, meaning that the KL divergence from P to Q is not equal to the KL divergence from Q to P. If P and Q are identical then $D_{KL} = 0$.

Figure 1 displays how predictions progress over time under three separate settings. The average KL divergence between the learned occupancy model and the true model is plotted for each setting. Note that the true model is not strictly ground truth occupancy, as it incorporates the variance in sensing for a more realistic comparison. It's worth noting that in a given week the learner will only sample each time period once, this means that over the course of a month the learner will have four bouts of experience to train its model. By grouping days into categories such as "working week" or "weekends", the learning time for occupancy prediction could be dramatically reduced as the information learned over a number of days could be aggregated. However we choose to treat each day as an independent event in order to present a clearer evaluation of the time it takes to learn an occupancy model. Figure 1 plots the course of the model learning process over 150 days. The first two curves detail the affect of the sensing error on the learning process, whilst the third demonstrates sensing error and model shift. Model

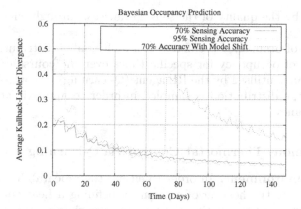

Fig. 1. Occupancy prediction evaluation with varying sensing error and model change

shift is intended to represent a scenario where the underlying occupancy pattern changes entirely, i.e. a user suddenly begins to work nights instead of days and their pattern of occupancy changes to reflect this.

When the sensor is only 70% accurate, the initial difference between the two distributions is less than when the sensor is 95% accurate. The reason for this is because the initial equiprobable binomial distribution is not too dissimilar to the true model as the true model has greater variance due to the sensing errors. For this reason when the accuracy of the sensor is switched to 95% the initial divergence is much higher as distributions differ by greater amounts. However, with the greater accuracy in sensing the learner is capable of better approximations of the true model, improving upon the 70% curve after approximately 40 days. It's clear that in both cases only a small amount of experience is required to make a good approximation of the underlying model. After 20 days the learner has built up a good predictive model of occupancy in both cases.

The last curve plots the effect of model shift on the approach. How a learning approach recovers from a shift in its underlying model is an important feature of online learning methods. Offline methods usually have to be retrained once such an event occurs but online methods should show adaptability to this type of behaviour. Model shifts are always challenging from a learning perspective because the agent has already significant past experience pointing to something which is no longer valid and how it adjusts its estimates determines its efficacy in the domain. If the agent simply disregards all the previous estimates in favour of the most recent, a temporary change could easily skew the predictive power of your solution.

After 70 days the occupancy behaviour of the user changes causing a jump in divergence. The key thing here is to note the recovery, i.e. within the space of 50 days the learner has returned to making good approximations of this new underlying model. The approach generally demonstrates good approximations without any prior knowledge, however the accuracy of the solution will always be constrained by the quality of the sensing devices, through which predictions are attained.

The output of occupancy prediction is a multinomial distribution governing the probability of occupancy for specific times over the course of a given day. This distribution is utilised by the RL agent (Q-learning) to control heating and cooling where the distribution is sampled in order to approximate a value for "time to occupancy".

5.2   Autonomous Thermostat Control via Q-learning

Simulation Environment. In order to ensure the repeatability of our experiments we simulate the heating of a room by defining a heat transfer rate, an input heating rate and calculating the temperature changes for each time step. Equation 6 describes the calculation of the heat transfer rate in Watts,

$$Heat_{transfer} = uValue \times surfaceArea \times (rt - ot) \tag{6}$$

where $rt - ot$ is the difference in temperature between the internal temperature and the external outside temperature. The u-value[2] is given in units of W/m^2K. By dividing the thickness in (m) of the materials (plaster, slab, screed, etc) by their manufacturer stated resistivity values one can compute an approximate u-value for the building/room. It is generally given as $1/totalResistance$. By measuring the total surface area of the room (m^2) one can work out the heat transfer rate i.e. the amount of heat energy in Watts leaving the room at any given moment.

To model the effects with respect to temperature changes we simulate using the following configuration. The specific heat of air is the amount of energy (Joules) required to raise the temperature of $1Kg$ of air by $1°K$ and works out to be approximately $718J/KgK$ given atmospheric pressure of $1\ atm$ and air density of $1.3Kg/m^3$. For simulation purposes we assume a resistive heater is heating the room and it's 100% efficient, meaning that if it's rated 1kW it is outputting 1kW of heat energy into the room.

We modelled the effects of heating on a perfectly uniform cubed shaped room which has a surface area of $54m^2$. We assume the ceiling, walls and floor are insulated with each having u-values of 0.4, 0.6, 0.5 respectively. If the outside temperature ot at time t is $10°C$ and the inside room temperature rt is $20°C$, then the temperature difference between inside and out is $10°C$. Using equation 6 one can compute the heat transfer for each component i.e. the heat escaping through the ceiling would be given by $0.4 \times 9 \times 10 = 36W$. Obviously rooms are often not entirely uniform but for simulation purposes it's a reasonable assumption. By aggregating the heat transfer of each component (ceiling, walls, floor) at time t we can determine the total heat transfer in Watts. We then subtract this value from the heat input to determine the net heat gain into the room. Say the simulated room has a heat transfer rate of $300W$, then 1 minute of

---

[2] Can also be known as the r-value in some countries.

heating by a $1kW$ heater into this room would result in a temperature increase of $(700 \times 60)/718/(1.3 \times 27) = 1.67°C$, where $27m^3$ is the room volume. This approach allows us to model the temperature changes in a repeatable and reproducible manner. Whilst we do not observe all room parameters such as the heat generated by individuals occupying the room or windows/doors being left open, the state space is sufficiently informative enough to ensure a good measure of control is possible.

In order to get a measure of the outside temperature ot for our simulations, we utilised data supplied from the weather station situated at the National University of Ireland, Galway. The University provided us with five months of environmental data dating from 1 January 2013 to 31 May 2013. The data was sampled every minute and consists of temperature, humidity, wind speed and atmospheric pressure. For experimental purposes we focus solely on room heating as the temperatures within Ireland are relatively moderate and cooling systems are generally not required in many environments such as the home. However from both a learning and control viewpoint the same principles will still apply.

The goal of this research is to produce a control solution which can effectively combine both occupant comfort with energy cost savings. For comparative purposes we focus on comparing the costs for the "Always On" and "Programmable Control" methods, ignoring the "Manual Control" method as it's not a realistic comparison with our proposed solution as from a cost perspective it cannot be optimised any further. From a comfort analysis we focus on comparing the affect of different learning rates on Q-learning and show how occupant comfort can be improved by adjusting the configuration settings on the learning approach.

Online Q-learning vs HVAC "Always On". Figure 2 plots a comparison between an online Q-learning approach and an "Always on" solution. Online learning with respect to Q-learning means that the agent is arbitrarily initialised in the beginning and has no prior knowledge of the domain. The "Always on" method means that the HVAC system operates 24/7. Many users operate their HVAC systems in this way, as they often cannot understand how to program their thermostat properly or if they are sick/elderly. Figure 2 plots the monetary costs of both solutions over the period from 1 January 2013 to 31 May 2013. From a cost perspective, it's clear that the online Q-learning method combined with occupancy prediction is capable of operating the heating of the room at more than half of what the "Always on" solution costs. The total costs for heating the room for the period under consideration were €152.55 vs €344.15. The results demonstrate the significant savings that are achievable using adaptive control via Q-learning and occupancy prediction when compared to "Always on".

However one of the significant advantages of constantly running your HVAC system is that you can always be sure of the comfort of the environment where the set point temperature is always maintained. The goal of the learning solution proposed by this paper is achieve significant cost reduction whilst concurrently optimising the comfort levels of the end user, so we need to make sure that this occurs.

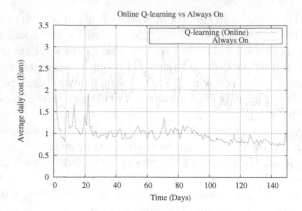

Fig. 2. Online Q-learning vs "Always on" control

Offline Q-learning Comfort Analysis. Offline learning involves an agent learning a good initial policy through simulation (offline) which can then be used when operating in the real world (online). It is commonly used to improve results over solely learning online. Since we interact with a simulator for our results we can utilise this method to demonstrate occupant comfort however online methods will still work, just more slowly.

Figure 3 plots the average amount of time in minutes when the temperature conditions were outside the setpoint temperature of $23°C$ with a threshold setting of plus or minus $1°C$. The results are carried out over multiple consecutive learning trials where the agent has no knowledge in the beginning, but carries forward its knowledge between trials. The graph considers two separate learning rates α with values of 0.1 and 0.5. The learning rate determines the amount by which the reinforcement learning agent backs up its value function estimates considering the new information presented to it. The higher the learning rate the shorter the amount of time it takes to learn a good policy, however too high a learning rate can lead to suboptimal policies where the approach takes too big of a step to correct the observed error in the estimate.

It's clear that after only a short number of learning trials the amount of time the comfort settings are not optimal has reduced to less than 40 minutes over the course of an entire day. Our results show that of these 40 minutes, 83% of these occur when the temperature is within $1°C$ of the threshold parameter. This means that whilst the environment is not optimal, the occupants would only experience mild discomfort. As the number of learning trials progresses this time reduces further. If the policy eventually turns to a completely greedy strategy over time then this should in theory drop to 0.

Offline Q-learning vs HVAC "Programmable Control". Continuing on from the previous section next we analyse the benefits of offline learning via simulation compared to a programmed schedule for operating the HVAC system. A

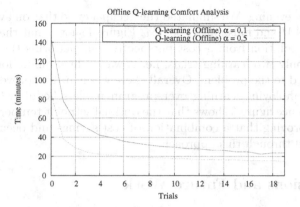

Fig. 3. Length of time in minutes when the setpoint temperature is not achieved but the room is occupied

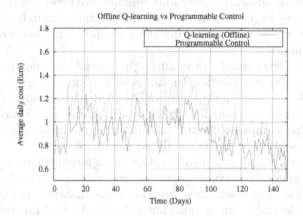

Fig. 4. Q-learning vs "Programmable Control"

number of other approaches could be utilised instead of offline training, i.e. function approximation would allow for generalisation over states and actions not yet visited over ones that have been if one were not able to perform offline training. In addition, parallel learning methods have also been proposed to achieve same where multiple independent thermostats could communicate in parallel in order to learn good policies. If one can simulate the environment, offline learning is a common technique where one can avoid the initial poor performance by yielding a good initial policy.

Figure 4 details the performance of offline learning against the programmed schedule. The schedule was designed by the facilities manager in the Schneider Electric Galway offices in accordance with how the building is currently operated. In the building, the HVAC systems are turned on at $7AM$ in the morning and go off at $8PM$ that evening. We simulated the occupancy so that on average, people begin at $8:30AM$ and finish at $6:30PM$. Figure 4 shows that the performance of the learning solution from a cost perspective out performs the programmed schedule with only two learning trials, i.e. it was trained offline for a single run and then applied to the problem. Overall there was a 10% improvement in costs as a result of employing learning over programmable schedules. Given enough learning experience figure 3 shows that the optimal setpoint temperatures can be achieved also proving that a combination of cost savings and occupant comfort can be achieved through this approach.

6   Conclusions and Future Work

This paper has demonstrated a reinforcement learning method combined with occupancy prediction capable of optimising the heating and cooling of a space autonomously with no prior information. Due to the limitations of our data set, our results focussed on heating only and demonstrated cost savings against two common strategies for controlling HVAC. In addition through offline learning via a simulator we demonstrated improved comfort and cost savings for the approaches in question.

In summary, if one carefully programs a thermostat and one's occupancy pattern is pretty regular, it's questionable how much energy savings can be achieved by a device such as a learning thermostat. The strategy is already optimal from a cost perspective. Thus we compared the approach against an "Always on" control method and "Programmable Control" method demonstrating cost reductions of 55% and 10% respectively in our simulated environments.

For future work the proposed state action space formalism could be extended further to give greater observation over the environment. In addition, methods from supervised learning such linear function approximation could be applied to generalise over the states and actions not yet visited based on those that have, reducing the time it takes to converge an optimal policy.

References

1. Fuzzy logic inference temp. controller for air conditioner, September 1, 1999. https://www.google.fr/patents/CN2336254Y?cl=en. cN Patent 2,336,254
2. Honeywell evohome, January 01, 2015. http://evohome.honeywell.com/
3. Nest thermostat, January 01, 2015. https://nest.com/thermostat/life-with-nest-thermostat
4. Ahmed, O.: Method and apparatus for determining a thermal setpoint in a hvac system, November 9, 2004. https://www.google.fr/patents/CA2289237C?cl=en. cA Patent 2,289,237
5. Barrett, E., Duggan, J., Howley, E.: A parallel framework for bayesian reinforcement learning. Connection Science **26**(1), 7–23 (2014)
6. Barrett, E., Howley, E., Duggan, J.: A learning architecture for scheduling workflow applications in the cloud. In: 2011 Ninth IEEE European Conference on Web Services (ECOWS), pp. 83–90. IEEE (2011)

7. Barrett, E., Howley, E., Duggan, J.: Applying reinforcement learning towards automating resource allocation and application scalability in the cloud. Concurrency and Computation: Practice and Experience (2012)
8. Choi, S., Yeung, D.Y.: Predictive q-routing: a memory-based reinforcement learning approach to adaptive tra c control. In: Advances in Neural Information Processing Systems 8, pp. 945–951 (1996)
9. Dage, G., Davis, L., Matteson, R., Sieja, T.: Method and system for controlling an automotive hvac system, July 22, 1998. https://www.google.fr/patents/EP0706682B1?cl=en. eP Patent 0,706,682
10. Dorigo, M., Gambardella, L.: Ant-q: a reinforcement learning approach to the traveling salesman problem. In: Proceedings of ML-95, Twelfth Intern. Conf. on Machine Learning, pp. 252–260 (2014)
11. Doshi, P., Goodwin, R., Akkiraju, R., Verma, K.: Dynamic workflow composition using markov decision processes. International Journal of Web Services Research **2**, 1–17 (2005)
12. Dutreilh, X., Kirgizov, S., Melekhova, O., Malenfant, J., Rivierre, N., Truck, I.: Using reinforcement learning for autonomic resource allocation in clouds: towards a fully automated workflow. In: The Seventh International Conference on Autonomic and Autonomous Systems, ICAS 2011, pp. 67–74 (2011)
13. Fadell, A., Rogers, M., Satterthwaite, E., Smith, I., Warren, D., Palmer, J., Honjo, S., Erickson, G., Dutra, J., Fiennes, H.: User-friendly, network connected learning thermostat and related systems and methods, July 4, 2013. https://www.google.fr/patents/US20130173064. uS Patent App. 13/656,189
14. Grzes, M., Kudenko, D.: Learning shaping rewards in model-based reinforcement learning. In: Proc. AAMAS 2009 Workshop on Adaptive Learning Agents, vol. 115 (2009)
15. Karray, F.O., De Silva, C.W.: Soft computing and intelligent systems design: theory, tools, and applications. Pearson Education (2004)
16. Nau, D., Ghallab, M., Traverso, P.: Automated Planning: Theory & Practice. Morgan Kaufmann Publishers Inc., San Francisco (2004)
17. Russell, S., Norvig, P., Canny, J., Malik, J., Edwards, D.: Artificial intelligence: a modern approach, vol. 2. Prentice hall Englewood Cliffs, NJ (1995)
18. Scott, J., Bernheim Brush, A., Krumm, J., Meyers, B., Hazas, M., Hodges, S., Villar, N.: Preheat: controlling home heating using occupancy prediction. In: Proceedings of the 13th International Conference on Ubiquitous Computing, pp. 281–290. ACM (2011)
19. Spiegelhalter, D.J., Dawid, A.P., Lauritzen, S.L., Cowell, R.G.: Bayesian analysis in expert systems. Statistical science, 219–247 (1993)
20. Strens, M.: A bayesian framework for reinforcement learning, pp. 943–950 (2000)
21. Tesauro, G.: Temporal difference learning and td-gammon. Communications of the ACM **38**(3), 58–68 (1995)
22. Tesauro, G., Kephart, J.O.: Pricing in agent economies using multi-agent q-learning. Autonomous Agents and Multi-Agent Systems **5**(3), 289–304 (2002)
23. Watkins, C.: Learning from Delayed Rewards. Ph.D. thesis, University of Cambridge, England (1989)
24. Wiering, M.: Multi-agent reinforcement learning for traffic light control. In: ICML, pp. 1151–1158 (2000)

Clustering by Intent: A Semi-Supervised Method to Discover Relevant Clusters Incrementally

George Forman[1]($\boxtimes$), Hila Nachlieli[2], and Renato Keshet[2]

[1] Hewlett-Packard Labs, Palo Alto, USA
george.forman@hp.com
[2] Hewlett-Packard Labs, Haifa, Israel

Abstract. Our business users have often been frustrated with clustering results that do not suit their purpose; when trying to discover clusters of product complaints, the algorithm may return clusters of product models instead. The fundamental issue is that complex text data can be clustered in many different ways, and, really, it is optimistic to expect relevant clusters from an *unsupervised* process, even with parameter tinkering.

We studied this problem in an interactive context and developed an effective solution that re-casts the problem formulation, radically different from traditional or semi-supervised clustering. Given training labels of some known classes, our method incrementally proposes complementary clusters. In tests on various business datasets, we consistently get relevant results and at interactive time scales. This paper describes the method and demonstrates its superior ability using publicly available datasets. For automated evaluation, we devised a unique cluster evaluation framework to match the business user's utility.

Keywords: Semi-supervised clustering · Class discovery · Topic detection

1 Introduction

Hewlett-Packard uses text mining techniques to help analyze customer surveys, customer support logs, engineer repair notes, system logs, etc. [11] Though clustering technologies are employed to discover important topics in the data, usually only a small fraction of the proposed clusters are relevant. This is expected by data mining practitioners, but can prove somewhat disappointing to business users. The fundamental issue is that such complex text data can be clustered in many different ways, and it is unlikely that an *unsupervised* algorithm stumbles upon the one that suits the user's current intent. We have often found they still fail to produce useful clusters even with repeated attempts at adjusting the various parameters by data mining experts.

Furthermore, once some initial large clusters are recognized and dealt with, the remaining data tends to produce decreasingly useful clusters. In fact, sometimes the removal of the known issues causes a shift to less relevant breakdowns

© Springer International Publishing Switzerland 2015
A. Bifet et al. (Eds.): ECML PKDD 2015, Part III, LNAI 9286, pp. 20–36, 2015.
DOI: 10.1007/978-3-319-23461-8_2

of the data, e.g., by setting aside some clusters of known laptop issues (old batteries or cracked displays), the remaining data may be more likely to cluster by product type or geography—frustrating the intent of the user.

One may think that semi-supervised clustering algorithms would provide the answer [2], but they do not. We explored using constrained clustering, a form of semi-supervised learning with must-link and cannot-link constraints [3,26], but we found its results mostly useless for our purposes (see Tables 1 and 2). Additionally, we considered constrained non-negative matrix factorization (CNMF) methods [8,18]. We tested three implementations, but found both their speed and their results unacceptable. (See the experiments in Section 3.) Fundamentally, most semi-supervised techniques are designed to improve classification, but instead we seek improved discovery of clusters by leveraging the known categories as partial supervision.

Besides the troublingly poor results, we find that clustering solutions tend to be slow.[1] We tried the research software of a half-dozen different publications that claimed to be 'fast'—such as for clustering web search results instantly as they are displayed—but none of them approached the speed needed for interactive use on our text datasets with tens or hundreds of thousands of rows. Research in semi-supervised clustering that involves pairwise constraints typically considers up to thousands of constraints. But once several hundred cases have been labeled for each of a dozen known categories, we end up with millions of pairwise constraints—not very scalable for interactive response times. Also, since clustering into too few clusters will mix different topics together, for our complex data we need to generate many clusters, resulting in linear slowdown for most algorithms. It is a poor interaction: the user waits and waits for the results, then hundreds of clusters appear for the user to examine one by one, the fixed results oblivious to the judgments the user makes as they peruse the voluminous output.

Clustering By Intent (CBI): By examining the practical needs of our interactive users, we reformulated the semi-supervised clustering problem as a substantially transformed data mining task with a distinct yet familiar character, which we shall call *Clustering By Intent*: As the user incrementally explores the dataset, they maintain a growing set of discovered, approved classes, each associated with labeled training cases (typically tens to thousands).[2] The user iteratively requests a cluster, which should be incrementally generated on demand with quick response time. The user may (a) reject the cluster (either being irrelevant or perhaps too impure), (b) accept it as a new class, or (c) merge it into

---

[1] Witness the large number of clustering publications with *fast*, *efficient*, or *scalable* in their titles, attesting to the problem.

[2] **Terminology:** Let us say that the underlying domain data consists of a set of generally non-overlapping ground-truth *topics* with respect to the user's current intent, e.g., different failure modes, or else product types, or else geographies—not all perspectives mixed together at once. The algorithm strives to return a *cluster* (list of cases) with high *purity*—the precision of the cluster with respect to the cluster's main topic (the most common topic among its cases). The user creates a *class* corresponding to one or more ground-truth topics of interest.

Table 1. Illustrative Comparison of Methods, Clustering By Sport: semi-supervised clustering of 28,166 Reuters news sports headlines, where the supervision given is a single class containing 986 headlines having the word 'baseball.' We show the first 18 outputs of each method, marking (✗) those that are repeats or not relevant. For Constrained K-Means, we report largest clusters first, showing the most distinguishing word of each cluster of documents. For CBI, we limited it to one word per cluster, but the method is more general.

Clustering By Intent	Constrained K-Means [26]
soccer	soccer
cricket	cricket
tennis	✗uk
rugby	✗africa
golf	✗first
racing	union (rugby)
skiing	tennis
athletics	racing
basketball	nhl (hockey)
hockey	✗tennis
cycling	✗spain
boxing	✗france
nfl (football)	✗sri (lanka)
swimming	golf
olympics	✗uk
rallying	✗cup
skating	skiing
motorcycling	athletics

an existing class. The algorithm should be responsive to previous user actions, including the most recent supervision.

A couple more points are in order. First, the *purity* of the returned cluster matters greatly to the domain expert. It is easier to recognize a topic if the cluster has high purity, ideally just a single topic. For our typical, complex text domain data, determining the meaning and worth of a proposed cluster can take the user awhile examining its cases. Thus, it is best to provide a manageable list of cases that are most typical or central to the cluster, rather than return a much larger set of cases that may include some other topics mixed in.[3]

Second, the *size* of the cluster topic matters to the user. Although the cluster may be described by a small list of cases, the underlying topic that it informs the user about may be large. We usually encounter complex datasets that have a long-tailed distribution of topic sizes. Users ordinarily prefer to discover the larger topics first, ideally working down the tail in order.[4]

For example, in the application of problem management one wants to discover the most common customer problems in order to address them first or with more

---

[3] It is useful to provide a symbolic description of the cluster as well, such as which query terms form the cluster or which keywords are most associated.

[4] In some business datasets, we have different priority considerations, but for the scope of this paper, we will use the number of cases in the underlying topic.

Table 2. Clustering the Same Data by Country Instead: Same as Table 1, but here the supervision given to the competing algorithms is a single class containing the 5401 headlines with the word 'UK.'

Clustering By Intent	Constrained K-Means
usa	✗soccer
france	✗division
south (africa)	✗tennis
spain	africa
germany	✗cup
italy	✗tennis
netherlands	✗nhl
zealand	✗union
switzerland	spain
republic (of china)	france
greece	✗standings
portugal	✗baseball
japan	✗cricket
canada	✗cricket
austria	✗golf
australian	✗alpine
indies	✗athletics
belgium	✗basketball

resources. The CBI task fits squarely with this application. Typically once many topics have been discovered, the user would ideally follow it with a period of active learning to expand the training set of the recently defined classes, and finish with a process of machine learning *quantification* [10] to estimate the true size or cost of each class.

The goal of the process is to gain insights from the dataset, and at no stage do we expect to achieve full dataset clustering, as real-world datasets are often not fully clusterable. We do, however, assume that the intent of the user is consistent and does not change viewpoint during the process.

Of course, the user may enact a separate analysis on the same dataset with a different perspective. We illustrate this briefly to show the major benefit of clustering *by intent*. Given a dataset of 28,166 news headlines about sports (from RCV1 [16]), we provide the supervision of a single class of 986 headlines containing the word 'baseball.' With no background knowledge or stopword lists, our CBI method (explained in section 2) iteratively generates the cluster queries shown on the left in Table 1, while the results on the right are generated by the well-known semi-supervised clustering method Constrained K-Means [26] using normalized cosine-similarity as its measure.[5] Alternatively, if the user instead gives the supervision of a single class of the 5401 headlines containing 'UK,' then we get the results in Table 2. The contrast in the CBI outputs for the

---

[5] We removed stopwords in order to assist Constrained K-Means, at the request of the reviewers; the results are substantially unimproved. (We avoided removing the common stopword 'us' to avoid masking the country 'US'. The ideal algorithm should not need tailored stopword lists in order to find the meaningful terms.)

two intents is night and day, whereas the contrast between the two sets of Constrained K-Means results is weak, and not apparently aligned in any meaningful way to the user's supervision.

The contributions of this paper include: (a) Distinguishing the *Clustering By Intent* data mining task—a new kind of semi-supervised learning. The supervision is given on the known classes and the goal is to discover large unknown topics that are relevant to the user's intent. (b) Detailing how CBI is different from the many recognized data mining tasks—Section 4. (c) Offering a specific CBI algorithm that excels for text domain datasets—Section 2. (d) Illustrating the effectiveness and directability of the method on an intuitive example dataset. (e) Providing a method of automated evaluation for this interactive task without a person in the loop—Section 3.1. (f) Using this method to quantitatively evaluate the algorithm and comparing it with other methods across a gamut of conditions drawn from six publicly available datasets—Section 3. (g) Identifying promising leads for future work—Section 5.

2   CBI Methods

The input to any *Clustering By Intent* method is a typical K-class training set T, plus a set of unlabeled examples U. Not only should one expect that U contains undiscovered classes, but also that some of these unlabeled examples belong to the K known classes. In practical business use, this is particularly the case for periodic analyses where additional unlabeled examples have accumulated for all classes (known and unknown) since the training set was previously developed. Notice that emerging, epidemic topics might have appeared in U since the previous analysis.

The output is an abstract sequence of clusters $C_{0,1,...}$ pulled by the user on demand. A cluster consists of a list of cases of U, and, optionally, a query or description of the topic being proposed. The user may volunteer or implicitly generate feedback on the disposition of any cluster to improve ensuing results. As soon as the training set T is changed, the algorithm retrains.

CBI: We begin by describing one of our leading CBI methods which is appropriate for sparse datasets, such as those that consist of text features in the common *bag of words* representation. (Note that categorical data fields can easily be represented as sparse key=value binary features.) We begin by training a base multi-class classifier that returns the confidence measure for each of its predictions: the margin between the highest scoring class and the runner-up class.[6] We select the low confidence cases of U as our 'residual' set R. The purpose is to avoid cases that are likely to belong to known classes.

Next, when a cluster is demanded, we select cases of the residual R according to the algorithm in Table 3, which also returns a descriptive query. After each cluster is returned, we remove its cases and query terms from R. The

---

[6] In the rare and short-lived circumstance when only a single class is known, a one-class classifier or a Positive-Unlabeled learner would be appropriate [17].

Table 3. CBI cluster & query construction algorithm

1: **INPUT:** training set T, residual set R from classifier, target cluster size
2: selected cases C = residual set of cases R
3: query = empty list
4: **loop**
5:    term = highest scoring term wrt C and T (see text)
6:    C' = cases of C containing term
7:    **if** $|C'| <$ target cluster size **then**
8:       **RETURN:** C, query
9:    **end if**
10:   C = C'
11:   query += term
12: **end loop**

algorithm iteratively appends terms to a conjunctive query until the resulting set would be below our target cluster size (25). At each iteration, we select the term with highest divergence with respect to C and the training set T. Here we have experimented with a variety of functions, including information gain, chi-squared, bi-normal separation, etc., with some variation. For this paper, we simply use the precision of the term in separating C from false positives matches in T, with a minimum floor of false positives (50).

We have tested various methods for selecting the residual. The experiments section shows results using three separate classifiers: Multinomial Naive Bayes (NB), Regularized Least Squares (RLS), and, as an upper-bound comparison, an oracle classifier (Oracle), which selects all the cases of the unknown topics.

KMeans: For each of these three classifiers independently, we also run the residual through Mini-Batch K-Means (K-Means++ initialization, batch size 400, K=10) [24], returning the clusters largest first. This represents a commonplace workflow: as one recognizes and removes known cases, he or she clusters the remainder to see what else can be found. We also try clustering the entire dataset, which may excel if the data has a propensity to cluster according to the hidden topics.

CNMF: Finally, we tried three different implementations of semi-supervised clustering via Constrained Non-negative Matrix Factorization [8,18]. The experiments show only the best of these.

After illustrating the weak results of Constrained K-Means [3,26] in the introduction and having faced its scalability problems on our larger datasets, we exclude it from the experiments. We have tried a panoply of ideas, but there is space to show only some representatives. Testing other ideas is left as an exercise for the reader. Our implementation leverages the *scikit-learn* package [22].

3   Experiments

There would be no laws and no cricket [without] substantial agreement
about what sort of thing cricket ought to be—if, for example, one party
thought of it as a species of steeplechase, while another considered it to
be something in the nature of a ritual dance...   -Dorothy Sayers

Clustering studies ordinarily measure the effectiveness of a method by how
well its clusters align with hidden ground-truth class labels in a benchmark
dataset, such as by the average purity of its generated clusters. This is philosoph-
ically problematic where one dataset may have multiple perspectives of hidden
class labels, such as by sports or by countries in our illustration.s[7] Against which
standard should an unsupervised clustering output be judged? Given the hidden
standard, it makes more sense to grade semi-supervised algorithms, where some
ground-truth labels are revealed to impart the desired breakdown.

Existing studies evaluate the set of produced clusters in entirety. What could
be wrong with that? It has long been recognized in information retrieval research
that it is useful for the objective function to mirror the practical point that a
user will need to review the results sequentially. They care much more about
the first results than the latter. For this reason, CBI changes the perspective to
judging results sequentially. The algorithm must produce a sequence of clusters,
not an (unordered) set as traditional methods. Within a single cluster, the user
will typically make their judgment about the proposal after reviewing a limited
number of cases. Finally, once a topic has been discovered by the user, no credit
is awarded for providing additional clusters on the same topic.

Research studies in semi-supervised clustering select training examples at
random. Their goal has been to see how much better the clustering results would
be if the user would provide just a bit of guidance, preferably applied uniformly.
But in our intended use case, our goal is to discover new clusters that are relevant
to the current intent. Thus, in CBI the training labels should be drawn from a
limited set of classes and credit should not be awarded for returning clusters
about the known classes. Furthermore, although we appreciate that obtaining
labeled data can be costly in practice, there should be no assumption on the
part of the method that the number of labeled cases will be small (it can be easy
enough to gather many similar training cases in some domains using a simple
binary classifier).

For these reasons and others, we developed a new experiment protocol suited
for evaluating CBI tasks. In fact, an important part of the work was to establish
an evaluation framework and a credible performance objective measurement in
order to then develop and test a wide variety of ideas.

3.1   Experiment Protocol

Our experiment protocol for evaluating CBI methods is shown in Table 4. Note
that, since popular topics are easy to discover, we assume the K known classes

---

[7] For a real, publicly available example, the MULAN dataset repository [25] has a
EUR-Lex dataset that has multiple distinct perspectives of labels [19].

Table 4. CBI Experiment Protocol

```
1:  for all benchmark dataset D with each case labeled with a ground-truth topic do
2:     for # of known classes K = 2, ..., 10 (taking largest topics first) do
3:        for labeled fraction P = 25%, ..., 100% do
4:           for all 100 random seeds do
5:              labeled = randomly select P% of each of the K known classes
6:              unlabeled = all unselected cases including all unknown dataset topics
7:              for all method M do
8:                 Train M on (labeled,unlabeled)
9:                 Get the first two cluster outputs of M on the unlabeled data
10:                Return the better score of the two clusters, scoring each cluster C as:
11:                score = purity(C.mode)² × topic_size(C.mode) / topic_size(largest)
12:             end for
13:          end for
14:       end for
15:    end for
16: end for
```

should always correspond to the K largest topics in the dataset, which is often so skewed that one can hardly fail to notice the first couple (see last column of Table 5). Note that the random sampling of labels is a only within each of the K known classes; no labels or class definitions are provided about the remainder of the dataset, not even the number of classes that might be expected. To vary the amount of supervision, one can vary both the number of known classes K as well as the percentage P of each topic's cases that are labeled. In our experiments, we vary one parameter while we hold the other fixed, and vice versa. We use the best score of only the first two clusters output, because we suppose the user is likely to change the training set in some way, and the model would be retrained before producing more outputs.

When it comes to assigning a score to a proposed cluster C, it depends on both size and purity factors. We first determine the cluster's most common represented ground-truth class, C.mode. The *topic size* of the cluster is the number of cases in the ground-truth topic C.mode; note that this is not the size of the cluster C itself. The final score should be directly proportional to the C.mode's topic size, as this is often proportional to real cost. Exceptionally, if C.mode represents a class that is already known in the training set, we give zero credit, in order to align scoring with our purpose of discovering new topics.

The *purity* of a proposed cluster C is evaluated by dividing the number of cases in C.mode by the size of the cluster. We have found that cluster purity matters to the user in a super-linear way: a cluster with, say, 50% purity is less than 50% likely to be understood by the user. Thus, the final score for a cluster

Table 5. Datasets. The last column characterizes the class skew by showing what percent of the dataset falls in the two largest classes.

Dataset	Rows	Features	Classes	K=2
eurlex-codes	16173	5000	20	53%
eurlex-subjects	5418	5000	113	26%
new3	9558	26832	44	13%
fbis	2463	2000	17	36%
re1 (Reuters)	1657	3758	25	42%
wap (web pages)	1560	8460	20	34%

is its purity squared times the topic size of its mode, normalized by the size of the largest unknown topic available to be found (so that finding it achieves 100% credit, rather than have the best possible score shrink as we increase K).

Table 5 shows the six benchmark datasets we use from the text classification domain. The last column characterizes the class skew of each dataset by showing the percent of the rows in the two largest classes. (We verified that the largest classes do not represent a catch-all 'none of the above' class.) The first two datasets are different breakdowns of the EUR-Lex[8] dataset of legal documents: the first by directory codes, the second by subject matter.[9] The remaining four text datasets have been used and described in a variety of other publications (e.g. [10]) and we have provided them in ARFF format at the WEKA dataset repository.[10]

For the first three datasets with >5000 rows, the methods select the residual as the 10% lowest confidence unlabeled cases. But for the three datasets with <2500 rows, 10% returns too few cases to mine. For example, dataset *re1* has 1657 rows and at K=2 already 42% of the dataset is in known classes. So, the true residual is the remaining 961 rows, and that is divided among the 15 remaining classes to be discovered. Selecting just 10% residual at P=75% yields fewer than 150 cases (distributed among all 17 classes)—not enough data. Thus, for the three small datasets, the methods use a threshold of 50%. (Our non-public business datasets usually have tens or hundreds of thousands of rows.)

Figure 1 shows (left) the head of the class distribution for each dataset, and (right) the classifyability of each respective class, as characterized by the F-measure of a NB classifier trying to discriminate that class vs. all others under 4-fold cross-validation. Whereas it is common knowledge that more training examples generally improve classification accuracy for a given class, clearly the difficulty of each individual class can have a larger effect.

---

[8] http://mulan.sourceforge.net/datasets-mlc.html

[9] In order to fit our experiment harness and reuse classification libraries equipped only for single-label datasets, we simply discarded any rows that actually had multiple labels (thus the number of cases differs).

[10] http://www.cs.waikato.ac.nz/ml/weka/datasets.html

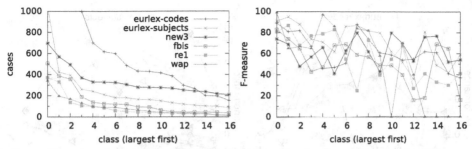

Fig. 1. Dataset characteristics. (left) Class size distribution, shown for the 17 largest classes of each dataset. (right) Classifiability of each class, respectively, as characterized by F-measure of one-vs-all classification by NB classifier under 4-fold cross-validation.

3.2   Results

We begin by comparing the average performance objective (§3.1) of the various methods as we vary the number of known classes K. We hold the percentage of training for each known class at P=75%; for K=2, this results in a supervision level of 10–40% of the dataset. Increasing K provides more training supervision which might benefit the classifier's accuracy, but concomitantly increasing K removes the largest classes from the remaining topics to be found, making the task harder. For different classes, the inherent classification difficulty and clusterability varies, of course. (Refer to Figure 1.) Thus, we expect substantial variation as we change K, not a diagonal trend.

Figure 2 shows the results. We see the semi-supervised method CNMF consistently gave poor results for this task (and this is the best of three implementations). Generally we see CBI methods (black) exceeded KMeans methods (blue), sometimes with the exception of KMeans in combination with the Oracle classifier, which is unachievable in practice. The Oracle classifier (×) did not always lead to the best performance. The classifiers may sometimes do a better job of isolating an interesting and cohesive subset of cases from which to discover a topic. Each of the two classifiers showed many situations where it substantially exceeded the other. In practice one can use cross-validation to select the best model for plain classification tasks, but the lack of training labels for the unknown topics would thwart cross-validation for the CBI objective.

To quantitatively summarize the results across the different datasets and values of K, we computed the average rank of each method, as shown in Figure 3. The red bar indicates those that are not statistically significantly different from the best ranked method, CBI-Oracle, according to the Friedman and Nemenyi tests at alpha=0.05, as prescribed by Demšar [9]. If we were to exclude the two Oracle methods for being impossible in practice, the two CBI methods are better than statistically significantly better than all other methods by a statistically significant margin.

Next we vary the percentage P of training labels provided for each known class, while we fix the number of known classes at K=4. Figure 4 shows these results. They have less variation than the previous results, because the known

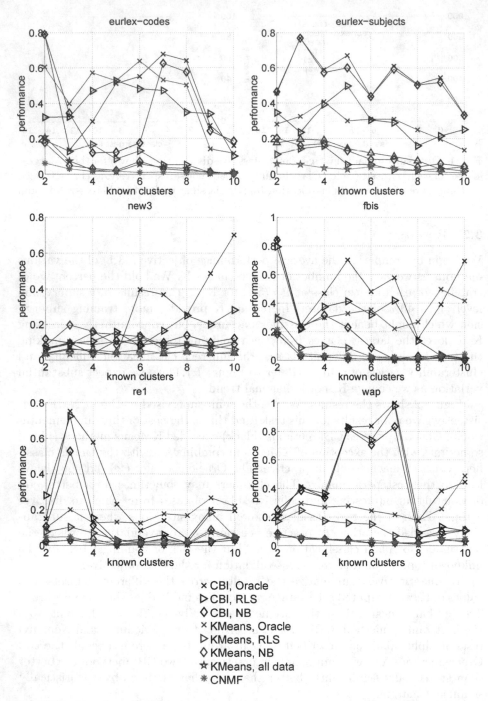

Fig. 2. Performance of all methods on each dataset as we vary the number of known classes K, fixing P=75% for training.

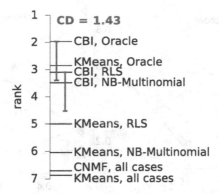

Fig. 3. Average rank and statistically significance differences. The red bar indicates that CBI-RLS is not statistically significantly worse than the oracle, and is significantly better than the state-of-the-art semi-supervised methods.

classes and unknown topics to be discovered are stable across the x-axis in each graph.

In contrast to learning curve studies in classification, it is most striking that increased supervision did not consistently lead to better performance for the CBI objective, though it often helped somewhat (even for CNMF). Labeling more cases removes them from the unlabeled set, reducing the risk for all methods of accidentally proposing a class that is already known. Separately we validated that, as the training level increases, the classifiers showed increasing precision in selecting a residual subset.[11]

Thus, clearly classifier accuracy is not the overriding priority for this task. Case in point, the Oracle returns a perfect residual regardless of training set, yet this does not ultimately lead to the best performance; the size and makeup of the training set affects the CBI method substantially and non-linearly. Increasing supervision benefits CBI-NB for datasets eurlex-subjects, fbis, re1, and wap.

Some of our CNMF results took 5–20 minutes to compute, and RLS classifier training for K=10 took minutes sometimes. This paper focuses on introducing the CBI task and on satisfying the performance objective; not on speed. That said, the CBI process ultimately needs to be put into an interactive loop in applications. For the most part we have not concerned ourselves with fast implementations, but we have prototyped a fast version of a text-based CBI method which clustered 40,000 rows of text data in under 100 ms—clearly suitable for interactive time scales. The CBI algorithm is linear in the number of documents and the number of frequent terms.

---

[11] Note that this is a different objective than their classification accuracy on the known classes, which is of little interest to us in this paper.

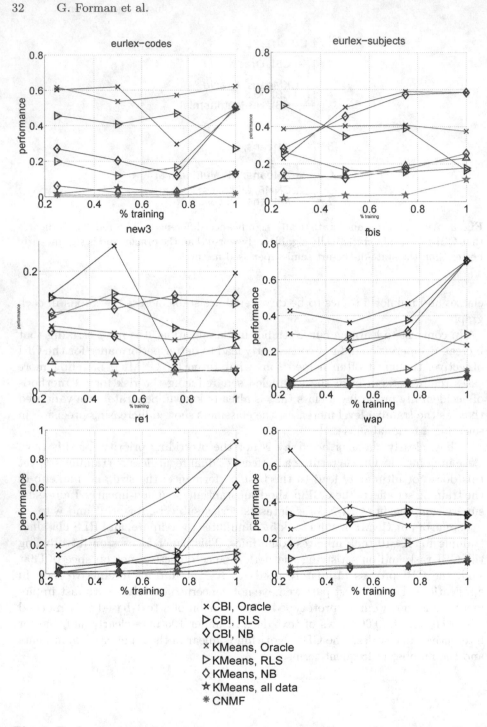

Fig. 4. Performance of all methods on each dataset as we vary the percentage P of training examples randomly selected from the K known classes, fixing K=4.

4   Related Work

The field of *topic detection* has similar goals to CBI: automatically finding new topically related material in streams of data [1]. The streams may be news feeds or Twitter streams [6]. Given their temporal impermanence, the clustering is expected to succeed without supervision, as it does well for Google News. In these domains, news articles are often copied verbatim and so they cluster very neatly. By contrast, CBI targets complex data that cannot be decomposed into user-relevant clusters without some guidance about the user's intent.

Semi-Supervised Learning (SSL) is a broad area, most of which augments labeled training sets with unlabeled data in order to improve the accuracy of classification [7,27]. The minority of research that focuses on semi-supervised *clustering* is described only somewhat differently: augmenting the unlabeled data with some few labels or, more typically, must-link and cannot-link pairwise constraints in order to improve the clustering. Examples include Constrained K-Means [3,26] and Constrained Non-negative Matrix Factorization [8,18]. Even so, researchers measure the clustering quality by its conformance to the (mostly) hidden ground-truth labels, e.g. classification accuracy, so it ends up being much like the SSL research for classification (e.g. [8]). Furthermore, the accepted experiment methodology uses random sampling to select the supervision. This leads to a high likelihood of covering most large classes, which have the greatest effect on the performance measurement. Thus, supervision is given for all the clusters that should be found, but in CBI we only have supervision for the clusters we already know about and not on the unknown clusters that need to be discovered.

Thus, none of this work is like our CBI framework, which might be said to use extremely 'skewed' supervision and does not hinge upon overall classification accuracy. Under CBI, the clustering algorithm gets no credit for returning clusters on classes that have training examples—this is the domain of classifiers (which can probably perform the job better). As we have seen, if the supervision is focused on only a few known classes, it does not seem to help CNMF or Constrained K-Means to adopt the perspective of the user's intent.

One-class classification, outlier detection, anomaly detection and *novelty detection* aim to recognize abnormal data points, generally with the assumption that they are rare events and not available in quantity at training time [23]. These methods are considered successful by recognizing individual test cases that are highly unusual; they do not attempt to cluster such cases into meaningful subgroups, as CBI.

Similarly, there are a variety of works in *novel class discovery* that attempt to seek individual cases that may stem from unknown classes [4,13,20]. These methods generally interact with a user via active learning and assume that the user can recognize a novel class when presented with an individual anomalous case. Thus, they evaluate their work as a learning curve showing the number of novel classes encountered over time. By contrast, in CBI there are generally many unknown classes, mostly very small ones in the long tail, and the goal is to find the larger ones. Further, we have the need to present a collection of similar cases to communicate the topic to the user.

The area of *subgroup discovery* sounds entirely appropriate to our goals, but it is actually an unsupervised task that attempts to find rules of interest associated with a feature of a dataset [14,15]. Gamberger et al. even use the title 'expert-guided subgroup discovery', but by this they mean 'the decision of which subgroups will be selected to form the final solution is left to the expert' [12]. *Contrast set mining* and *emerging pattern mining* are variants that seek rules that find significant differences in databases, such as old and new datasets [21]. None of these methods take in the multi-class supervised data of CBI, although perhaps it would be interesting future work to see whether they could be adapted to produce useful results for CBI tasks.

Finally, there is the idea of *meta-clustering*, which takes many different unsupervised clustering results and produces clusters of them which the user can select among [5]. Ideally one could imagine that different meta-clusters would correspond to different user intents. We have not tried it, but without supervision it seems unlikely to produce results nearly as relevant as CBI, even if the user could determine the most appropriate meta-cluster.

As we mentioned earlier, the area of *PU Learning*—learning from positive labeled cases and unlabeled cases, some of which may actually be positives—may be pertinent in the classification stage of CBI at first when the user has only identified a single class. PU Learning addresses binary classification problems, and sometimes considered streams that may have concept drift over time [17], which is ultimately also an issue of concern for the real-world business user.

5 Conclusions and Future Work

We have labored—and made our internal business users to labor—under poor clustering results for years when seeking to discover new clusters relevant to a particular purpose. We found semi-supervised clustering methods intellectually promising, but, unfortunately, we saw little benefit in practice. By stepping back from our assumptions and recasting the task substantially, we have been able to crack a variety of business datasets, and with a natural user-interaction that is quick. This paper elucidates the task, a suitable method, and its evaluation with a novel protocol devised to work easily with any publicly available multi-class benchmark dataset.

We expect future work in this area to compare additional methods, improve on these results, relax assumptions, and remove limitations. In particular, though our focus has been on the text domain, this could be broadened. In order to perform model selection and tuning in practice, future work could develop a form of cross-validation for CBI tasks—a challenge, having no labels available for the long tail. Next, although in this paper the value of discovering a topic was posited to be proportional to its size, in some datasets, we have a cost indicator associated with each case—such as parts and labor costs to resolve each case. In these settings, the total cost of the topic is a more appropriate indicator of value than simply the number of cases in the topic. Cost quantification techniques [10] could also be applied to prioritize probabilistic cluster definitions. Finally, future work may add emerging topics and/or concept drift to the evaluation with methods to handle them.

Lastly, a philosophical remark. While classification excels at the 'more like this' task, and clustering could be said to excel at the 'find various topics' task, CBI provides a qualitatively new capability: 'find topics different from those, yet alike in some important way.' Even so, it has no higher level concept of how those things are alike. For example, in the sports illustration of Table 1, we know these are sports, but there is no runtime representation of this meta-information.

References

1. Allan, J. (eds.): Topic Detection and Tracking, The Information Retrieval Series, vol. 12 Springer (2002)
2. Bair, E.: Semi-supervised clustering methods. Wiley Interdisciplinary Reviews: Computational Statistics **5**(5), 349–361 (2013)
3. Basu, S., Bilenko, M., Mooney, R.J.: A probabilistic framework for semi-supervised clustering. In: KDD 2004, pp. 59–68 (2004)
4. Bouveyron, C.: Adaptive mixture discriminant analysis for supervised learning with unobserved classes. J. Classif. **31**(1), 49–84 (2014)
5. Caruana, R., Elhawary, M., Nguyen, N., Smith, C.: Meta clustering. In: ICDM 2006, pp. 107–118 (2006)
6. Cataldi, M., Di Caro, L., Schifanella, C.: Emerging topic detection on twitter based on temporaland social terms. In: MDMKDD 2010, pp. 4:1–4:10 (2010)
7. Chapelle, O., Schölkopf, B., Zien, A.: Semi-supervised Learning. Adaptive computation and machine learning. MIT Press (2006)
8. Chen, Y., Rege, M., Dong, M., Hua, J.: Non-negative matrix factorization for semi-supervised data clustering. KAIS **17**, 355–379 (2008)
9. Demšar, J.: Statistical comparisons of classifiers over multiple data sets. JMLR **7**, 1–30 (2006)
10. Forman, G.: Quantifying trends accurately despite classifier error and class imbalance. In: KDD 2006, pp. 157–166 (2006)
11. Forman, G., Kirshenbaum, E., Suermondt, J.: Pragmatic text mining: minimizing human effort to quantify many issues in call logs. In: KDD 2006, pp. 852–861 (2006)
12. Gamberger, D., Lavrac, N.: Expert-guided subgroup discovery: Methodology and application. J. AI Research **17**(1), 501–527 (2002)
13. Haines, T.S., Xiang, T.: Active rare class discovery and classification using dirichlet processes. Int. J. Computer Vision **106**(3), 315–331 (2014)
14. Herrera, F., et al.: An overview on subgroup discovery: Foundations and applications. Knowledge and Information Systems **29**(3), 495–525 (2011)
15. Lavrač, N., Kavšek, B., Flach, P., Todorovski, L.: Subgroup discovery with CN2-SD. JMLR **5**, 153–188 (2004)
16. Lewis, D., et al.: RCV1: A new benchmark collection for text categorization research. JMLR **5**, 361–397 (2004)
17. Li, X., Yu, P.S., Liu, B., Ng, S.: Positive unlabeled learning for data stream classification. In: SIAM 2009, pp. 259–270 (2009)
18. Liu, H., Wu, Z.: Non-negative matrix factorization with constraints. In: AAAI 2010, pp. 506–511 (2010)

19. Mencía, E.L., Fürnkranz, J.: Efficient pairwise multilabel classification for large-scale problems in the legal domain. In: ECML/PKDD 2008, pp. 50–65 (2008)

20. Miller, D.J., Browning, J.: A mixture model and em-based algorithm for class discovery, robust classification, and outlier rejection in mixed labeled/unlabeled data sets. IEEE Trans. Pattern Anal. Mach. Intell. **25**(11), 1468–1483 (2003)

21. Novak, P.K., Lavrač, N., Webb, G.I.: Supervised descriptive rule discovery: A unifying survey of contrast set, emerging pattern and subgroup mining. JMLR **10**, 377–403 (2009)

22. Pedregosa, F., et al.: Scikit-learn: Machine learning in Python. JMLR **12**, 2825–2830 (2011)

23. Pimentel, M.A., Clifton, D.A., Clifton, L., Tarassenko, L.: A review of novelty detection. Signal Processing **99**, 215–249 (2014)

24. Sculley, D.: Web-scale K-means clustering. In: WWW 2010, pp. 1177–1178 (2010)

25. Tsoumakas, G., Katakis, I., Vlahavas, I.: Mining Multi-label Data. Data Mining and Knowledge Discovery Handbook

26. Wagstaff, K., Cardie, C., Rogers, S., Schroedl, S.: Constrained K-means clustering with background knowledge. In: ICML 2001, pp. 577–584 (2001)

27. Zhu, X.: Semi-supervised learning literature survey. Technical Report 1530, Computer Sciences, University of Wisconsin-Madison (2005)

Country-Scale Exploratory Analysis of Call Detail Records Through the Lens of Data Grid Models

Romain Guigourès[1], Dominique Gay[2], Marc Boullé[2](✉),
Fabrice Clérot[2], and Fabrice Rossi[3]

[1] Zalando, Berlin, Germany
[2] Orange Labs Lannion, Lannion, France
[3] SAMM EA 4543, Univeristé Paris 1, Paris, France
marc.boulle@orange.com

Abstract. Call Detail Records (CDRs) are data recorded by telecommunications companies, consisting of basic informations related to several dimensions of the calls made through the network: the source, destination, date and time of calls. CDRs data analysis has received much attention in the recent years since it might reveal valuable information about human behavior. It has shown high added value in many application domains like e.g., communities analysis or network planning.

In this paper, we suggest a generic methodology based on data grid models for summarizing information contained in CDRs data. The method is based on a parameter-free estimation of the joint distribution of the variables that describe the calls. We also suggest several well-founded criteria that allows one to browse the summary at various granularities and to explore the summary by means of insightful visualizations. The method handles network graph data, temporal sequence data as well as user mobility data stemming from original CDRs data. We show the relevance of our methodology on real-world CDRs data from Ivory Coast for various case studies, like network planning strategy and yield management pricing strategy.

Keywords: Classification rule · Bayes theory · Minimum description length

1 Introduction

Telco operators' activities generate massive volume of data, mainly from three sources: networks, service platforms and customers data bases. Particularly, the use of mobile phones generates the so called Call Detail Records (CDRs), containing information about end-point antenna stations, date, time and duration of the calls (the content of the calls is excluded). While this data is initially stored for billing purpose, useful information and knowledge (related to human

---

Romain Guigourès was with Orange Labs when this work began.

© Springer International Publishing Switzerland 2015
A. Bifet et al. (Eds.): ECML PKDD 2015, Part III, LNAI 9286, pp. 37–52, 2015.
DOI: 10.1007/978-3-319-23461-8_3

mobility [1,23], social interactions [22] and economic activities) might be derived from the large sets of CDRs collected by the operators.

Recent studies have shown the potential added-value of analyzing such data for several application domains: United Nations Global Pulse [21] sums up some recent research works on how analysis of CDRs can provide valuable information for humanitarian and development purposes, e.g., for disaster response in Haiti, combating H1N1 flu in Mexico, etc. Also, leveraging country-scale sets of CDRs in Ivory Coast, the recent Orange D4D challenge (Data For Development [5]) has given rise to many investigations in several application domains [4] such as health improvement, analysis of economic indicators and population statistics, communities understanding, city and transport planning, tourism and events analysis, emergency, alerting and preventing management, mobile network infrastructure monitoring. Thus, the added-value of analysis of CDRs data does not need to be proved any longer.

Various classical data mining techniques [4] have been applied on CDRs data depending on the features and the task considered: e.g., considering network graphs from (source antenna, destination antenna) data or temporal sequences from (source antenna, date) data appeals for different clustering techniques for summarizing information in the data.

Contribution: in this paper, we suggest an efficient and generic methodology for summarizing CDRs data whatever the features are retained in the analysis. The method is based on data grid models [6], a parameter-free joint distribution estimation technique that simultaneously partitions sets of values taken by each variable describing the data (numerical variables are discretized into intervals while the categories of categorical variables are grouped into clusters). The resulting data grid – that can be seen as a coclustering – constitutes the summary of the data. The method is thus able to summarize various types of data stemming from CDRs: network graph data, temporal sequence data as well as user mobility data. We also suggest several criteria *(i)* to exploit the resulting data grid at various granularities depending on the needs of analysis and *(ii)* to interpret the results through meaningful visualizations. The whole methodology aims at demonstrating strong impacts on two key points on economic strategy: network planning and pricing strategy.

Outline: in the next section, we discuss further recent work related to CDRs and mobile phone trace data analysis as well as data grid models. In Section 3, we summarize the impacts of the various case studies on the economic development strategy related to the specific context of telecommunications in Ivory Coast. A brief description of the CDRs data characteritics is also given. Section 4 recalls the main principles of data grid models and introduces the tools for exploiting the resulting data grid. In section 5, we report the experimental results on the various case studies.

2   Related Work

CDRs data have received much attention in recent years. Famous applications of CDRs data analysis are for the benefit of social good: e.g., in the transportation domain, [2] suggest a system for public transport optimization; in the health domain, e.g., [8] suggest a model for epidemic spread.

Mobile phones may also provide other types of data (e.g., the Nokia Mobile Data Challenge [15]), like applications events, WLAN connection data, etc. For instance, [13] pre-processed phone activities of one million users to obtain information about their approximative temporal location, then mined daily motifs from the spatio-temporal data to infer human activities. Finally, smart phones are or will be equipped with accelerometers and/or gyroscopes providing data about physical activities of users: [16] suggest a complete system of activity recognition based on smartphone accelerometers with potential application to health monitoring.

Research work related to data grid models: We are *not* coclustering data (objects × attributes) like pioneering work of Hartigan [12]. Data grid models are related to the work of Dhillon et al. [7] who have proposed an information-theoretic coclustering approach for two discrete random variables Y_1 and Y_2: the loss in Mutual Information $MI(Y_1, Y_2) - MI(Y_1^M, Y_2^M)$ is minimized to obtain a locally-optimal grid with a user-defined number of clusters for each dimension. This is limited to two variables and requires to choose the number of clusters per variable. Going beyond 2D matrices, recent significant progress has been done in multi-way tensor analysis [14,19]. Dealing with k-adic data, (also known as co-occurrence data, like contingency table), [17] suggest a coclustering method for social network and temporal sequence (with pre-discretization of time). The Information Bottleneck (IB) method [20] stems from another information-theoretic paradigm: given the joint probability $P(X, Y)$, IB aims at grouping X into clusters T in order to both compress X and keep as much information as possible about Y. IB also minimizes a difference in Mutual Information: $MI(T, X) - \beta MI(T, Y)$, where β is a positive Lagrange multiplier. Wang et al. [24] build upon IB and suggest a coclustering method for two categorical variables. Extending IB for more than two categorical variables, Slonim et al. [18] have suggested the agglomerative multivariate IB that allows constructing several interacting systems of clusters simultaneously; the interactions among variables are specified using a Bayesian network structure.

To the best of our knowledge, our summarization approach is the only one to combine the following advantages: it is parameter-free, scalable and can be applied to mixed-type attributes (categorical, numerical, thus multiple types of time dimensions without pre-processing). Therefore, the same generic method can be used to analyze network graph, temporal sequence and mobility data.

3   Impacts on Economic Strategy

Besides the high-level knowledge extracted from country-scale data and confirmed by local sociologists from the University of Bouaké in Ivory Coast, these studies have also a strong impact on future economic development strategy, mainly in two identified branches:

- *Network planning strategy:* In 2014, there are around 20M inhabitants in Ivory Coast and the mobile service penetration rate is $\simeq$ 84% – with a still growing mobile phone market in a context of demographic growth. The analyses of the first two case studies and the resulting map projections (that can be seen as the network of calls available at various granularities, see Sections 5.1 and 5.2) are considered as an additional input for network planning and investment; for instance to help network designer in answering questions about how many and where the next antennas have to be set while preserving the quality of service at a reasonable cost.
- *Yield management pricing strategy:* a part of the pricing policy, called *Bonus Zone*, established in Ivory Coast offers discount prices (from 10% to 90%) to calling users depending on the location and hour of the emitting call. Maps and calendars resulting from the last two case studies on temporal distribution of output calls (see Section 5.3) and on mobility data (see Section 5.4) that are available at various granularities, provide valuable information to economic analysts in order to design optimized spatio-temporal pricing policy in the context of *Bonus Zone*.

Data Description and Studies. The CDRs data under study come from the Orange D4D challenge[1] (Data For Development [5]). We consider several case studies on two anonymized CDRs data sets from Ivory Coast, namely communication data and mobility data:

Case studies on communication data. Communication data consists in 471 millions mobile calls and covers a 5-month period (from 2011, December 1st to 2012, April 28th). The records are described by the four following variables: *emitting antenna* (1214 categorical values); *receiving antenna* (1216 categorical values); *time of call* (with hour precision); *date of call* (from 2011/12/01 to 2012/04/28). From this data set, we consider three subsets for:

1. *Analysis of call network between antennas.* Considering emitting antennas, receiving antennas and the calls made between antennas, the data set can be seen as a directed multigraph where nodes are antennas and links are the calls between antennas.
2. *Analysis of output traffic w.r.t. date of call.* We consider emitting antennas and the number of days for each call from referral to first day of recording. This data set can be considered as a temporal event sequence spanning over the whole observation period, where the time is the number of days passed and the events are the emitting antenna IDs.
3. *Analysis of output traffic w.r.t. week day and hour of call.* We consider emitting antennas, the day of the week (stemming from the date and considered as a numerical variable) and the hour of the day for each call. Here the time dimension is represented by two variables and the data of the whole period are folded up to week day and hour.

---

[1] http://d4d.orange.com/en/home

Case studies on mobility data. Mobility data consists in mobility traces of 50000 users over a 2-week period (from 2012 December 12th to 2012 December 24th), i.e. approximatively 55 millions records. The records are described by the four following variables: *anonymized user ID* (50000 categorical values); *connexion antenna* (1214 categorical values); *time of call* (minute precision); *date fo call* (from 2012/12/12 to 2012/12/24).

From this data, we consider the user trajectories (identified by user ID) inside the network for the following analysis:

1. *Analysis of user mobility w.r.t. week day and hour.* We consider the user ID, antennas, week day and hour. This data set can be considered as a set of spatio-temporal footprints, where each user ID is associated with a sequence of antenna usage over the time dimension. Here again, the time dimension is represented by two variables and the data of the whole period is folded up to week day and hour.

4   Exploratory Analysis through Data Grid Models

Data grid models aim at estimating the joint distribution between K variables of mixed-types (categorical as well as numerical). The main principle is to simultaneously partition the values taken by the variables, into groups/clusters of categories for categorical variables and into intervals for numerical variables. The result is a multidimensional (K-d) data grid whose cells are defined by a part of each partitioned variable value set. Notice that in all rigor, we are working only with partitions of variable value sets. However, to simplify the discussion we will sometime use a slightly incorrect formulation by mentioning a "partition of a variable" and a "partitioned variable".

In order to choose the "best" data grid model M^* (given the data) from the model space $\mathcal{M}$, we use a Bayesian Maximum A Posteriori (MAP) approach. We explore the model space while minimizing a Bayesian criterion, called cost. The cost criterion implements a trade-off between the accuracy and the robustness of the model and is defined as follows:

$$cost(M) = -\log(\underbrace{p(M \mid D)}_{\text{posterior}}) \propto -\log(\underbrace{p(M)}_{\text{prior}} \times \underbrace{p(D \mid M)}_{\text{likelihood}})$$

Thus, the optimal grid M^* is the most probable one (maximum a posteriori) given the data. Due to space limitation, the details about the *cost* criterion and the optimization algorithm (called KHC) are available in appendix of [11]. Hereafter, we focus on the tools for exploiting the grid and their applications on large-scale CDRs data. The key features to keep in mind are: *(i)* KHC is parameter-free, i.e., there is no need for setting the number of clusters/intervals per dimension; *(ii)* KHC provides an effective locally-optimal solution to the data grid model construction efficiently, in sub-quadratic time complexity ($O(N\sqrt{N}\log N)$ where N is the number of data points). Figure 1 illustrates the input data and output results of KHC on an examplary mobility data stemming from CDRs.

IdUser	Date	Time	O-Antenna	D-Antenna
u1	14/07/2014	07:25:02	A1	A25
u1	14/07/2014	07:57:22	A2	A2
u1	14/07/2014	09:05:57	A2	A16
u1	14/07/2014	10:32:16	A2	A25
...	...	...	...	...
u4	11/11/2014	22:52:32	A11	A17

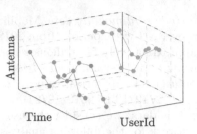

(a) Toy example of CDRs data. Here, mobility data (grey columns): UserId, Time, OriginAntenna.

(b) Resulting 3D data grid with 2 clusters of users, 4 time intervals and 3 clusters of antennas.

Fig. 1. From 3D mobility data, stemming from CDRs data, to data grid.

4.1   Data Grid Exploitation and Visualization

Because of the very large number observations in CDRs data, the optimal grid M^* computed by KHC can be made of hundreds of parts per dimension, i.e., millions of cells, which is difficult to exploit and interpret. To alleviate this issue, we suggest a grid simplification method together with several criteria that allow us to choose the granularity of the grid for further analysis, to rank values in clusters and to gain insights in the data through meaningful visualizations.

Dissimilarity Index and Grid Structure Simplification. We suggest a simplification method of the grid structure that iteratively merge clusters or adjacent intervals – choosing the merge generating the least degradation of the grid quality. To this end, we introduce a dissimilarity index between clusters or intervals which characterize the impact of the merge on the *cost* criterion.

Definition 1 (Dissimilarity index). *Let $c_{.1}$ and $c_{.2}$ be two parts of a variable partition of a grid model M. Let $M_{c_{.1} \cup c_{.2}}$ be the grid after merging $c_{.1}$ and $c_{.2}$. The dissimilarity $\Delta(c_{.1}, c_{.2})$ between the two parts $c_{.1}$ and $c_{.2}$ is defined as the difference of cost before and after the merge:*

$$\Delta(c_{.1}, c_{.2}) = cost(M_{c_{.1} \cup c_{.2}}) - cost(M) \tag{1}$$

When merging clusters that minimize Δ, we obtain the sub-optimal grid M' (with a coarser grain, i.e. simplified) with minimal *cost* degradation, thus with minimal information loss w.r.t. the grid M before merging. Performing the best merges w.r.t. Δ iteratively over the K variables without distinction, starting from M^* until the null model $M_\emptyset$, K agglomerative hierarchies are built and the end-user can stop at the chosen granularity that is necessary for the analysis while controlling either the number of clusters/cells or the information ratio kept in the model. The information ratio of the grid M' is defined as follows:

$$IR(M') = \frac{cost(M') - cost(M_\emptyset)}{cost(M^*) - cost(M_\emptyset)} \tag{2}$$

where $M_\emptyset$ is the null model (the grid with a single cell).

Typicality for Ranking Categorical Values in a Cluster. When the grid is coarsen during the hierarchical agglomerative process, the number of clusters per categorical dimension decreases and the number of values per cluster increases. It could be useful to focus on the most representative values among thousands of values of a cluster. In order to rank values in a cluster, we define the typicality of a value as follows.

Definition 2 (Typical values in a cluster). *For a value v in a cluster c of the partition Y^M of dimension Y given the grid model M, the typicality of v is defined as:*

$$\tau(v,c) = \frac{1}{1 - P_{YM}(c)} \times \sum_{\substack{c_j \in Y^M \\ c_j \neq c}} P_{YM}(c_j)(cost(M|c \setminus v, c_j \cup v) - cost(M)) \quad (3)$$

where $P_{YM}(c)$ is the probability of having a point with a value in cluster c, $c \setminus v$ is the cluster c from which we have removed value v, $c_j \cup v$ is the cluster c_j to which we add value v and $M|c \setminus v, c_j \cup v$ the grid model M after the aforementioned modifications.

Intuitively, the typicality evaluates the average impact in terms of *cost* on the grid model quality of removing a value v from its cluster c and reassigning it to another cluster $c_j \neq c$. Thus, a value v is representative (say typical) of a cluster c if v is "close" to c and "different in average" from other clusters $c_j \neq c$. Notice that this measure does not introduce any numerical encoding of the categories of the categorical variable under study.

Insightful Visualizations with Mutual Information. It is common to visualize 2D coclustering results using 2D frequency matrix or heat map. For KD coclustering, it is useful to visualize the frequency matrix of two variables while selecting a part of interest for each of $K - 2$ other variables. We also suggest an insightful measure for co-clusters to be visualized, namely, the Contribution to Mutual Information (CMI) – providing additional valuable visual information inaccessible with only frequency representation. Notice that such visualizations are also valid whatever the variable of interest.

Definition 3 (Contribution to mutual information). *Given the $K - 2$ selected parts $c_{i_3...i_K}$, the mutual information between two partitioned variables Y_1^M and Y_2^M (from the partition M of Y_1 and Y_2 variables induced by the grid model M) is defined as:*

$$MI(Y_1^M; Y_2^M) = \sum_{i_1=1}^{J_1} \sum_{i_2=1}^{J_2} MI_{i_1 i_2} \text{ where } MI_{i_1 i_2} = p(c_{i_1 i_2}) \log \frac{p(c_{i_1 i_2})}{p(c_{i_1 .}) p(c_{. i_2})} \quad (4)$$

where $MI_{i_1 i_2}$ represent the contribution of cell $c_{i_1 i_2}$ to the mutual information, $p(c_{i_1 i_2})$ is the observed joint probability of points in cell $c_{i_1 i_2}$ and $p(c_{i_1 .}) p(c_{. i_2})$ is the expected probability in case of independence, i.e., the product of marginal probabilities.

Thus, if $MI_{i_1 i_2} > 0$ then $p(c_{i_1 i_2}) > p(c_{i_1 .})p(c_{. i_2})$ and we observe an excess interaction between $c_{i_1 .}$ and $c_{. i_2}$ located in cell $c_{i_1 i_2}$ defined by parts i_1 of Y_1^M and i_2 of Y_2^M. Conversely, if $MI_{i_1 i_2} < 0$, then $p(c_{i_1 i_2}) < p(c_{i_1 .})p(c_{. i_2})$, and we observe a deficit of interactions in cell $c_{i_1 i_2}$. Finally, if $MI_{i_1 i_2} = 0$, then either $p(c_{i_1 i_2}) = 0$ in which case the contribution to MI and there is no interaction or $p(c_{i_1 i_2}) = p(c_{i_1 .})p(c_{. i_2})$ and the quantity of interactions in $c_{i_1 i_2}$ is that expected in case of independence between the partitioned variables.

The visualization of cells' CMI highlight valuable information that is local to the $K - 2$ selected parts and bring complementary insights to exploit the summary provided by the grid.

5   Exploration Results

Each application of KHC (available at http://www.khiops.com) for the various case studies data is achieved within a day of computation on a commodity computer – which confirms the efficiency of the method.

5.1   Analysis of Call Network between Antennas

The application of data grid models on the CDRs provides a segmentation with 1150 clusters, that corresponds to nearly one antenna per cluster. This is due to the large amount of data – 471 millions CDRs. Indeed, the number of calls is so high for each antenna that the distribution of calls originating from (resp. terminating to) each antenna can be distinguished from each other. In order to

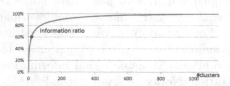

Fig. 2. Evolution of the information kept in the data grid model w.r.t. the number of clusters using the ascending hierarchical post-processing – from optimal data grid M^* (100%) to the null model $M_\emptyset$ (0%).

obtain a more interpretable segmentation, we apply the post-processing introduced in the Section 4.1. Figure 2 shows the information ratio (see definition 1) versus the number of clusters for all intermediate models obtained during the ascending hierarchical post-processing. Interestingly, the resulting Pareto curve shows that very informative models are obtained with few clusters. In our study, we decrease the number of clusters until keeping 60% of the model informativity – corresponding to 20 clusters, an admissible number for the interpretation. Throughout the simplification process, both partitions of source and target antennas stay identical. Thus we consider only the partition of source antennas for the rest of the study. Those clusters are projected on a map of Ivory Coast in Figure 3. Antennas are identified using dots, which color matches with the cluster they belong to.

The first observation is the strong correlation between the clusters and the geography of the country. Indeed, antennas from a same cluster are close to each other. The size of the clusters is almost the same in terms of area and match with the administrative zones of the country, except for Abidjan, the economic capital, which is split into four clusters. This is due to the high concentration of antennas in the city (32% of the ivorian antennas) and the dense phone traffic (34% of the calls).

We use the typicality (see definition 2) to rank the antennas of each cluster. The place, where the antenna with the highest typicality is located, is used to label the cluster. On the map in Figure 3, the size of the dots are proportional to the antenna typicality. Most typical antennas are located in the main cities of Ivory Coast. This phenomenon has already been observed in [3] and [10]: the discovered clusters match with the area of influence of the main cities of a country. We observe few exceptions: the cluster of the city of Sassandra contains the antennas of the city of Divo, while Divo is almost 4 times bigger than Sassandra (population wise) and is the sixth Ivorian city. Antennas in Divo are 40% less typical than the ones in Sassandra, meaning that allocating them to another cluster would be less costly for the criterion. Actually, calls emitted from Divo are significant in direction to other regions of Ivory Coast whereas calls from Sassandra are more inter-

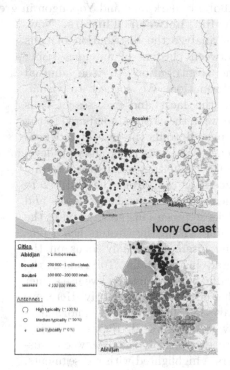

Fig. 3. Twenty clusters displayed on Ivory Coast map. There is one color per cluster.

nal to its region. In more formal terms, the calls distributions of the antennas in Divo are closer to the marginal distribution than to its cluster's distribution. This observation is not really surprising because Divo has experienced a recent growth of its population, due to migrations within the country [9]. Divo is also located in an area specialized in the intensive farming, that attracts seasonal workers from other parts of Ivory Coast.

Now, focusing on the segmentation of Abidjan: the city is divided into four parts with a strong socioeconomic correlation. The first cluster – in red in Figure 3 – covers central Abidjan, including the Central Business District (le Plateau), the transport hub (Adjamé) and the embassies and upper class area (Cocody). The second cluster – in light green – is located in the South of the city. The covered neighborhoods are mainly residential areas and ports. Note that this cluster and the previous one are separated by a strip of sea, except for its North part that is included in the previous cluster. This very localized neighborhood matches with the party area of Abidjan. Finally, the last two clusters group antennas located in two areas with a similar profile: these are lower class neighborhoods. These clusters are separated not only because they are located in different parts of the city but especially because their call distribution differs: Abobo in dark blue and Yopougon in grey in the Figure 3.

Traffic between Clusters. Now, we analyze the distribution of calls between clusters of antennas using the contribution to the mutual information. We suggest to visualize the lacks and excesses of calls between the clusters, compared to the expected traffic in case of independence. Whatever the granularity level of the clustering, we observe a strong excess of calls from the clusters to themselves and weaker excesses and lacks between clusters. Studying the traffic within the clusters has a limited interest. We only focus on the inter-clusters traffic. To visualize the traffic between clusters, we use a finer clustering than previously. Here, we have 355 clusters for 95% informativity (see Figure 2). Figure 4 depicts the excesses of traffic between clusters – highlighted with red segments. The end points of the segments are drawn at the positions of the most representative antennas of the associated clusters (i.e with the highest typicalities). The opacity of a segment is

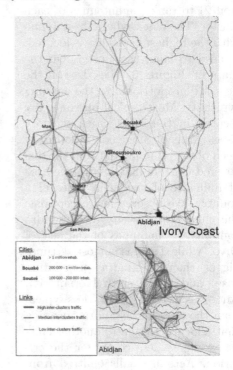

Fig. 4. Analysis of Excess of calls between clusters of antennas

proportional to the value of the contribution to mutual information and its width is proportional to the number of calls between clusters. The biggest cities – like Bouaké, San Pedro and Man – are clearly marked on the map: they are regional capitals, a fact that is confirmed and highlighted by the call traffic visualization. The case of Bouaké is particularly interesting: although it is not the country capital, its national influence seems bigger than the one of Yamoussoukro, the actual capital. Yamoussoukro is twice smaller than Bouaké (population wise) and is a quite recent city where there is no major economical activity, contrary to Bouaké. This fact can explain our observation.

We also observe that excess of traffic between major cities is a rare phenomenon. Cities are more like phone hubs, except in the West of the country around Soubré. This area is not a densely populated area but corresponds to a region with important migration flows. Finally, in Abidjan, we observe important excesses of traffic within neighborhoods, but not between neighborhoods.

5.2 Temporal Analysis of the Calls Distribution

From previous section, we learnt that the correlation between source and destination antennas is very high. The evolution of the calls distribution over time might be the same for both sets of antennas. Therefore, to track the evolution of traffic over time, we only study the evolution of the originating calls: one call is described by the emitting antenna and a day count (stemming from the date).

Again, the clustering of antennas resulting from the optimal data grid is also too fine for an easy interpretation (1051 clusters of antennas and 140 intervals for the day count). We coarse the grain of the grid with our hierarchical post-processing so that the informativity of the model is 80%, with ten clusters of antenna and twenty time segments. Since, missing values are abundant in this data, i.e., some antennas emitted no call during some time periods, consequently, we obtain time segments that are strongly correlated with missing data. For the same reason, antennas are grouped together because they experienced an absence of calls during one or several similar periods. In the Figure 5, the colored antennas belong to clusters hav-

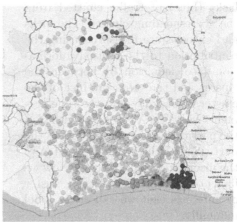

Fig. 5. Antennas activity clusters projected on Ivory Coast map. Colored clusters show inactivity periods while grey clusters indicate antennas whose traffic is complete over the period.

ing experienced simultaneous absences of calls. We observe that the green, orange, light blue and purple clusters are located in localized area. The missing data appear during short periods for these clusters. This grouping might be due to localized technical issues on the network. The antennas of the yellow cluster are spread over the country. These antennas are grouped because they have been activated at the same date. This use case provides a better understanding the dysfunctions in the network over the year.

5.3 Analysis of Output Communications w.r.t. Week Day and Hour

Our objective is to build simultaneously a partition of the antennas, a partition of the week days and a discretization of the hour, i.e., a triclustering. For the same reasons as previously, we only keep the emitting antennas.

At the finest level, we obtain a triclustering with 806 clusters of emitting antennas, 7 clusters of days and 22 time segments. Again, these results must be simplified to ease the interpretation. However, we fix the numbers of clusters of days and time segments, since they are acceptable for the analysis and we only reduce the number of clusters of antennas. With four clusters of antennas, we keep 51% of the informativity of the model.

Antennas are displayed on the map of Figure 6. We also build a calendar (see Figure 7) for each cluster with days in columns and time segments in lines. The color of the cells indicates the excesses (red) or the lacks (blue) of traffic emitted from the corresponding cluster. The lacks and excesses are measured using the contribution to the mutual information (see definition 3) between the cluster and the cross product of the cluster of weekday and the time segment: $MI(X_1^M; X_2^M \times X_3^M)$, with X_1^M the partition of the antennas, X_2^M the partitions of the weekdays and X_3^M the discretization of the time. Now we focus on the analysis of each cluster of antennas that we can easily label manually:

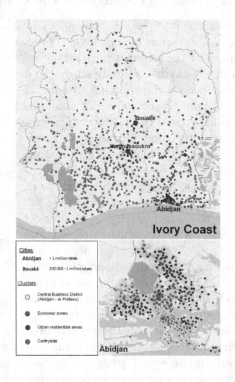

Fig. 6. Clusters on the map of Ivory Coast. Dots are antennas. There is one color per cluster.

Abidjan - Le Plateau (yellow). This cluster covers exactly the Central Business District of Abidjan. In the calendar of Figure 7, we observe an excess of calls from the Monday to the Friday, between 8-9am and 4-5pm. The rest of the time, there is a low lack of traffic emitted from this area. In other words, during the office hours, the phone traffic is higher than expected and lower the rest of the time. This is expected and representative of this type of area: a non-residential business district.

Economic Zones (red). The antennas of this cluster are located either in the commercial areas of the cities or in areas with a strong economic activity, like plantations or mines. In Abidjan, these antennas are located in industrial zones (South and North-West), the shopping districts (North of the business district) and the universities and embassies neighborhood (East). The traffic in these areas is mainly in excess from the Monday to the Saturday between 9 am and

5 pm. The correlation is very strong between the working hours and the calls traffic on these areas.

Urban Residential Areas (blue). The antennas belonging to this cluster are mainly located in the cities like Abidjan, Bouaké and Yamoussoukro. If we focus on Abidjan, we observe that the cluster covers the residential neighborhood located in the West and in the North-East of the city. At a finer level of partition of the antennas, this cluster would be split according to the socioeconomic class of the neighborhood: the upper class neighborhood in the East of the city is separated from the lower class neighborhoods, located in the North and the West. The calendar shows lacks of calls during the office hours and excesses the weekend, the night and the early morning during the week. This is correlated with the presence of people in residential areas. Note that the excesses of calls start around 8 pm, while it stops around 5 pm in the Central Business district or in economic areas. This time lag is due to the cheaper price of calls after 8 pm.

The Countryside (green). The antennas of this cluster are spread over the country, except in Abidjan and other cities in general. The calendar for this cluster is quite similar to the one of the urban residential areas, except that the excess periods are limited to the early evening and the whole Sunday.

5.4 User Mobility Analysis w.r.t. Week Day and Hour

Among the 50000 anonymized users, we focus on mobile users characterized by a frequent use of a large set of distinct antennas: after filtering, 6894 users are under study. For these 4-d data (user, antenna, week day and hour), KHC operates a tetra-clustering: as a result, users with the same mobility profile are grouped together, i.e., users who have connected to similar groups of antennas, on similar days of the weeks at similar time periods.

At the finest grain, we obtain 237 clusters of users, 218 clusters of antennas and three time segments ,while week days remain as singletons. Again, the granularity prevent us from an easy interpretation, and we simplify the model. We keep 50% of informativity, that enables a reduction of the numbers of clusters of users and antennas to 40, and the numbers of groups of week days and hour segments to two. The week is divided in two parts: the working days and the weekend. For the hour dimension, the split occurs around 6 pm. The intervals are 0 am - 6 pm and 6 pm - 12 am. Note that the bound at midnight is artificial, because the day start as this time. The cut at 6 pm is the last in the hierarchy of the time segmentation. Then it would have been more relevant to consider a day from 6 pm to 6 pm the next day. Nevertheless, it is easier to have an interpretations on a "usual" time period between 0 am and 12 pm. Therefore we keep the following segmentation: 0 am - 6 pm, 6 pm - 12 pm.

To illustrate the characterization of users' behaviors in terms of mobility provided by the grid, we focus on a group of users. The maps of Figure 8 shows the excesses and lacks of traffic in Abidjan during the week, for both periods of the day and for the selected group of users. The colors correspond to the mutual information $MI(X_1^M; X_2^M \times X_3^M \times X_4^M)$ where X_1^M is the partition of antennas;

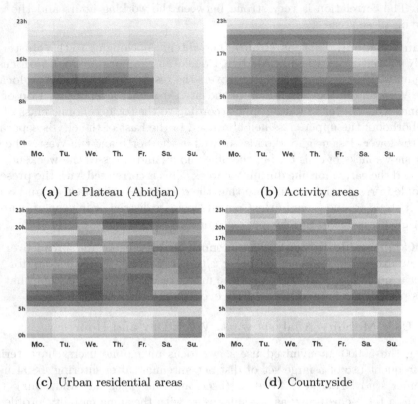

(a) Le Plateau (Abidjan)   (b) Activity areas

(c) Urban residential areas   (d) Countryside

Fig. 7. Calendars of excesses (red) and lacks of calls emitted from each of the four clusters of antennas, in function of the weekday and the daytime.

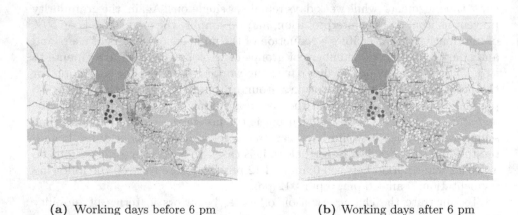

(a) Working days before 6 pm   (b) Working days after 6 pm

Fig. 8. For a group of user, excesses and lacks of uses of antennas according to the day of the week and the time of the day. Focus on Abidjan.

X_2^M, the partition of the weekdays; X_3^M the discretization of the daytime; and X_4^M, the selected partition of the users.

The selected group of users mainly connects to the antennas located in the East of Abidjan after 6 pm during the working days, while they rarely connect to the same antennas before 6 pm the same days. Then, it can be assumed that the selected cluster of users is composed people living in the same area. This hypothesis is reinforced by the socioeconomic nature of this part of Abidjan: it is a residential area. The contributions to mutual information of the other clusters of antennas are smaller. Three areas experience excesses of traffic before 6 pm and lacks after 6 pm. They correspond to the business district (Le Plateau), the embassies and universities neighborhood and the industrial zone located in the West of the city. The common feature of all these areas is their economic activity during the day. To sum up, we can assume that the users of the selected cluster are similar in that they live in the same area and work during the week in three localized area of Abidjan. Similar observations stand for several other clusters of users – thus we are able to summarize users' mobility behavior.

6   Conclusion

Motivated by two key points of economic development strategy of a telco in emerging countries, we have instantiated a generic methodology for exploratory analysis of CDRs data. Our method is based on a joint distribution estimation technique providing the user analyst with a summary of the data in a parameter-free way. We have also suggested several tools for exploring and exploiting the summary at various granularities and highlighting its relevant components. We have demonstrated the applicability of the method on graph data, temporal sequence data as well as user mobility data stemming from country-scale CDRs data. The results of the exploratory analysis are currently considered as valuable additional input to improve network planning strategy and pricing strategy.

References

1. Becker, R.A., Cáceres, R., Hanson, K., Isaacman, S., Loh, J.M., Martonosi, M., Rowland, J., Urbanek, S., Varshavsky, A., Volinsky, C.: Human mobility characterization from cellular network data. Commun. ACM **56**(1), 74–82 (2013)
2. Berlingerio, M., Calabrese, F., Di Lorenzo, G., Nair, R., Pinelli, F., Sbodio, M.L.: AllAboard: a system for exploring urban mobility and optimizing public transport using cellphone data. In: Blockeel, H., Kersting, K., Nijssen, S., Železný, F. (eds.) ECML PKDD 2013, Part III. LNCS, vol. 8190, pp. 663–666. Springer, Heidelberg (2013)
3. Blondel, V., Krings, G., Thomas, I.: Regions and borders of mobile telephony in Belgium and in the Brussels metropolitan zone. Brussels Studies **42** (2010)
4. Blondel, V., de Cordes, N., Decuyper, A., Deville, P., Raguenez, J., Smoreda, Z.: Mobile phone data for development - analysis of mobile phone datasets for the development of ivory coast (2013). http://perso.uclouvain.be/vincent.blondel/netmob/2013/D4D-book.pdf

5. Blondel, V.D., Esch, M., Chan, C., Clérot, F., Deville, P., Huens, E., Morlot, F., Smoreda, Z., Ziemlicki, C.: Data for development: the D4D challenge on mobile phone data. CoRR abs/1210.0137 (2012)
6. Boullé, M.: Data grid models for preparation and modeling in supervised learning. In: Guyon, I., Cawley, G., Dror, G., Saffari, A. (eds.) Hands-On Pattern Recognition: Challenges in Machine Learning, vol. 1, pp. 99–130. Microtome (2011)
7. Dhillon, I.S., Mallela, S., Modha, D.S.: Information-theoretic co-clustering. In: KDD, pp. 89–98 (2003)
8. Frías-Martínez, E., Williamson, G., Frías-Martínez, V.: An agent-based model of epidemic spread using human mobility and social network information. In: Social-Com/PASSAT, pp. 57–64 (2011)
9. Gnabéli, R.: La production d'une identité autochtone en Côte d'Ivoire. Journal des anthropologues. Association française des anthropologues 114–115, 247–275 (2008)
10. Guigourès, R., Boullé, M.: Segmentation of towns using call detail records. In: NetMob Workshop at IEEE SocialCom (2011)
11. Guigourès, R., Gay, D., Boullé, M., Clérot, F., Rossi, F.: Country-scale exploratory analysis of call detail records through the lens of data grid models (2015). http://arxiv.org/abs/1503.06060
12. Hartigan, J.A.: Direct clustering of a data matrix. Journal of the American Statistical Association 67, 123–129 (1972)
13. Jiang, S., Fiore, G.A., Yang, Y., Ferreira Jr., J., Frazzoli, E., Gonzàlez, M.C.: A review of urban computing for mobile phone traces: current methods, challenges and opportunities. In: UrbComp@KDD (2013)
14. Kolda, T.G., Sun, J.: Scalable tensor decompositions for multi-aspect data mining. In: ICDM, pp. 363–372 (2008)
15. Laurila, J.K., Gatica-Perez, D., Aad, I., Blom, J., Bornet, O., Do, T.M.T., Dousse, O., Eberle, J., Miettinen, M.: From big smartphone data to worldwide research: The mobile data challenge. Pervasive and Mobile Computing 9(6), 752–771 (2013)
16. Lockhart, J.W., Weiss, G.M.: The benefits of personalized smartphone-based activity recognition models. In: SDM, pp. 614–622 (2014)
17. Peng, W., Li, T.: Temporal relation co-clustering on directional social network and author-topic evolution. Knowledge and Information Systems 26(3), 467–486 (2011)
18. Slonim, N., Friedman, N., Tishby, N.: Agglomerative multivariate information bottleneck. In: NIPS, pp. 929–936 (2001)
19. Sun, J., Tao, D., Faloutsos, C.: Beyond streams and graphs: dynamic tensor analysis. In: KDD 2006, pp. 374–383 (2006)
20. Tishby, N., Pereira, O.C., Bialek, W.: The information bottleneck method. In: Allerton Conference on Communication, Control and Computing (1999)
21. United Nations Global Pulse: Mobile phone network data for development (2013). www.unglobalpulse.org/Mobile_Phone_Network_Dat_for_Dev
22. Vieira, M.R., Frías-Martínez, V., Oliver, N., Frías-Martínez, E.: Characterizing dense urban areas from mobile phone-call data: discovery and social dynamics. In: SocialCom/PASSAT, pp. 241–248 (2010)
23. Wang, D., Pedreschi, D., Song, C., Giannotti, F., Barabási, A.L.: Human mobility, social ties, and link prediction. In: KDD, pp. 1100–1108 (2011)
24. Wang, P., Domeniconi, C., Laskey, K.B.: Information bottleneck co-clustering. In: Workshop TextMining@SIAM DM 2010 (2010)

Early Detection of Fraud Storms in the Cloud

Hani Neuvirth[1](✉), Yehuda Finkelstein[1], Amit Hilbuch[1], Shai Nahum[1], Daniel Alon[1], and Elad Yom-Tov[2]

[1] Azure Cyber-Security Group, Microsoft, Herzelia, Israel
{haneuvir,t-yehudf,amithi,snahum,dalon}@microsoft.com
[2] Microsoft Research, Herzelia, Israel
eladyt@microsoft.com

Abstract. Cloud computing resources are sometimes hijacked for fraudulent use. While some fraudulent use manifests as a small-scale resource consumption, a more serious type of fraud is that of fraud storms, which are events of large-scale fraudulent use. These events begin when fraudulent users discover new vulnerabilities in the sign up process, which they then exploit in mass. The ability to perform early detection of these storms is a critical component of any cloud-based public computing system.

In this work we analyze telemetry data from Microsoft Azure to detect fraud storms and raise early alerts on sudden increases in fraudulent use. The use of machine learning approaches to identify such anomalous events involves two inherent challenges: the scarcity of these events, and at the same time, the high frequency of anomalous events in cloud systems.

We compare the performance of a supervised approach to the one achieved by an unsupervised, multivariate anomaly detection framework. We further evaluate the system performance taking into account practical considerations of robustness in the presence of missing values, and minimization of the model's data collection period.

This paper describes the system, as well as the underlying machine learning algorithms applied. A beta version of the system is deployed and used to continuously control fraud levels in Azure.

1   Introduction

The adoption of the public cloud as an agile model for computational resources consumption is continuously increasing. The high scalability of these services offer many opportunities, as well as new challenges. Examples include failure detection [1, 2], resources optimization [3–5], and security [6, 7]. However, a common challenge to all is the efficient and effective analysis of large quantities of data that is continuously accumulated by such cloud platforms.

A significant portion of the data collected at the cloud is in the form of time series data, e.g., signals from the monitoring of continuous resource use. Therefore, machine learning algorithms performing time series analysis and forecasting are commonly applied. Numerous algorithms have been developed for this purpose over the years [8]. The most established ones are auto-regressive models, integrated models and moving average

© Springer International Publishing Switzerland 2015
A. Bifet et al. (Eds.): ECML PKDD 2015, Part III, LNAI 9286, pp. 53–67, 2015.
DOI: DOI: 10.1007/978-3-319-23461-8_4

models, modeling the signal value at a certain time-point using linear dependencies on preceding data points. Various extensions to these algorithm for multivariate analysis where the underlying entity is a vector of co-evolving signals also exist [9].

A common analysis setting of these time series in the cloud is anomaly detection [10]. Wang et al. developed EbAT, an entropy based anomaly testing for detecting anomalies in cloud utilization patterns [11]. In a more recent paper they describe a lightweight online algorithm for this task based on the Tukey and Relative Entropy statistics [12]. Dean et al. created the Unsupervised Behavior Learning (UBL) algorithm which leverages self-organizing maps to detect performance anomalies [13]. Vallis et al. focused on long-term anomaly detection analyzing data from twitter [14].

With the cloud environment being noisy and dynamic, short periods when reliable data is only partially available often occur. Therefore, in order to be applicable in practice, the algorithms must be able to operate in the presence of missing values. For univariate time series, linear and spline interpolation methods are a common approach [15]. An advanced approach is provided by the DynaMMo algorithm, which aims to enhance the information from the correlations among multiple dimensions, through modeling co-evolving time series through the use of latent variables [15]. Wellenzohn et al suggest a method imputing the missing values from the k most similar patterns in historical data [16]. Xie et al. developed the MOUSSE algorithm which is based on submanifold approximation, which can handle some amount of missing data, and demonstrated it on theft detection and solar flare detection on video streams [17].

Among the many challenges in the cloud, of primary importance is ongoing services availability. In parallel to the genuine users, these services attract fraudsters trying to utilize the resources afforded for more nefarious means. In order to avoid decreases in service quality due to fraudulent activity, cloud providers have to either maintain excessive resources or detect, as soon as possible, consumption by fraudsters and take relevant action.

Here we address the early detection of significant peaks in fraudulent activity in cloud systems, which we term "fraud storms" (FS). In order to avoid a reduction in the quality of service (QoS) as a result of this kind of indirect attacks [18], the service provider has to maintain large amount of excessive resources, which has significant cost implications. We examine two different possible approaches for the problem – a supervised approach aiming to predict the continuous fraud levels, and an unsupervised approach searching for anomalies in multivariate capacity related signals. The requirement for prompt detection poses a constraint on the resolution of data available for the analysis, which is collected at a datacenter level, as opposed to subscription level data, which is too big to collect within the required time constraints.

2   Methods

2.1   The Problem Setting

Microsoft Azure is one of the largest public-cloud platforms available, spanning several large data centers around the world. The size of each data center varies depending on local needs, economic considerations, etc. From a fraud protection perspective, this means that the effect of fraud peaks on a certain data center depends on its size.

In addition, Azure has a wide variety of offer types through which users can subscribe to the system. For example, free trial accounts that offer limited resources to new subscribers, and Pay-as-you-go offers allowing users to pay at the end of the billing period for specific resources consumed during this time. Each of these offer types requires a different verification process on registration to the system, designed to address the associated fraud risk.

The final true labeling of a certain subscription as fraudulent is based on post-usage information, for example, through information that certain credit cards were reported stolen. These labels necessarily arrive much later than the fraudulent use, typically in the order of a month. However, if a specific subscription is suspected to be fraudulent, the cloud operator has the ability to manually investigate the user further, and if found to be fraudulent, the subscription is closed. In the case where a group of fraudulent subscriptions is identified, a bulk shutdown process can be employed.

The type of resources consumed by fraudsters highly depends on the specific fraud type. Bitcoin mining would usually require high compute resources, while spamming and click-frauds might have a greater effect on the bandwidth use.

To achieve a high-resolution control over fraud storms, supplying adequate reliability for the different data center regions and offer types, similar offer types have been grouped together by a domain expert, and the analysis is performed independently for each pair of region and offer group, which we term here a sample-set. Effectively, in the available data there is some correlation between FS events on different regions, however, in this study, the analysis of each sample-set is independent of the others.

2.2   The Labels

The objective of this work is to detect fraudulent peaks affecting compute resources. To obtain FS labels, the ongoing number of fraudulent cores in use was extracted, and a domain expert used this information to assign labels to hourly data on an eight months period, spanning from January 6 to September 10. Note that this procedure was performed on past events, once the full information about fraud was available. The labels indicate the start date of the FS event, the date of the first bulk-shutdown of fraudulent subscriptions employed to discard the fraudulent accounts, and the date in which the fraudulent cores consumption returned to its limited regular level.

2.3   The Analyzed Signals

We use two signals sampled at hourly intervals in this version of the FS detection component: average of the total number of cores used in the past hour in a region and offer group, and the total number of new subscriptions belonging to a certain offer group. These signals are extracted at the level of the datacenter, and thus available at a short latency. Here we use data gathered between January 6 and September 10, 2014. Some of the regions and offer groups became operational only later within this period, and thus their analysis sequence is shorter. Note that the breakdown of the new subscriptions signal to the different regions is not available, and thus the exact same signal is shared across all regions. The design of the algorithms is such that new signals can be easily added.

2.4   Learning Approaches

Two very different learning approaches can be utilized for this problem. One is a two-layer supervised approach, analogous to the labeling process, where the first layer predicts the continuous signal of the number of fraudulent cores that was used to derive the classification, and the second layer uses a classification algorithm over this signal, to predict the binary FS labeling. An alternative approach follows the observation that each FS is an outlier of the system and hence applies an anomaly detection approach.

The Supervised Analysis Approach

This analysis includes two independent layers: using regression to predict the number of fraudulent cores (FC) at each time-point from the available new-subscriptions and total-core-usage signals, followed by a classification of this prediction to retrieve the binary FS prediction.

The input to the present version of the system are two signals, sampled hourly: the number of cores in use, and the total number of new subscriptions created for this offer group. The number of cores at each region and offer group was first processed to define their analysis starting point, discarding data with no sufficient variation, by requiring a minimum standard deviation of 3 on a 7-days window. This is due to the fact that data starts to flow from the data centers and offer types already at their testing phase, before they actually become operational.

The regression analysis in the first layer was performed independently for each sample-set (region and offer group). The features used include the values at the past 1,2,4 and 8 hours, 1, 2, 4, 7 and 14 days, and averages of the signal in windows of 2, 4, 8, 12, 24 and 48 hours, as well as their polynomial products of degree 2. These features were first filtered based on low variance ($<10^{-10}$). Then, they were further filtered using a false discovery rate (FDR) procedure on the p-values obtained from the correlation coefficient of each feature with the outcome setting alpha=0.05. Finally, they were standardized, and fed into a ridge regression model counting over ridge values ranging from 1 to 10^{12} at a logarithmic scale. Model selection was performed using generalized cross validation with the python scikit-learn package.

In the training phase, the regression, predicting the number of fraudulent cores is performed in 5-fold cross-validation in order to obtain signal with the noise resulting from the prediction model. Another regression model is trained on the full training data to be used on the test set. Three features were extracted from this noisy signal for each time-point t, relative to the preceding time point (t-1): their difference (diff=FC(t)-FC(t-1)), their relative difference (diff/mean(FC)), and their ratio (FC(t)/FC(t-1). These features coming from all the sample-sets are then fed into a logistic regression model to train the classification. For the labeling of the classification layer, two schemes were explored: a strict labeling considering only the time period until the bulk shutdown as FS, named "bulk shutdown" (see section 2.2), and a looser definition representing the whole FS period as positive.

The Outlier Detection Approach

The outlier detection algorithm processes the time series of the same selected signals. It is composed of two parts: prediction of the value at the next time point, and evaluation of the error. Our signals show typical seasonality patterns (An example is shown in Figure 6), and thus, an analysis similar to the one employed by Bay et al. was used [19]. Specifically, we utilized a sliding window approach, in which the values of both signals in a recent time window are used to predict the value of each signal at the next timepoint. The selected features include the values at the past 1,2,4, and 8 hours, 1, 2, 4, 7 and 14 days, and averages of the signal in windows of 2, 4, 8, 12, 24 and 48 hours.

Several regression models have been explored for the prediction, including random forest regression, k-nearest neighbors and SVM with polynomial and RBF kernels, however all gave a similar performance to the one obtained by a ridge regression model, thus, this simpler model was selected. The ridge in this model was selected using cross validation over a log-scale set of values between 1 and 10^{12}.

In the training phase, prediction was performed using a sliding window starting from a 60 days history, and in gaps of 30 days. The history length used for training each window was set to a minimum of 60 days, and a maximum of 100 days, in order to have data that is both sizable and relevant. For each window, we used linear regression on the features listed above to predict the value at the next time-point in 5-fold cross validation (CV). One additional training is performed on the full data window.

The CV data was used to estimate the error distribution. A multivariate Gaussian distribution is assumed, and the mean vector and covariance matrix are calculated using a robust estimation procedure available in the python scikit-learn package [20]. These are then used to calculate a p-value for each time-point. There exist two possible definitions for the p-value of the multivariate normal distribution (MVN). The first is analogous to the one-dimensional counterpart, and is based on the Mahalanobis distance [21]. The second considers the distribution in a vectorized manner, and estimates the weight of the cumulative distribution function in a rectangle of values higher than the observed values [22]. The latter demonstrated better performance (data not shown). This observations is aligned with the fact that FS by definition refers to anomalies at the higher cores and subscription values, and thus this p-value estimation method was selected. The p-value bound was optimized by counting over a few selected values.

There are two sources for randomization in this approach: in the sampling of the CV partition, and in the robust covariance estimation [20]. All the experiments were repeated three times, with different seeds, to estimate the associated standard deviation. Unless indicated otherwise, the standard deviations of all the points in the figures in this paper fell below 1%.

2.5   Performance Evaluation

To compare the binary predictions with the true labels, all time-points identified as a FS that are less than 36 hours apart were clustered. Following, each cluster was mapped to the earliest overlapping FS, if any.

Two measures of interest were calculated over these clusters in order to compare the models: the number of FS events detected, and the median time to detect in hours, which is the difference between the start time of the predicted cluster to the start time of the FS label. This measure may take negative values in case the predicted cluster starts before the manually labeled FS start date.

In addition, we calculated FPDays, the fraction of days in which a false FS alert was raised for the sample-set, and a summary measure indicating the fraction of sample-sets having FPDays < 0.05. Note, that high values of this measure correspond to low FP rates. Naturally, the aim is to maximize the number of FS events detected and the fraction of sample-sets having a low FP rate, while minimizing the detection time.

3   Results

This section begins with comparing the different analysis approaches, showing that in our setting the anomaly detection approach is superior to the supervised analysis. Then, we compare the performance of our algorithm, named FraudStormML, with the Seasonal Hybrid ESD algorithm, a state of the art anomaly detection algorithm [14]. Further, we address two key challenges that stem from the practical nature of the problem: making the analysis robust to missing and corrupted data, and enabling fast utilization of the system on newly available sample-sets.

3.1   Analysis Approaches Comparison

Two main analysis approaches have been explored in this study. The input to both is the two hourly signals providing the core usage and count of new subscriptions per region and offer group. Recall that the signal used to derive the manual labels is the number of fraudulent cores. The first approach performs supervised analysis to predict the number of fraudulent cores. Following, a binary classification stage is used to map the continuous prediction to the binary FS label. The second approach performs multivariate anomaly detection on the two input signals.

In the initial analysis using the supervised approach, no FS has been detected. To further investigate this, we explored the errors of each of the layers independently. Figure 1 shows the performance of the classification layer that is based on the true number of fraudulent cores that was used for the manual labeling (gray line with circle markers). These are extremely good results, with a false-positive rate lower than 5% in nearly all sample sets while providing full detection of all FSs. This implies that despite the scarcity of the labels, the logistic regression model can easily capture the manual labeling process, and the failure is related to the regression layer or to the combination of the two layers. Note that these results were obtained only with the "bulk shutdown" classification labeling scheme. When the full FS duration was labeled as positive, no FS has been detected. This can be ascribed to the use of a linear model for the classification, however, the overall good performance suggests that the steep rise in the number of fraudulent cores was the main factor in the manual classification.

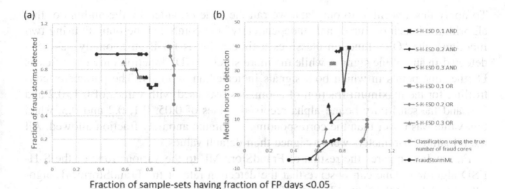

Fig. 1. Performance comparison of the fraud storm detection approaches

Investigating the error of the regression layer, we observed good prediction power in general, but high RMSE values specifically in regions where a FS occurred (Figure 2, Pearson correlation coefficient=0.53, p-value=4e-4). This suggests that the number of fraudulent cores cannot be easily modeled in our context, since the behavior during FS is significantly different from normal, and thus an anomaly detection scheme would be more appropriate.

(a)

Offer Groups

	A	B	C	D	E
1	0	0.037	0	0	0.037
2	0	0.037	0.074	0	0.037
3	0	0	0.037	0	0.037
4	0	0	0	0	0.074
5	0	0.111	0	0	0.074
6	0	0.074	0	0	0.037
7	0	0	0	0	0.074
8	0	0	0	0	0.074
9	0	0	0	0	0.074
10	0	0	0	0	0.111

regions

(b)

Offer Groups

	A	B	C	D	E
1	8.22	1066.85	50.34	25.98	1738.46
2	7.76	3121.29	3391.17	29.52	1035.49
3	16.95	535.71	1656.73	110.33	446.35
5	12.07	965.95	38.24	26.11	392.26
6	43.34	868.81	471.86	10.06	523.73
7	2.77	0	0	2.39	180.85
8	0.67	55.69	0	0.56	473.42
9	4.61	633.47	44.76	19.64	487.6

regions

Fig. 2. (a) The distribution of FS over regions and offer groups. (b) The RMSE obtained on the regression analysis for each region and offer group. A green-to-red color scale is used to represent the values ranges depicted in the boxes. The correlation between the figures is evident (Pearson correlation coefficient=0.53, p-value=4e-4), proving the regression model fails on FS events, which suggests that an anomaly detection approach might be more suitable. The data for regions 4 and 10 was not available early enough for them to qualify for this analysis.

Turning to the anomaly detection approach, we compare our algorithm, named FraudStormML, to the Seasonal Hybrid ESD (S-H-ESD) algorithm, a state of the art algorithm recently published for anomaly detection in the cloud, specifically designed for time-series data [14]. This algorithm is similar to the one we employed in that both algorithms account for the daily and weekly seasonality in the data. It differs from our algorithm in the specific statistical analysis applied, utilizing the generalized Extreme Studentized Deviate test, and by the fact that the analysis is a univariate one.

To apply this algorithm to our data, we ran the code provided by the authors of this algorithm on each of our signals independently, and combined the outputs using two functions: the "OR" function classifies a time point as FS if an anomaly has been detected in any of the signals, while the more strict "AND" function only classifies as FS the time points in which both signals indicated an anomaly. The parameter controlling for the maximum fraction of anomalies was used with values of 0.1, 0.2 and 0.3, and the sensitivity bound alpha received values of 0.05, 0.1, 0.2 and 0.3, when this value was lower than the corresponding maximum anomaly fraction allowed. All the remaining parameters were assigned their default values.

Figure 1a compares the results of FraudStormML to the various runs of the S-H-ESD algorithm. One can observe that the detection rate of the FraudStormML algorithm is higher at the same FP rates (Figure 1a), and the median time-to-detect is lower (Figure 1b).

3.2   Handling Missing and Corrupted Data

With the cloud environment being a highly dynamic one, it is not uncommon for data to become temporarily corrupted or unavailable. From our experience, those vulnerable times are often associated with fraud storms. Hence, allowing a fast recovery of the system from such events is highly important. Furthermore, when training an anomaly detection model, it would be wise to discard past FS events, as these may reduce the sensitivity of the model. Therefore, an important aspect of a FS detection system is its ability to be trained and applied in the presence of missing data.

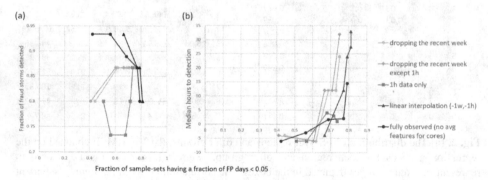

Fig. 3. Performance comparison of the anomaly detection in the presence of missing values

In this analysis we assume a broad missing data model in which the cores-usage data is missing for a period of one week where the same hour at the previous week is the most recent time-point that is available. For the simplicity of this analysis, the features calculating past averages were completely dropped for this signal. The missingness pattern was applied at the feature extraction stage for all the data points in the sample set, and the prediction performance was evaluated.

Figure 3 presents a comparison of five different models: a fully observed model (excluding the averaging features), a model with the most recent week missing, a model where the most recent week except the preceding hour is missing, a model that is based only on the preceding hour, and a model deriving all the features from a linear interpolation between the value at the preceding week and the value at the preceding hour. We can observe that dropping the recent week has a significant effect of about 10% on the FP rate, and a similar effect on the fraction of FS detected. The effect is somewhat reduced in the model which adds in the preceding hour (1h) feature, simulating the case when the information on the recent hour has been accumulated, however, not to a significant level. Dropping the longer history features, and counting on the 1h feature alone provides similar performance, yet also a noisier one. In contrast, the linear interpolation model manages to recover the performance to a level very similar to the fully observed model.

3.3   The Effect of the Data Accumulation Period Length

As the cloud keeps evolving, new offer groups are created from time to time. Therefore, it is important to be able to estimate the minimal time required for the system to become effective. Note, that as a measure for the system performance on new offer types, this is merely an approximation, since new offer types might have different behavior from the longstanding offers for which we have data.

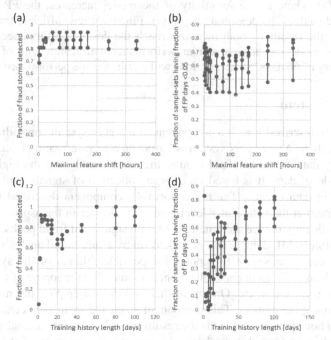

Fig. 4. The performance achieved as a function of the data accumulation period.

The length of the data accumulation period is a sum of two parameters in our model: the longest shift used to derive features, and the amount of data required for training. Figure 4a presents the effect of reducing the extent of the features used. It plots the detection achieved for different feature shifts ranging over values in [2, 4, 8, 16,

24, 48, 72, 96, 120, 144, 168, 240, 336] hours, where in each experiment a single shift was added to all the smaller ones. All dots connected with a line are the results of the same model configuration, with different p-value thresholds ranging from 10^{-5} to 10^{-15}. The experiments for this plot were performed with full training data as was finally used by the FraudStormML system, requiring a minimum of 60 days and at most 100 days. As before, each dot is the mean of three experiments with different randomization seeds. Figure 4a presents a significant decrease in the sensitivity of the model below 48 hours. Then, the detection rate is stabilized. For feature shifts of 1–2 weeks there is some improvement in the noise of the model, making it less sensitive to the p-value threshold. The FP rates are stable for all values, which might be related to the large training set used (Figure 4b).

We evaluate the performance of the model given different history lengths of data used for training. Figures 4b, 4c show the performance as a function of the history length, ranging between 1 to 100 days, with the sliding window gaps set to the history length in case it is smaller than 30, and 30 days otherwise (as in the final model). Experiments have been repeated with three different seeds. The standard deviation was lower or equal to 1% for FP, and lower or equal to 4% for the fraction of FSs detected. Figure 4c shows initial good performance, which is misleading, because, as Figure 4d shows, it is associated with very high FP rates (low fraction of the sample-sets having FP rate < 5%). Then, as the sensitivity of the model increases, the FP rates on (d) decrease, and so does the detection rate on (c). Finally, when sufficient training data is accumulated, the results on both measures reach the desired performance. The results imply that the model can achieve similar performance with feature shift of 1 week, and training history of 60 days. Shorter periods will result in some performance degradation.

4   The System

The algorithm described here is a component of a bigger system controlling the ongoing fraud levels in Azure. This system is based on a map-reduce architecture collecting telemetry data from the cloud around the world, analyzing the high-resolution, subscription-level data. It is based on a large collection of signals, and thus suffers from delays in data latency. The FraudStormML component described here aims to control the fraud at the data center level, and to be thinner in the aspect of signals, in order to significantly reduce the time-to-detect on significant fraud peaks.

The FraudStormML component, is backed-up by a rule-based analysis component. This component is important to support for possible failures of the system, which might result from failures in data collection, for example in mapping of new subscriptions to their offer types, etc. On top of this system, we employ the more sensitive machine learning component. This component scans the main data centers and offer groups on an hourly basis, and feeds the resulting evaluation into a database. A Power BI dashboard reads from this database and provides an overview of the fraud storm state of the system. A snapshot of this system is presented in Figure 5. A detailed figure for each of the regions and offer groups suspected to be affected by a FS is also produced, to allow a deeper manual investigation. The model for this component is trained weekly. Time periods where data was corrupted or missing are being manually fed into a database, and are interpolated during training.

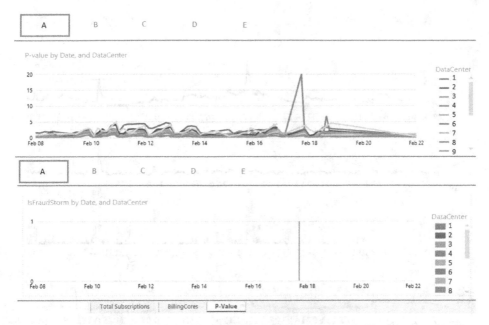

Fig. 5. A snapshot of the FraudStormML component dashboard.

5   Discussion

This paper presents a machine leaning based fraud storm detection system. We examined the performance of a supervised approach, and compared it to an unsupervised anomaly detection approach. Selecting between the two approaches a-priori is not straight forward, as each of them has its advantages. On one hand, the cloud environment is a very noisy one supplying many outliers that are not necessarily FS events. On the other hand, a supervised approach is challenging due the scarcity of FS events. Moreover, we observed that the regression prediction accuracy is especially low at regions where FS occurred, as they are outliers of the model. The supervised approach might gain from use of more sophisticated, non-linear models. However, in a big data system designed to continuously monitor the cloud, the advantage of light-weight models is evident. With an underlying such model, the anomaly detection approach performed significantly better than the supervised alternative.

There are many sources for false positive detections in this system. First is the use of manual labeling, which is a common practice in many machine learning domains, albeit a noisy one. Figure 6 presents a few examples for FP detections. All the panels on Figure 6 share the same x-axis, indicating the timeline. The first (top) panel shows the core-usage signal, the second panel shows the new subscriptions signal, the third panel plots the p-values obtained by the FraudStormML system, with red bars indicating a detection. The panel on the bottom is the continuous signal used for labeling, i.e. the number of fraudulent cores. The purple background indicates regions that were manually labeled as a FS.

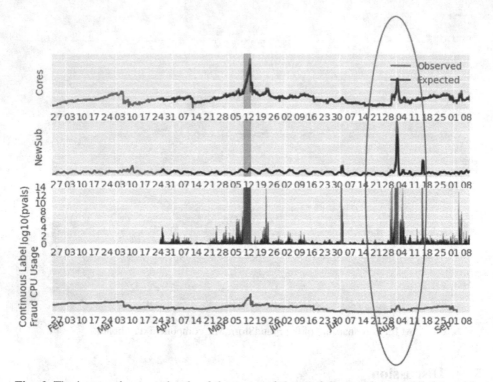

Fig. 6. The input and output signals of the system for one of the sample sets. All the panels share the same x-axis indicating the timeline. The two top panels present the values of the analyzed signals (cores and new subscriptions). In these panels, the green curve presents the observed values, while the black curve represents FraudStormML's predicted values. The third panel presents the $-\log_{10}$(p-value) of the prediction obtained by the FraudStormML component, with red bars indicating a detection. The bottom panel presents the continuous label, namely, number of fraudulent cores in use. The time window that was manually labeled as a FS is depicted in purple background at the top three panels. There appear to be a few anomalies that are FP detections of FS. The FP anomaly marked with the circle is also apparent at the continuous label panel at the bottom, thus representing a limited but true event.

For this region and offer group we observe a single FS event labeled. This event was successfully detected. However, there are a few other anomalies detected by the system. The one designated by the red circle demonstrates one FP detection. Even though this event was not manually labeled as a FS, the corresponding small anomaly in the number of fraudulent cores used (bottom panel) is evident.

Other anomalies detected are true anomalies that are not FS. In practice, the information about their cause is not always available on past data. Since the system became operational, we encountered several such events resulting from normal but rare events such as switching the default region of the system. Each event leads to a manual investigation, which would eventually allow us to identify more relevant signals that could be added to the system in order to reduce the FP rate. Another way to reduce FP rate is to use information coming from different regions, as FSs often affect more than one region, however, we observed that the migration of a FS between regions is slow, and utilizing this comes at the expense of the detection time of the system.

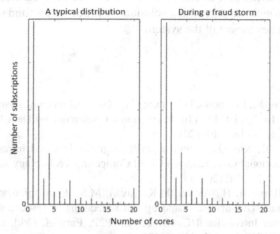

Fig. 7. A qualitative comparison of the distributions of the number of cores used typically, and during a specific fraud storm event on one of the data centers. Typical users aim to minimize their costs, and thus they mostly utilize the minimal required compute power, while fraudsters in this case aimed to maximize their utilization of the system, resulting in a peak at the high cores during a fraud storm.

Modeling fraud is a challenging task as fraudsters constantly keep learning the detection systems, and searching for new ways to bypass them. Features that distinguish between fraudulent and genuine activity can be divided to two types: those monitoring the resource consumption, and those monitoring the fraudsters' breaking-in method. While both are valuable, the latter is not necessarily steady and could very easily change. Conversely, the fraudsters' resources utilization is steadier, usually changing only with the specific fraud purpose, and thus basing a system on these features, as in our analysis, would result in a system with a longer life expectancy. Figure 7 presents an example for such feature, by comparing the typical distribution of the core usage per subscription to the one observed during a FS in one of the data centers. While the genuine user aims to minimize their cost though minimizing the resources consumption, a fraudster that managed to break into the system usually aims to maximize its utilization as much as possible. Therefore, on the right panel we observe a large peak at the high cores consumption during a FS. This information can be modeled into our system through a few possible features: the total cores consumption (as present in the above analysis), the mean core consumption per subscription, with the high values truncated to reduce the noise, and the entropy of the cores-per-subscription histogram. Note that the entropy feature has the unsteadiness disadvantage described above. In our analysis, both alternatives did not improve the prediction over the total number of cores.

Ongoing improvement of the systems is an important aspect of future research. The FraudStormML algorithm can be easily enhanced with new signals, like bandwidth and disk usage. This is important, since the typical fraudster's motivation is dynamic and changes over time, which affects the type of resources they consume. In the course of system operation, some anomalies will be detected, those that would turn out to be FP can provide new insights that would further be integrated into the system. Having a design which supports this evolutionary process is a fundamental challenge in this field, and a key aspect of the system.

References

1. Bhaduri, K., Das, K., Matthews, B.L.: Detecting abnormal machine characteristics in cloud infrastructures. In: 2011 IEEE 11th International Conference on Data Mining Workshops (ICDMW), pp. 137–144. IEEE (2011)
2. Zhu, Q., Tung, T., Xie, Q.: Automatic fault diagnosis in cloud infrastructure. In: 2013 IEEE 5th International Conference on Cloud Computing Technology and Science (Cloud-Com), pp. 467–474. IEEE (2013)
3. Hormozi, E., Hormozi, H., Akbari, M.K., Javan, M.S.: Using of machine learning into cloud environment (A Survey): managing and scheduling of resources in cloud systems. In: 2012 Seventh International Conference on P2P, Parallel, Grid, Cloud and Internet Computing (3PGCIC), pp. 363–368 (2012)
4. Beloglazov, A., Buyya, R.: Energy efficient resource management in virtualized cloud data centers. In: Proceedings of the 2010 10th IEEE/ACM International Conference on Cluster, Cloud and Grid Computing, pp. 826–831. IEEE Computer Society (2010)
5. Beloglazov, A., Abawajy, J., Buyya, R.: Energy-aware resource allocation heuristics for efficient management of data centers for cloud computing. Future Gener. Comput. Syst. **28**, 755–768 (2012)
6. Hashizume, K., Rosado, D.G., Fernández-Medina, E., Fernandez, E.B.: An analysis of security issues for cloud computing. J. Internet Serv. Appl. **4**, 1–13 (2013)
7. Fernandes, D.A., Soares, L.F., Gomes, J.V., Freire, M.M., Inácio, P.R.: Security issues in cloud environments: a survey. Int. J. Inf. Secur. **13**, 113–170 (2014)
8. Bontempi, G., Ben Taieb, S., Le Borgne, Y.-A.: Machine learning strategies for time series forecasting. In: Aufaure, M.-A., Zimányi, E. (eds.) eBISS 2012. LNBIP, vol. 138, pp. 62–77. Springer, Heidelberg (2013)
9. Aggarwal, C.C.: Outlier analysis. Springer Science & Business Media (2013)
10. Chandola, V., Banerjee, A., Kumar, V.: Anomaly detection: A survey. ACM Comput. Surv. CSUR. **41**, 15 (2009)
11. Wang, C., Talwar, V., Schwan, K., Ranganathan, P.: Online detection of utility cloud anomalies using metric distributions. In: 2010 IEEE Network Operations and Management Symposium (NOMS), pp. 96–103. IEEE (2010)
12. Wang, C., Viswanathan, K., Choudur, L., Talwar, V., Satterfield, W., Schwan, K.: Statistical techniques for online anomaly detection in data centers. In: 2011 IFIP/IEEE International Symposium on Integrated Network Management (IM), pp. 385–392 (2011)
13. Dean, D.J., Nguyen, H., Gu, X.: Ubl: unsupervised behavior learning for predicting performance anomalies in virtualized cloud systems. In: Proceedings of the 9th International Conference on Autonomic Computing, pp. 191–200. ACM (2012)
14. Vallis, O., Hochenbaum, J., Kejariwal, A.: A novel technique for long-term anomaly detection in the cloud. In: Proceedings of the 6th USENIX Conference on Hot Topics in Cloud Computing, USENIX Association, Berkeley, CA, USA, pp. 15–15 (2014)

15. Li, L., McCann, J., Pollard, N.S., Faloutsos, C.: Dynammo: Mining and summarization of coevolving sequences with missing values. In: Proceedings of the 15th ACM SIGKDD International Conference on Knowledge Discovery and Data Mining, pp. 507–516. ACM (2009)
16. Wellenzohn, K., Mitterer, H., Gamper, J., Böhlen, M.H., Khayati, M.: Missing Value Imputation in Time Series using Top-k Case Matching
17. Xie, Y., Huang, J., Willett, R.: Changepoint detection for high-dimensional time series with missing data. ArXiv Prepr. ArXiv12085062 (2012)
18. Modi, C., Patel, D., Borisaniya, B., Patel, H., Patel, A., Rajarajan, M.: A survey of intrusion detection techniques in Cloud. J. Netw. Comput. Appl. **36**, 42–57 (2013)
19. Bay, S., Saito, K., Ueda, N., Langley, P.: A framework for discovering anomalous regimes in multivariate time-series data with local models. In: Symposium on Machine Learning for Anomaly Detection, Stanford, USA (2004)
20. Rousseeuw, P.J., Driessen, K.V.: A fast algorithm for the minimum covariance determinant estimator. Technometrics **41**, 212–223 (1999)
21. Multivariate normal distribution (2015). http://en.wikipedia.org/w/index.php?title=Multivariate_normal_distribution&oldid=651587942
22. Genz, A., Bretz, F.: Computation of multivariate normal and t probabilities. Springer Science & Business Media (2009)

Flexible Sliding Windows for Kernel Regression Based Bus Arrival Time Prediction

Hoang Thanh Lam$^{(\boxtimes)}$ and Eric Bouillet

IBM Research, Building 3, IBM Technology Campus, Damastown, Dublin 15, Ireland
{t.l.hoang,bouillet}@ie.ibm.com

Abstract. Given a set of historical bus trajectories D and a partially observed bus trajectory S up to position l on the bus route, kernel regression (KR) is a non-parametric approach which predicts the arrival time of the bus at location $l+h$ $(h > 0)$ by averaging the arrival times observed at same location in the past. The KR method does not weights the historical data equally but it gives more preference to the more similar trajectories in the historical data. This method has been shown to outperform the baseline methods such as linear regression or k-nearest neighbour algorithms for bus arrival time prediction problems [9]. However, the performance of the KR approach is very sensitive to the method of evaluating similarity between trajectories. General kernel regression algorithm looks back to the entire trajectory for evaluating similarity. In the case of bus arrival time prediction, this approach does not work well when outdated part of the trajectories does not reflect the most recent behaviour of the buses. In order to solve this issue, we propose an approach that considers only recent part of the trajectories in a sliding window for evaluating the similarity between them. The approach introduces a set of parameters corresponding to the window lengths at every position along the bus route determining how long we should look back into the past for evaluating the similarity between trajectories. These parameters are automatically learned from training data. Nevertheless, parameter learning is a time-consuming process given large training data (at least quadratic in the training size). Therefore, we proposed an approximation algorithm with guarantees on error bounds to learn the parameters efficiently. The approximation algorithm is an order of magnitude faster than the exact algorithm. In an experiment with a real-world application deployed for Dublin city, our approach significantly reduced the prediction error compared to the state of the art kernel regression algorithm.

1 Introduction

Recently, bus arrival time prediction using GPS data has become an important problem attracting many researchers from both academia and industry labs [9,11]. Beside having important application in urban transportation management systems, GPS data providing accurate real-time locations of buses is very cheap and easy to collect. In such system, GPS devices equipped on-board continuously update real-time locations of the buses in a reasonable fine-grained

© Springer International Publishing Switzerland 2015
A. Bifet et al. (Eds.): ECML PKDD 2015, Part III, LNAI 9286, pp. 68–84, 2015.
DOI: 10.1007/978-3-319-23461-8_5

time resolution (from seconds to minutes). Updated locations of the buses are used to predict bus arrival time at any location of the trajectory. Prediction can be used to provide urban citizens with real-time information about bus arrival time at any bus stop or to estimate the likelihood of bus bunching from which bus operators can direct bus drivers to avoid those unexpected events.

Literature on bus arrival time prediction problem is very rich [2-3,6-14]. Depending on type of data used for prediction, different methods were proposed. However, when only GPS data is available, the *state-of-the-art* algorithm is relied on the *kernel regression* (KR) method [9]. The KR algorithm exploits the similarity of the currently observed trajectory of the bus and the historical trajectories to make prediction. Since behaviour of buses travelling on a fixed bus route is highly repeated, KR approach has been shown to outperform the baseline approaches such as linear regression or k-nearest neighbour [9].

Example 1 (Kernel regression). In order to illustrate the intuition behind the KR method we show an example with 4 different historical spatio-temporal trajectories A,B,C,D and a partially observed trajectory S at location l in Figure 1. The y-axis shows the time offsets since the time when the buses start and the x-axis shows the distance (in meters) from the departure stop.

Prediction of the arrival time of the bus at location $l + h$ $(h > 0)$ denoted as $\hat{S}(l + h)$ is done by averaging the arrival time of the bus at location $l + h$ observed in the historical data:

$$\hat{S}(l + h) = \alpha_A * A(l + h) + \alpha_B * B(l + h) + \alpha_C * C(l + h) + \alpha_D * D(l + h)$$

Where the weight values $\alpha_A, \alpha_B, \alpha_C, \alpha_D$ summing up to 1 and can be calculated as follows:

$$\alpha_A = \frac{sim(S,A)}{sim(S,A)+sim(S,B)+sim(S,C)+sim(S,D)}$$

$$\alpha_B = \frac{sim(S,B)}{sim(S,A)+sim(S,B)+sim(S,C)+sim(S,D)}$$

$$\alpha_C = \frac{sim(S,C)}{sim(S,A)+sim(S,B)+sim(S,C)+sim(S,D)}$$

$$\alpha_D = \frac{sim(S,D)}{sim(S,A)+sim(S,B)+sim(S,C)+sim(S,D)}$$

Function $sim(S, A)$ denotes the similarity between two trajectories S and A till position l. The larger the value of $sim(S, A)$ the more similar the two trajectories are.

An benefit of using KR in practice is that there is no need to build a model for every location along the bus route as does with parametric approaches such as linear regression. This property is a big plus in an industrial setting when fast deliver of solution is required. However, KR is sensitive to the choice of the similarity function which usually considers the whole trajectory for similarity evaluation [9]. Under the context of bus arrival time prediction application, bus journeys might be influenced by hidden spatial or temporal contextual factors such as changes from crowded to less traffic locations or unplanned events like accidents. In such cases, the entire trajectory is no longer relevant for making prediction because most of the information is out of date and does not reflect

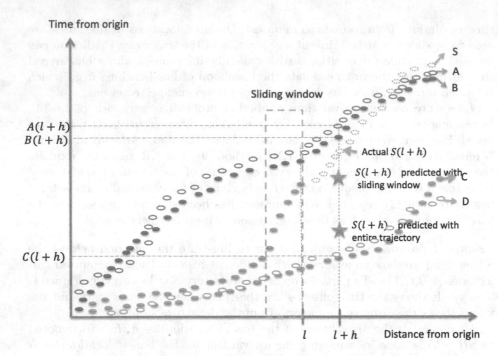

Fig. 1. Four trajectories A, B, C and D in a historical log and a partially observed (orange solid points) bus trajectory S up to location l (the orange dotted points are future positions). As we can observe on the figure, the bus was moving very close to C and D from the beginning but due to some unplanned events happening at location l the bus was delayed and started moving closer to A and B during the last part of the journey. At location l, considering the whole trajectory, S is much closer to C and D than to A and B. On the other hand, in the sliding window with only recent data, S is much closer to A and B. Therefore, making prediction based on recent data is more accurate as it captures recent behaviours of the bus.

the recent behaviour of the buses. The following example shows a situation in which the KR method is sensitive to the choice of similarity evaluation methods.

Example 2 (Sensitivity to similarity function). Figure 1 shows four historical trajectories A, B, C and D and a partially observed (orange solid points) trajectory S at location l (the orange dotted points are future positions). Prediction of bus arrival time at location $l + h$ is made by copying information about arrival times of the bus at location $l + h$ from similar historical trajectories.

Different predictions (marked with star-like symbols) of bus arrival time $\hat{S}(l+h)$ at location $l + h$ are made using either the entire trajectory or only recent data in a sliding window with size $w > 0$. As we can observe on the figure, the bus on the journey S was moving very close to C and D from the beginning but due to some reasons, e.g. an unplanned event happening at location l, the bus was delayed and started moving closer to A and B during the last part of the journey. At location l, considering the whole trajectory, S is much closer to

C and D than to A and B. On the other hand, in the sliding window (marked with a dotted orange box) with only recent data, S is much closer to A and B. Therefore, making prediction based on recent data in that window is more accurate as it captures the recent behaviour of the bus.

The key assumption in Example 2 is that each location along the bus route is associated with a hidden spatial or temporal context that determines how long we should look back to the past for predicting the future bus arrival times. An important question to ask is: how do we discover relevant window size for each location along the bus route to make prediction better? In this work, we try to answer this question by proposing a method that automatically learns appropriate window sizes from training data. The key technical contributions of this work can be summarized as follows:

- A method to optimize the kernel regression based prediction algorithm for a real-world application relied on flexible sliding window length learning.
- An approximation algorithm (with error bound guarantees) for speeding up the parameter learning process when the training size is large.
- Our method reduced the prediction error from 40-60 %.
- The approximation algorithm proposed for the window size learning task achieved a speeds up of 15-20x while preserving high accuracy of the prediction as the brute-force learning method does.
- Our approach speeds up real-time evaluations of the similarity between trajectories as only short parts of the trajectories are considered for evaluation.

2   Problem Definition

In this section, we first recall the kernel regression algorithm, and then formally define the problem. All the notations are shown in Table 2.

Notation	Meaning
$S = t_1 t_2, \cdots, t_n$	A trajectory with arrival times at n locations
$S(l)$	Arrival time at location l
$\hat{S}(l)$	Predicted arrival time at location l of trajectory S
$\hat{S}^w(l)$	Predicted arrival time when window length is w
$S(l, h)$	A sequence of arrival times from location l to h
$D = \{A_1, A_2, \cdots, A_m\}$	A reference set with m historical trajectories
$Sim(A, B)$	A similarity function between two sequences A and B
$e_l(w)$	Prediction error at location l when window size is w
$E[e_l(w)]$	Expectation of prediction error when window size is w
w_l^*	The best window length at location l

2.1   Kernel Regression

A bus trajectory is a sequence of pairs $S = (p_1, t_1), (p_2, t_2), \cdots, (p_n, t_n)$ where p_i is a location on the bus route and t_n is the arrival time at that location counting from the beginning of the journey. In our bus arrival time prediction application, bus route is fixed beforehand and GPS-based locations are interpolated such that the trajectory contains arrival time for every location on the route at the distance one meter. Therefore, a bus trajectory can be considered as a sequence of arrival times $S = t_1 t_2, \cdots, t_n$ because the corresponding location can be implicitly inferred from the context with the index of the arrival time.

A historical dataset is a collection of bus trajectories $D = \{A_1, A_2, \cdots, A_m\}$ each has length equal to n. Given a trajectory A_i, denote $A_i(l)$ as the bus arrival time at location l and $A_i(l, h)$ $(h > l)$ as the sequence of arrival times starting from location l to location h: $A_i(l+1)A_i(l+2) \cdots, A_i(h)$. Let denote $Sim(A, B)$ as a similarity function between two sequences with the same length A and B. The larger value of $Sim(A, B)$ is the more similar the sequences are.

Let S be a partially observed trajectory up to location l, from now on we call S the *target trajectory* while the trajectories in historical data are called as *reference trajectory*. Assume that our main goal is to predict the arrival time at location $l + h$ for a prediction horizon $h > 0$ of the target trajectory, i.e predict the unobserved value of $S(l + h)$. The kernel regression algorithm estimates the value $\hat{S}(l + h)$ with the help of the historical dataset D as follows:

$$\hat{S}(l + h) = \frac{\sum_{i=1}^m Sim(S(0, l), A_i(0, l)) * A_i(l + h)}{\sum_{i=1}^m Sim(S(0, l), A_i(0, l))} \tag{1}$$

The prediction is made by averaging observations of arrival time at location $l + h$ in the historical data. Every observation $A_i(l + h)$ is weighted by the similarity between the current trace S and the historical trace A_i. For the bus arrival time prediction application, Sinn et al. [9] suggested to use the *Gaussian kernel* $Sim(A, B) = e^{-\sum_{i=0}^l \frac{1}{b} \frac{(A(i) - B(i))^2}{\sigma_i}}$ where σ_i denotes the variance of arrival time at location i and b is a bandwidth parameter.

2.2   Problem Definition

As we have discussed earlier, the KR algorithm is sensitive to the way we evaluate the similarity between the target trajectory and the reference trajectories. The state-of-the-art algorithm considers the entire trajectory for evaluating the similarity score. In this work, we introduce a method that uses recent data for evaluating the similarity score. Let w_l $(w_l \leq l)$ denote the length of the sliding window which we use to calculate the similarity score at position l. In particular, the $Sim(S(0, l), A_i(0, l))$ function in the kernel regression method is replaced by $Sim(S(l - w_l, l), A_i(l - w_l, l))$. Therefore, for a bus route with length n we have a vector of n window lengths $W = (w_1, w_2, \cdots, w_n)$

An important question is how we determine an appropriate window length w_l for every location l along the bus route. Finding optimal window lengths can

---

Algorithm 1. A brute-force algorithm

1: **Input**: a training set $D = \{A_1, A_2, \cdots, A_m\}$ and a prediction horizon h
2: **Output**: A vector of window lengths $W = \{w_1^*, w_2^*, \cdots, w_n^*\}$
3: **for** l=1 to n **do**
4:     **for** w=1 to l **do**
5:         **for** i=1 to m **do**
6:             $e_l^i(w) \leftarrow (A_i(l+h) - \hat{A}^{w_i}(l+h)^2$
7:             $e_l(w) \leftarrow e_l(w) + e_l^i(w)$
8:         **end for**
9:         $e_l(w) \leftarrow \frac{e_l(w)}{m}$
10:     **end for**
11:     $w_l^* = argmin_w e_l(w)$
12: **end for**

---

be formulated as an optimization problem as follows. We use the set of historical trajectories $D = \{A_1, A_2, \cdots, A_m\}$ as a training set and evaluate the set of window lengths through a leave-one-out cross-validation process.

In particular, let h be a horizon for prediction, l be a location. We pick A_i from the training set and consider it as a target trajectory while the set $D \setminus \{A_i\}$ is considered as a reference set. The prediction error at location l when the window length is set to w can be calculated as follows:

$$e_l^i(w) = (A_i(l+h) - \hat{A}_i^w(l+h))^2 \tag{2}$$

Where $\hat{A}_i^w(l+h)$ denotes the predicted arrival time at location $l+h$ when the sliding window length is set to w. The leave-one-out cross-validation average prediction error over the entire training set D can be estimated as : $e_l(w) = \frac{\sum_{i=1}^m e_l^i(w)}{m}$. The problem of learning the optimal window length at location l can be formulated as follows:

Problem 1 (Window length learning). For every location l, find the window length w_l^* which minimizes the expectation of the leave-one-out cross-validation average prediction error $e_l(w)$, i.e. $w_l^* = argmin_w E[e_l(w)]$

3 A Brute-Force Algorithm

In this section, we introduce a brute-force search approach that solves Problem 1. It evaluates all possible values of the window length at every location in the bus route. At location l, since $0 < w < l$ we simply calculate $e_l(w)$ for every possible value of w and choose w_l^* that minimizes the accumulative prediction error $e_l(w)$. The same task can be repeated for every location $1 \leq l \leq n$ to obtain a complete set of window lengths for every location along the bus route.

Algorithm 1 describes the main steps of the brute-force algorithm. It iterates over every location l along the bus route (lines 3-12) and evaluate the accumulative prediction error $e_l(w)$ for every candidate window length w (lines 4-10). Evaluation of the accumulative prediction error is done by summing up the prediction error made when each trajectory A_i is picked as a target and the remaining trajectories are combined to a reference set (lines 5-8). In line 11, the best window length w_l^* for location l is selected as the one that minimizes the accumulative prediction error.

The complexity of Algorithm 1 is $O(m^2n^2)$ because of the three outer loops and the computation of the prediction error in lines 5-8 (requiring another loop over the training set). It is important to notice that the calculation of the prediction error at location l in lines 5-8 requires a full pass over the training data which incurs a complexity of $O(mn)$. However, this number can be reduced to $O(m)$ when the computation only needs to update from the result of the prior step at location $l - 1$. In general, the brute-force algorithm is not scalable especially when the training size is large because of the quadratic complexity. In the next subsection, we will discuss an approximation algorithm that speeds up the brute-force method significantly.

4   An Approximation Algorithm

The complexity of Algorithm 1 is $O(m^2n^2)$, where m is the training size and n is the trajectory length. Since trajectory length is usually fixed, it can be considered as a large constant number. The training size m is equal to the number of historical trajectories collected so far. The training size increases as long as data is collected everyday, so we propose a method called FLOW (flexible sliding window for kernel regression) which approximates the solution of Problem 1 by optimizing the expensive computation caused by the training size m.

4.1   Approximation Algorithm

Recall that in Algorithm 1, for every candidate window length w, the prediction error $e_l(w)$ is accumulated by looping over every trace A_i in the training set. Our key intuition is that at a certain iteration i where i is large enough, we can estimate how likely a window length w is the optimal window length for the given location. For instance, if we observe that the value of $e_l(w_1)$ is significantly less than $e_l(w_2)$ then we can conclude that w_2 has a very low chance of being the optimal window length. In that case, we can stop evaluating $e_l(w_2)$ as long as the accuracy of the learning process is not too much sacrificed.

Algorithm 2 is not much different from Algorithm 1. It iterates over every location l on the bus route and find the best window length among a set of candidates stored in a list L (lines 3-4). Different from the brute-force algorithm, in each iteration of the leave-one-out cross-validation (lines 5-15), FLOW checks if the remaining candidates in the list L have very low probability (less than a user defined parameter ϵ) of being the best window (line 11). If the answer is yes then the candidates are removed from the list (line 12) and never be considered in the following iterations. In experiments, we will show that Algorithm 2 is much more efficient than the brute-force algorithm because it

---

Algorithm 2. FLOW algorithm

1: **Input**: a training set $D = \{A_1, A_2, \cdots, A_m\}$, a parameter ϵ and a prediction horizon h
2: **Output**: A vector of window lengths $W = \{w_1^*, w_2^*, \cdots, w_n^*\}$
3: **for** l=1 to n **do**
4:     $L \longleftarrow= \{1, 2, \cdots, l\}$
5:     **for** i=1 to m **do**
6:         **for** each w in L **do**
7:             $e_l(w) \leftarrow \frac{e_l(w)*(i-1)+(A_i(l+h)-\hat{A}_i^w(l+h))^2}{i}$
8:         **end for**
9:         $w^* \leftarrow argmin_w e_l(w)$
10:        **for** each w in L **do**
11:            **if** $e_l(w) > e_l(w^*) + \Delta(i, \epsilon)$ **then**
12:                L.remove(w)
13:            **end if**
14:        **end for**
15:    **end for**
16:    $w_l^* = argmin_w e_l(w)$
17: **end for**

---

prunes the computation significantly. An important point need to explain is the condition used for pruning a candidate w:

$$e_l(w) > e_l(w^*) + \Delta(i, \epsilon) \tag{3}$$

Where w^* is the current best candidate (up to the current iteration of the loop in line 5) with the smallest accumulative prediction error, i.e. $w^* \leftarrow argmin_w e_l(w)$, and $\Delta(i, \epsilon)$ is a function that depends on the tolerance parameter ϵ and the current iteration i.

The tolerance $\epsilon << 1$ is given as an input parameter which tells the program the desired bounded probability of missing the best window length during the search process. In the next subsection we will show how to derive the value of the function $\Delta(i, \epsilon)$ in each iteration. The key meaning of the pruning condition is: if the accumulative error of the candidate w is far deviated from the accumulative prediction error of the best candidate by a large number $\Delta(i, \epsilon)$ then it is safe to prune w from the list of candidates with a small chance of missing the optimal candidate (less than ϵ).

4.2 Theoretical Analysis

Recall that at iteration i, $e_l(w) = \frac{\sum_{k=1}^i e_l^k(w)}{i}$, where $e_l^k(w)$ is the prediction error when the A_k trajectory is picked as a target. Our main assumption in the analysis is that each $e_l^k(w)$ is a random variable with the same mean value and they are independent to each other. Therefore, from the definition of $e_l^k(w)$ for any k: $E[e_l^k(w)] = E[e_l(w)]$.

Let us denote the maximum and the minimum value of the arrival time at location $l+h$ as max and min. The following theorem shows how to calculate the function $\Delta(i, \epsilon)$ to guarantee that the probability of missing the optimal window length is less than ϵ:

Theorem 1 (Error Bound). *In Algorithm 2, if in each iteration we choose* $\Delta(i, \epsilon) = \frac{\sqrt{2}(max-min)^2}{\sqrt{i}} \log \frac{2}{\epsilon}$ *then the probability that FLOW misses the optimal candidate is upper-bounded by ϵ.*

Proof. Because of space limit, please refer to the link[1] to see the proof.

Theorem 1 shows that the value of function $\Delta(i, \epsilon)$ decays linearly with the value of $\sqrt{i}$. Therefore, when i is large enough FLOW can prune the computation substantially.

4.3   Further Optimization

In order to find the best window length for any location l along the bus route, we need to evaluate all values of window length $1 \leq w \leq l$. Our observation with the bus dataset shows that the prediction error in the leave-one-out cross-validation process is a smooth function of window lengths. Its value does not change significantly when the window length is slightly modified.

Therefore, instead of considering all values of w, we limit the search for the best window length to a subset with maximum $\log(l)$ values: $\{l, \lceil \frac{l}{2} \rceil, \lceil \frac{l}{2^2} \rceil, \cdots, 1\}$. This well-known technique in data stream called as pyramidal time frame [1] usually used for reducing the search space with preference to recent part of data. With this minor change to the FLOW and the brute-force algorithm we can reduce the worst-case complexity to $m^2 n \log(n)$. This techniques allows us to speed up the search algorithm especially for the case when n is large. In experiments, the results were reported when both brute-force and FLOW algorithm used this technique to reduce the search space.

5   Experiments

We will first start with a subsection that describes detail information about the dataset used in the paper and the experiment settings. Subsequently, we will report the empirical results including the prediction accuracy and the effectiveness of the approximation method.

The original implementation of the KR algorithm was chosen for comparison because it is considered as the state-of-the-art algorithm for this application. Finally, we performed several analyses on the distribution of the optimal window lengths chosen by our learning method for every location along the bus route to understand the hidden contexts existing in the data.

---

[1] https://drive.google.com/file/d/0BwWtvZfA5UCSUHpfS0d3a2pMVTQ/view?usp=sharing

46A bus trajectories

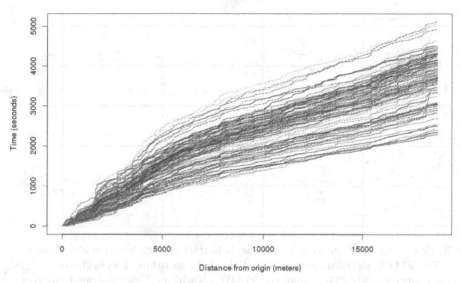

Fig. 2. One hundred bus trajectories sampled from the 46A bus dataset. The length of the journey is about 18 Km, in average, buses need one hour to complete the journey.

5.1   Dataset and Experiment Settings

The dataset used for empirical study consists of 1500 bus traces from the bus route 46A (outbound) in Dublin city in a period of one month. This route was chosen because it is the most frequent route in the city. The dataset is available for download at Dublinked[2]. Each bus trace was created from GPS data updated every 20-30 seconds. The GPS trajectories were projected to the known bus route and the positions were interpolated at one-meter long resolution.

All traces can be considered as sequences with the same length with 18587 data points corresponding to a 18 Km long bus route. More detail about the dataset and the preprocessing steps can be found in [4,9]. All the source codes were written in Java and the program was run on a Linux machine with 4GB of RAM. The implementation of the KR algorithm for comparison follows the description in Sinn et al. [9]. In that implementation, a parameter needs to set concerning the bandwidth of prediction. We fixed that value to 1 as suggested in the original work to ensure a fair comparison [9].

Recall that the FLOW algorithm uses the tolerance parameter ϵ to control the early pruning strategy. In our experiments, the tolerance parameter ϵ was set to 0.01 which limits the probability of missing the optimal window length to less than one percent. In addition to that, the Max and Min values in the

---

[2] http://www.dublinked.ie/

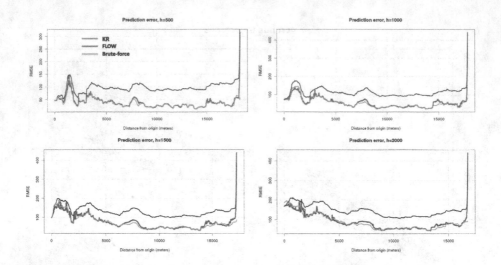

Fig. 3. Prediction error (in seconds, smaller is better) at each location along the bus route. The FLOW algorithm outperformed the KR algorithm. The brute-force algorithm was only slightly better than the FLOW algorithm. This confirmed our theoretical analysis in section 2 which shows that we didn't loose a lot of accuracy when approximation algorithm is used instead of the brute-force search.

formula calculating the pruning factor $\Delta(i, \epsilon)$ were set to the maximum and the minimum value of the arrival time observed in the training data.

5.2   Prediction Accuracy

In this subsection, we show the prediction accuracy calculated as rooted mean square error (RMSE smaller is better) at every location along the bus route. Given a test set $T = \{T_1, T_2, \cdots, T_M\}$ and a prediction horizon value h, the root mean square prediction error at location l is calculated as follows:

$$RMSE_l = \sqrt{\sum_{i=1}^{M}(T_i(l + h) - \hat{T}_i^{w^i}(l + h))^2} \qquad (4)$$

The experimental settings and the dataset were kept to the same as described in [9]. Since Sinn et al. has shown that the KR method outperformed the linear regression and the k-nearest neighbour algorithm, so in this work we just need to compare FLOW directly with to the original implementation of the kernel regression algorithm denoted as KR.

Figure 3 shows the RMSE (in seconds on the y-axis) at every location along the bus route (x-axis) when the prediction horizon h was set to 500, 1000, 1500

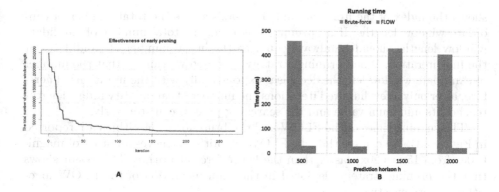

Fig. 4. (A). With the early pruning strategy, FLOW can prune a substantial amount of computation as the number of window length candidates decreases exponentially with the number of iterations. (B). Comparison of running time between the FLOW and brute-force algorithms when the prediction horizon is varied. FLOW achieves about an order of magnitude (15-20x) faster than the brute-force algorithm.

and 2000 meters respectively. The RMSE were reported via a ten-fold cross-validation. The first impression from Figure 3 is that the FLOW algorithm out-performed the KR algorithm with 40-60 % reduction in prediction error. The results are stable across different locations and with various prediction horizons.

Moreover, The difference between the FLOW and the brute-force algorithm is negligible. In fact, as we can see from the plots, the brute-force algorithm was only slightly better than the FLOW algorithm. This empirical results confirmed our theoretical analysis in section 3 that we didn't loose much accuracy when the approximation algorithm was used instead of the brute-force algorithm.

At the locations near to the end of the route, we can see that the errors increase significantly. The reason is that in the set of trajectories there are a few outliers on which the bus needs about more than four hours to complete the journey instead of one hour in average as usual. For those outlier traces, delay happened close to the end of the journey which explains why we see a peak in prediction error at the end of the trajectories.

5.3 Effectiveness of Approximation

In subsection 5.2, we have shown that the prediction accuracy of the approxima-tion algorithm is very similar to the prediction error of the brute-force algorithm. In this subsection, we will show that the FLOW algorithm is significantly more efficient than the brute-force algorithm.

First, recall that the approximation algorithm works by early pruning the set of candidate window lengths that with high confidence cannot be the opti-mal window length. In order to evaluate how effective the pruning strategy is in practice we plotted the number of candidate window lengths observed in each iteration (lines 5-15) of Algorithm 2 in Figure 4.A. In that figure, the x-axis shows the index of the iteration and the y-axis shows the total number of candidate window lengths. If no pruning was used, the total number of candidate window lengths should be always equal to the initial number of candidates at the first iteration. When pruning strategy was used we can see that the number of candidate window lengths decreases exponentially with the iterations. Therefore, after only a few hundred iterations the number of candidate window lengths reaches its minimum value and the searching process can stop early.

The running times of the FLOW and the brute-force algorithms are reported in Figure 4.B. We can see that the FLOW algorithm achieves an order of magnitude (from 15x to 20x) faster than the brute-force algorithm. This result shows that the pruning strategy deployed in the implementation of the FLOW algorithm is very effective.

5.4 Interpretation of the Results

The distribution of the best window length at different locations along the bus route is shown in Figure 5. As we can observe, in most location, FLOW only picks a few recent data points. Thanks to this, evaluation of similarity between the target and the reference trajectories is very efficient because the window length is very short. Interestingly, there are several locations in which the window length suddenly increases. This may concern a hidden spatio-temporal context that causes the change. These shifting contexts might provide bus operators with deep insights about the data.

6   Related Work

Recently, bus arrival time prediction problem attracts a lot of attention because of its useful application in public transport management systems. The most popular methods were relied on artificial neutral networks (ANNs) [2,3,8]. The issue with an ANN is that it is very sensitive to the network structure design and easily overfits data [15]. Besides ANNs, methods based on Kalman filters are also very popular. For instance, Wall et al. [12], Son et al. [10] and Yang et al. [14] combined data from automatic vehicle location services and historical data to make bus arrival time prediction.

Other machine learning approaches have also been proposed for this problem. Li et al. [7] used linear regression model with fused data from different sources such as GPS sensors, wired loop sensors and red radio radar etc. for bus arrival time prediction. Zhou et al. [16] used mobile sensing data from participating users for making prediction. Bin et al. [15] used support vector machine (SVM) using different features extracted from weather condition, type and time of the date, travel time in the previous segments. The SVM method has been shown to be superior to the methods based on ANN.

Under the context of the bus arrival time prediction problem, no method works well for all applications because each of them requires a specific type of

Average window lengths

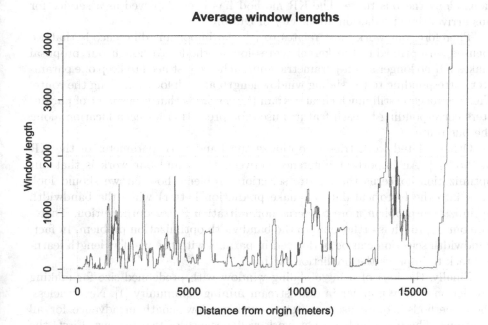

Fig. 5. Distribution of window length (y-axis) as a function of the location (x-axis). We can see that in most case FLOW only picks a few recent data. Interestingly, there are several locations in which the window length suddenly increases. This may concern a hidden spatio-temporal context that causes the change. These shifting contexts might provide bus operators with deep insights about the data.

data used for prediction. In practice, not always different types of data are available, e.g. mobile sensing data is only owned by telco companies while GPS data is collected by bus operator companies. When only GPS data is available, there are several approaches [9,11,13,17]. Nevertheless, the state-of-the-art algorithm for the bus arrival time prediction problem is relied on the kernel regression algorithm [9] in which Sinn et al. showed that the kernel regression approach outperformed the other methods based on linear regression models and k-nearest neighbour prediction algorithms.

Another reason that makes kernel regression attractive is that it is a non-parametric approach. Therefore, we don't need to learn different predictive models for every location along the bus route. It doesn't require intensive human efforts for feature extraction and selection. This property enables us to deploy scalable online prediction algorithms for large-scale applications because it does not require expensive training tasks and bookkeeping a model for every locations along the bus route. The KR method has been deployed as a service for bus arrival time prediction at the city of Dublin[3].

Therefore, our work most relates to [9]. An important difference is that we focused on optimizing the kernel regression methods. Although our proposal makes KR no longer a non-parametric approach, we just need to keep one parameter corresponding to the sliding window length at each location along the route. This approach is still much cheaper than the methods that keeps a set of parameters corresponding to each features used for prediction for each location along the bus route.

Other related work tries to optimize the bandwidth parameter of the KR method [5]. An important difference between those and our work is that our optimization concerns the feature selection problem (how far we should look back into the historical data to make prediction better) while the bandwidth optimization problem more concerns normalization factor optimization. Therefore our approach is orthogonal to the bandwidth optimization problem. In fact, bandwidth selection can be performed in parallel with the window length learning task to improve the prediction further.

Finally, the idea of using a sliding window with predefined size for making prediction is very popular in data stream mining community [1]. Nevertheless, these methods require users to set a fixed window length in advance for all locations. These methods do not work well because of two reasons. First, the users do not know which value they should choose for the window size. Second, the optimal window lengths as observed in Figure 5 vary a lot depending on the location along the bus. Different from those work, our method can automatically learn appropriate window lengths for every location along the bus route.

7   Conclusion and Future Work

In this work, we exploited the implicit spatial or temporal contexts to improve the prediction accuracy of the state of the art prediction algorithm for bus arrival time prediction problem with GPS data. Our algorithm searches for relevant data at each location that needs to use for improving the prediction. The results with a real-world dataset show that our method can improve the prediction accuracy significantly (from 40-60 % reduction in RMSE). Since the learning algorithm is time consuming, we proposed an approximation algorithm for the learning process which reduces the learning time significantly (15x-20x faster).

There are several possibilities to extend the current work. For instance, our algorithm was proposed for static data. It is interesting to discover and search for relevant data in an online settings to capture the real-time effect of several unplanned events such as accidents. Another important problem needs to solve is how to associate the discovered window lengths with the hidden spatial contextual information in order to better understand the behaviour of the bus on each segment of the route.

---

[3] http://dublinbus.ie/

Acknowledgments. We would like to thank Dublin City Council (DCC) for providing us with the Dublin bus dataset. We also thank Dr. Mathieu Sinn for his useful comments during the development of the work.

References

1. Aggarwal, C.C. (ed.): Data Streams - Models and Algorithms, Advances in Database Systems, vol. 31. Springer (2007)
2. Chen, M., Liu, X., Xia, J., Chien, S.I.: A dynamic bus-arrival time prediction model based on APC data. In: Computer-Aided Civil and Infrastructure Engineering, pp. 364–376, July 2004
3. Chien, S., Ding, Y., Wei, C.: Dynamic bus arrival time prediction with artificial neural networks. Journal of Transportation Engineering **128**(5), 429–438 (2002)
4. Coffey, C., Pozdnoukhov, A., Calabrese, F.: Time of arrival predictability horizons for public bus routes. In: Proceedings of the 4th ACM SIGSPATIAL International Workshop on Computational Transportation Science, CTS 2011, pp. 1–5. ACM, New York (2011)
5. Hardle, W., Marron, J.S.: Optimal bandwidth selection in nonparametric regression function estimation. The Annals of Statistics, 1465–1481 (1985)
6. Hoeffding, W.: Probability inequalities for sums of bounded random variables. Journal of the American Statistical Association **58**(301), 13–30 (1963)
7. Li, F., Yu, Y., Lin, H., Min, W.: Public bus arrival time prediction based on traffic information management system. In: IEEE International Conference on Service Operations, Logistics, and Informatics (SOLI), pp. 336–341, July 2011
8. Mazloumia, E., Rosea, G., Curriea, G., Sarvia, M.: An integrated framework to predict bus travel time and its variability using traffic flow data. Journal of Intelligent Transportation Systems: Technology, Planning, and Operations (2011)
9. Sinn, M., Yoon, J.W., Calabrese, F., Bouillet, E.: Predicting arrival times of buses using real-time gps measurements. In: 15th International IEEE Conference on Intelligent Transportation Systems (ITSC), pp. 1227–1232, September 2012
10. Son, B., Kim, H.-J., Shin, C.-H., Lee, S.-K.: Bus arrival time prediction method for ITS application. In: Negoita, M.G., Howlett, R.J., Jain, L.C. (eds.) KES 2004. LNCS (LNAI), vol. 3215, pp. 88–94. Springer, Heidelberg (2004)
11. Sun, D., Luo, H., Fu, L., Liu, W., Liao, X., Zhao, M.: Predicting bus arrival time on the basis of global positioning system data. Transportation Research Record: Journal of the Transportation Research Boardg (2007)
12. Wall, Z., Dailey, D.J.: An algorithm for predicting the arrival time of mass transit vehicles using automatic vehicle location data. In: 78th Anual Meeting of the Transportation Research Board (1999)
13. Xinghaoa, S., Jinga, T., Guojuna, C., Qichongb, S.: Predicting bus real-time travel time basing on both GPS and RFID data. In: 13th COTA International Conference of Transportation Professionals (CICTP 2013), pp. 2287–2299, November 2013
14. Yang, J.-S.: Travel time prediction using the GPS test vehicle and kalman filtering techniques. In: Proceedings of the 2005 American Control Conference, vol. 3, pp. 2128–2133, June 2005

15. Yu, B., Yang, Z., Yao, B.: Bus arrival time prediction using support vector machines. Journal of Intelligent Transportation Systems, 151–158, July 2006
16. Zhou, P., Zheng, Y., Li, M.: How long to wait?: Predicting bus arrival time with mobile phone based participatory sensing (2013)
17. Zhu, T., Ma, F., Ma, T., Li, C.: The prediction of bus arrival time using global positioning system data and dynamic traffic information. In: Wireless and Mobile Networking Conference, pp. 1–5, October 2011

Learning Detector of Malicious Network Traffic from Weak Labels

Vojtech Franc[1,2]($\boxtimes$), Michal Sofka[1], and Karel Bartos[1]

[1] Cisco Systems, Prague, Czech Republic
xfrancv@cmp.felk.cvut.cz
[2] Faculty of Electrical Engineering, Department of Cybernetics,
Czech Technical University in Prague, Prague, Czech Republic

Abstract. We address the problem of learning a detector of malicious behavior in network traffic. The malicious behavior is detected based on the analysis of network proxy logs that capture malware communication between client and server computers. The conceptual problem in using the standard supervised learning methods is the lack of sufficiently representative training set containing examples of malicious and legitimate communication. Annotation of individual proxy logs is an expensive process involving security experts and does not scale with constantly evolving malware. However, weak supervision can be achieved on the level of properly defined bags of proxy logs by leveraging internet domain black lists, security reports, and sandboxing analysis. We demonstrate that an accurate detector can be obtained from the collected security intelligence data by using a Multiple Instance Learning algorithm tailored to the Neyman-Pearson problem. We provide a thorough experimental evaluation on a large corpus of network communications collected from various company network environments.

Keywords: Computer security · Malware detection · Multiple-instance learning · Support vector machines

1 Introduction

Recent report has revealed that 100 percent of all investigated corporate networks had malicious traffic going to web sites that host malware [1]. Detecting malware infections is challenging since malware is rapidly evolving, the complexity of the attacks is increasing, and the number of different variants is rising. This security challenge creates a need for an automated analysis and detection of malware. We propose a learning-based system leveraging publicly available blacklists of malware domains to train a detector that automatically finds malicious communication.

Traditional approaches to network analysis rely on extracting communication patterns from HTTP proxy logs that are distinctive for malware [11]. The pattern matching is fast but it is difficult to keep up with constantly changing malware. Behavioral techniques extract features from the proxy log fields and build a

© Springer International Publishing Switzerland 2015
A. Bifet et al. (Eds.): ECML PKDD 2015, Part III, LNAI 9286, pp. 85–99, 2015.
DOI: 10.1007/978-3-319-23461-8_6

detector that generalizes to the particular malware family exhibiting the targeted behavior [10]. Previous algorithms were engineered to specific types of malicious activity such as spam [6], phishing [4], or communication with a command and control server of the attacker [9].

Our system extracts a number of generic features from proxy logs of HTTP requests and trains a detector of malicious communication using publicly-available blacklists of malware domains [7]. Since the blacklists contain labeling only at the level of domains while the detector operates on richer proxy logs with a full target web site URL, the labeled domains only provide weak supervison for training. We propose a variant of the Multiple Instance Learning algorithm (MIL) [5] that uses bags of proxy logs describing communication of users to the black-listed domains instead of manually labeled positive examples of network communication. The proposed MIL algorithm seeks for the Neyman-Pearson detector with a very low false positive rate that is necessary in the real-life deployment of the system. The resulting non-convex learning problem is solved by Averaged Stochastic Gradient Descent [13].

The algorithm results in a generic system that can recognize malicious traffic by learning from weak annotations. This simplifies the update process based on newly discovered threats since the detector of malicious HTTP requests can be reliably trained from domain blacklists. We show on an extensive dataset of traffic logs obtained from a large corporate network that the algorithm reliably detects new malware while keeping low false positive rates.

The paper is organized as follows. The malware detection problem is briefly described in Section 2. Formulation of the learning problem and its solution is presented in Section 3. Section 4 details the used database of the network traffic and the data representation. Experiments are given in Section 5 and Section 6 concludes the paper.

2   Malware Detection

The initial computer infection with a malware begins by executing a malicious program, for example by clicking on a link embedded in a website or an email. The reasons behind the infections are different (usually prompted by financial gain through ex-filtrating and abusing sensitive data) but in all cases the attacker (and the malware) needs to communicate over the network connection. The communication can involve scanning for potential targets, initial download of the malicious binary or library, or connection to the command and control server maintained by the attacker.

In a network analysis system, the HTTP communication is captured by network proxy logs (also called flows) that contain several fields specific to the HTTP protocol. Since the logs are recorded for every elementary action on the network (e.g. click on a link and downloading parts of a website), they only contain basic information about the data transfer and do not include the target website content or the malicious binary file. Despite their simplicity, the logs can be used to detect communication corresponding to malicious behavior.

The detection of malicious communication is done based on features extracted from the proxy log fields, such as the URL, flow duration, number of bytes transferred from client to server and from server to client, user agent, referer (address of the previous web page that was followed to this page), MIME-Type, and HTTP status. The URL is the most informative and has been traditionally represented by n-grams and their statistics. The resulting n-dimensional feature vector represents each proxy log and is used to discriminate between malicious and legitimate HTTP request. In this work, we train a binary classifier and process the logs individually. Although this ignores important high level information since the malicious communication is sometimes periodic and could therefore benefit from temporal features, we will show that our low level model can already detect malware reliably.

The problem of supervised training in network security is the lack of sufficiently large and representative dataset of labeled malicious and legitimate samples. The labels are expensive to obtain since the process involves forensic analysis performed by security experts. Sometimes, the labels are not even possible to assign, especially if the context of the network communication is small or unknown and the assignment is desired at a proxy-log level. Furthermore, the labeled dataset becomes obsolete quite quickly, as a matter of weeks or months, due to constantly evolving malware. As a compromise, domain-level labeling has been frequently adopted by compiling blacklists of malicious domains registered by the attackers. The domain blacklists can then be used to block network communication based on the domain of the destination URL in the proxy log. However, the malicious domains typically change frequently as a basic detection evasion technique. Even though the domains might change, the other parts of the HTTP request (and the behavior of the malware) remain the same or similar. This is exploited in our behavioral model of malicious traffic.

Our semi-supervised training takes the advantage of security intelligence captured in the form of domain blacklists collected from various security reports, data feeds, and sandboxing analysis. Since our goal is to detect malicious behavior at the level of individual flows, the domain blacklists offer a weak supervision in the classification task. The proxy logs originating at a particular user machine are grouped into bags based on the domains in the URL. Therefore, the bags are constructed for each user machine and all flows to a domain visited in a particular time window (24 hours). The bags are labeled according to the domain: if the domain was marked as positive in any of the blacklists, the bag has a positive label. Otherwise, the bag is labeled as negative.

Leveraging the labels at the level of bags has the advantage that publicly available sources of domain blacklists can be used for the classifier training. The problem is then formulated as weakly supervised learning since the bag labels are used to train a classifier of individual flows. We propose to solve the problem by Multiple Instance Learning (MIL) which we describe next.

3   Multiple Instance Learning of Neyman Pearson Detector

We start with defining a statistical model of the data. A flow is described by a vector of features $x \in \mathcal{X} \subseteq \mathbb{R}^d$ and a label $y \in \mathcal{Y} = \{+1, -1\}$ where $y = +1$ means the malicious and $y = -1$ the legitimate flow, respectively. The network traffic monitored in a given period is fully described by the *completely annotated data* $\mathcal{D}_{\text{cmp}} = \{(x_1, y_1), \ldots, (x_m, y_m)\} \in (\mathcal{X} \times \mathcal{Y})^m$ assumed to be generated from i.i.d. random variables with an unknown distribution $p(x, y)$. Obtaining the complete annotation is expensive hence we collect a weaker annotation by assigning labels to bags of flows. The *weakly annotated data* $\mathcal{D}_{\text{bag}} = \{x_1, \ldots, x_m, (\mathcal{B}_1, z_1), \ldots, (\mathcal{B}_n, z_n)\}$ are composed of the flow features $\{x_1, \ldots, x_m\} \in \mathcal{X}^m$ along with their assignment to labeled bags

$$\{(\mathcal{B}_1, z_1), \ldots, (\mathcal{B}_n, z_n)\} \in (\mathcal{P} \times \mathcal{Y})^n$$

where $\mathcal{P}$ is a set of all partitions [1] of indices $\{1, \ldots, m\}$. The i-th bag is a set of flow features $\{x_j \mid j \in \mathcal{B}_i\}$ label by $z_i \in \mathcal{Y}$. The weakly annotated data $\mathcal{D}_{\text{bag}}$ carry a partial information about the completely annotated data $\mathcal{D}_{\text{cmp}}$. In particular, we assume that:

1. The flow features $\{x_1, \ldots, x_m\}$ in $\mathcal{D}_{\text{cmp}}$ and $\mathcal{D}_{\text{bag}}$ are the same.
2. The negative bags contain just a single instance and its label is correct, that is, $z_i = -1$ implies $|\mathcal{B}_i| = 1$ and $y_i = -1$.
3. The positive bags have a variable size and at least one instance is positive, that is, $z_i = +1$ implies $\exists j \in \mathcal{B}_i$ such that $y_j = +1$.

Our ultimate goal is to construct the Neyman-Pearson detector (see e.g. [12]) $h^* \in \mathcal{H} \subseteq \mathcal{Y}^{\mathcal{X}}$ which attains the minimal false negative rate

$$\text{FN}(h) = \mathbb{E}_{p(x|y=+1)}[h(x) = -1]$$

among all detectors with the false positive rate

$$\text{FP}(h) = \mathbb{E}_{p(x|y=-1)}[h(x) = +1]$$

not higher than a prescribed threshold $\beta > 0$. That is, we want to find h^* such that $\text{FP}(h^*) \leq \beta$ and

$$\text{FN}(h^*) = \inf_{h \in \mathcal{H}} \text{FN}(h) \quad \text{s.t.} \quad \text{FP}(h) \leq \beta. \tag{1}$$

In practice it can be more convenient to solve an equivalent problem $\inf_{h \in \mathcal{H}} [\text{FN}(h) + \beta' \text{FP}(h)]$ where β' is the Lagrange multiplier of (1) whose value depends on β.

The Neyman-Pearson problem (1) cannot be solved directly since the distribution $p(x, y)$ is unknown and hence also the key quantities $\text{FN}(h)$ and $\text{FP}(h)$

---

[1] A partition of $\{1, \ldots, m\}$ is a sequence of sets $\mathcal{B}_1, \ldots, \mathcal{B}_n$ such that $\cup_{i=1}^n \mathcal{B}_i = \{1, \ldots, m\}$, $\cap_{i=1}^n \mathcal{B} = \emptyset$ and $\mathcal{B}_i \neq \emptyset$, $\forall i \in \{1, \ldots, n\}$.

cannot be computed. We use the weakly annotated data $\mathcal{D}_{\text{bag}}$ to solve the problem approximately via the empirical risk minimization approach. To this end, we have to concretize the hypothesis space $\mathcal{H}$ and to approximate $\text{FN}(h)$ and $\text{FP}(h)$ by empirical estimates computed from the weakly annotated data $\mathcal{D}_{\text{bag}}$:

- We consider the hypothesis space $\mathcal{H}$ composed of the linear decision rules

$$h(\boldsymbol{x}; \boldsymbol{w}, w_0) = \begin{cases} +1 \text{ if } \langle \boldsymbol{x}, \boldsymbol{w} \rangle + w_0 \geq 0, \\ -1 \text{ if } \langle \boldsymbol{x}, \boldsymbol{w} \rangle + w_0 < 0, \end{cases} \tag{2}$$

parametrized by $\boldsymbol{w} \in \mathbb{R}^d$ and $w_0 \in \mathbb{R}$.
- The number of false positives is approximated by

$$\overline{\text{FP}}(\boldsymbol{w}, w_0) = \sum_{i \in \mathcal{I}_-} \max \left\{ 0, 1 + \langle \boldsymbol{w}, \boldsymbol{x}_{\mathcal{B}_i} \rangle + w_0 \right\}$$

where $\mathcal{I}_- = \{i \in \{1, \ldots, n\} \mid z_i = -1\}$ are indices of negatively labeled bags. Recall that the negative bags contain just a single instance, hence $\boldsymbol{x}_{\mathcal{B}_i}$ denotes the single $\boldsymbol{x}_j$, $j \in \mathcal{B}_i$. It is seen that $\overline{\text{FP}}(\boldsymbol{w}, w_0)$ is a convex upper bound of the number of false positives which the linear rule with parameters $(\boldsymbol{w}, w_0)$ makes on the completely annotated data $\mathcal{D}_{\text{cmp}}$.
- The number of false negatives is approximated by

$$\overline{\text{FN}}(\boldsymbol{w}, w_0) = \sum_{i \in \mathcal{I}_+} \max \left\{ 0, 1 - w_0 - \max_{j \in \mathcal{B}_i} \langle \boldsymbol{w}, \boldsymbol{x}_j \rangle \right\} \tag{3}$$

where $\mathcal{I}_+ = \{i \in \{1, \ldots, n\} \mid z_i = +1\}$ are the indices of the positive bags. Let

$$\overline{\text{FN}}_{\text{optim}}(\boldsymbol{w}, w_0) = \sum_{i \in \mathcal{I}_+} [\![h(\boldsymbol{x}_j; \boldsymbol{w}, w_0) = -1, \forall j \in \mathcal{B}_i]\!]$$

be the most optimistic estimate (that is, the minimal possible) of the number of false negatives made by the linear rule with the parameters $(\boldsymbol{w}, w_0)$ on the completely annotated data $\mathcal{D}_{\text{cmp}}$. It is seen that $\overline{\text{FN}}(\boldsymbol{w}, w_0)$ is an upper bound of $\overline{\text{FN}}_{\text{optim}}(\boldsymbol{w}, w_0)$ obtained by replacing the step-function with the hinge loss.

With these approximations, we formulate learning of the Neyman-Pearson detector as the following optimization problem

$$(\boldsymbol{w}^*, w_0^*) = \operatorname*{argmin}_{\boldsymbol{w} \in \mathbb{R}^d, w_0 \in \mathbb{R}} \left[\alpha \cdot \overline{\text{FP}}(\boldsymbol{w}, w_0) + (1 - \alpha) \cdot \overline{\text{FN}}(\boldsymbol{w}, w_0) \right], \tag{4}$$

where $\alpha \in \mathbb{R}_{++}$ is a cost factor used to tune the trade-off between the number of false negatives and false positives. The optimization problem (4) is not convex due to the term $\overline{\text{FP}}(\boldsymbol{w}, w_0)$. We solve the problem (4) approximately by the averaged stochastic gradient descent described in the next section.

Note that the task (4) is a straightforward modification of the Multiple-Instance Support Vector Machines (mi-SVM) algorithm [5]. The original mi-SVM optimizes the classification error (meaning that α is fixed to $\frac{1}{2}$), the objective function contains an additional regularization term $\frac{\lambda}{2}\|\boldsymbol{w}\|^2$ used to prevent overfitting and, finaly, the negative bags can contain more than a single instance. We dropped the regularization because the chance of over-fitting is very low in our case. In particular, the ratio of the number of examples and the parameters to be learned is $m/d > 400$.

3.1   Averaged Stochastic Gradient Descent

We formulated learning of the Neyman-Pearson detector as a non-convex optimization problem (4). In this section we describe a simple solver based on the Averaged Stochastic Gradient Descent algorithm [13]. For the sake of simplicity, we rewrite (4) as

$$\boldsymbol{v}^* = \operatorname*{argmin}_{\boldsymbol{v}\in\mathbb{R}^{d+1}} \sum_{i=1}^{n} r_i(\boldsymbol{v}) \tag{5}$$

where

$$r_i(\boldsymbol{v}) = \begin{cases} (1-\alpha)\max\left\{0, 1 - \max_{j\in\mathcal{B}_i}\langle\boldsymbol{v}, \tilde{\boldsymbol{x}}_j\rangle\right\} & \text{if } z_i = +1\,, \\ \alpha\max\left\{0, 1 + \langle\boldsymbol{v}, \tilde{\boldsymbol{x}}_{\mathcal{B}_i}\rangle\right\} & \text{if } z_i = -1\,, \end{cases}$$

$\boldsymbol{v} = (\boldsymbol{w}, w_0)$ and $\tilde{\boldsymbol{x}}_i = (\boldsymbol{x}_i, 1)$. The SGD algorithm approximates the sub-gradient of the objective of the task (5) by the sub-gradient of a randomly selected term $r_i(\boldsymbol{v})$. A sub-gradient of $r_i(\boldsymbol{v})$ can be computed as $r_i'(\boldsymbol{v}) = \lambda_i\tilde{\boldsymbol{x}}_{i^*}$ where $i^* = \operatorname{argmax}_{j\in\mathcal{B}_i}\langle\boldsymbol{w}, \tilde{\boldsymbol{x}}_j\rangle$ and

$$\lambda_i = \begin{cases} \alpha - 1 & \text{if } z_i = +1 \text{ and } \langle\boldsymbol{w}, \tilde{\boldsymbol{x}}_{i^*}\rangle + w_0 < 1\,, \\ \alpha & \text{if } z_i = -1 \text{ and } \langle\boldsymbol{w}, \tilde{\boldsymbol{x}}_{i^*}\rangle + w_0 > -1\,, \\ 0 & \text{otherwise.} \end{cases}$$

A pseudo-code of the Averaged SGD solver is summarized in Algorithm 1. The ASGD solver has two free parameters: the fixed step-size γ and the number of epochs E. There is no generic theory how to set these parameters optimally. Hence, their values have to be tuned on particular data.

3.2   Baseline SVM Detector

We use the standard binary SVM classifier as the baseline solution [14]. A simple workaround when dealing with the weak annotations is to consider all instances in the positive bags to be positive. Given the training bags $\mathcal{D}_{\text{bag}}$ we define the sets $\mathcal{J}_+ = \cup_{j\in\mathcal{I}_+}\mathcal{B}_j$ and $\mathcal{J}_- = \cup_{j\in\mathcal{I}_-}\mathcal{B}_j$ containing indices of all positive and all negative flows, respectively. We learn the SVM detector by solving the following convex program

---

Algorithm 1. Averaged SGD

Require: cost factor $\alpha > 0$, number of epochs $E > 0$, step-size $\gamma > 0$
Ensure: vector $\overline{v}^t$ approximately solving the problem (5)
1: randomly set $v^0 \in \mathbb{R}^d$ and $\overline{v}^0 \leftarrow v^0$, $t \leftarrow 0$
2: **for** epoch in $\{1, \ldots, E\}$ **do**
3:     **for** i in randperm(n) **do**
4:         $v^{t+1} \leftarrow v^t - \gamma r'_i(v^t)$
5:         $\overline{v}^{t+1} \leftarrow \frac{t-1}{t}\overline{v}^t + \frac{1}{t}v^{t+1}$
6:         $t \leftarrow t + 1$
7:     **end for**
8: **end for**

$$(w^*, w_0^*) = \operatorname*{argmin}_{w \in \mathbb{R}^d, w_0 \in \mathbb{R}} \left[\frac{1}{2}\|w\|^2 + C \cdot (1 - \alpha) \sum_{i \in \mathcal{J}_+} \max\left\{0, 1 - \langle w, x_i\rangle - w_0\right\} \right.$$
$$\left. + C \cdot \alpha \sum_{i \in \mathcal{J}_-} \max\left\{0, 1 + \langle w, x_i\rangle + w_0\right\} \right]$$

(6)

where $\alpha \in \mathbb{R}_{++}$ is a cost factor used to tune the trade-off between the number of the false negatives and the false positives. The constant $C \in \mathbb{R}_{++}$ steers the strength of the regularization.

4 Specification of the Datasets

This Section provides detailed description of the datasets and features we used in the experimental evaluation. The datasets are divided into three parts: training, validation, and testing. All datasets represent real network traffic from 14 days of a large international company (80,000 seats) in form of proxy logs. These logs capture HTTP/HTTPS network communication and contain flows, where one flow represents one communication between a user and a server. More specifically, one flow is a group of packets with the same source and destination IP address, source and destination port, and protocol. As flows from the proxy logs are bidirectional, both directions of a communication are included into each flow.

A flow consists of the following fields: user name, source IP address, destination IP address, source port, destination port, protocol, number of bytes transferred from client to server and from server to client, flow duration, timestamp, user agent, URL, referer, MIME-Type, and HTTP status. The most informative field is URL, which can be decomposed further into 7 parts as illustrated in Figure 1. From the flow fields mentioned above, we extracted more than 317 features listed in Table 1. Features from the right column are extracted from all URL parts, including URL itself and referer.

Flows were grouped into bags, where each bag contains flows with the same user (or source IP) and the same second-level domain. Thus, each bag represents communication of a user with a particular domain. As the datasets were

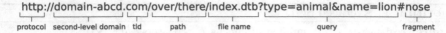

Fig. 1. URL decomposition into seven parts.

Table 1. List of features extracted from proxy logs. Features from the right column are extracted from all URL parts, including URL itself and referer.

Features	Features extracted from all URL parts & referer
duration	length
HTTP status	digit ratio
is URL encrypted	lower case ratio
is protocol HTTPS	upper case ratio
number of bytes up	vowel changes ratio
number of bytes down	has repetition of '&' and '='
is URL in ASCII	starts with number
client port number	number of non-base64 characters
server port number	has a special character (one feature per character)
user agent length	max length of consonant stream
MIME-Type length	max length of vowel stream
number of '/' in path	max length of lower case stream
number of '/' in query	max length of upper case stream
number of '/' in referer	max length of digit stream
is second-level domain rawIP	ratio of a character with max occurrence

originally unlabeled, we used available blacklists and other malware feeds from Collective Intelligence Framework (CIF) [7] to add positive labels to the training dataset. All bags with domains marked as malicious in CIF were labeled as positive. Negative samples were acquired from the list of popular domains (Alexa top 250,000 domains [2]). Bags with domains that were not in CIF nor in Alexa were discarded from the training set. Note that some popular domains are used very frequently, which may outweigh the importance of rarely-used domains. For this reason, each domain has at most 1000 flows in the training set. Flows with domains which appear in the positive training samples are removed from the testing dataset. We also removed all flows with second-level domains in IP (4-tuple digit) format (e.g. 192.168.0.0) due to the lack of negative samples of this type. Summary of the datasets is described in Table 2.

Each flow is described by 317 real valued features described above. The linear decision rule (2) operates on a binary representation of these features which is created as follows. Range of values of each feature observed in the training data is split into $p = 8$ bins of equal size. Each real valued feature is then represented as a binary vector with 8 elements all set to zero but one encoding the active bin. Stacking the binary vectors for all 317 features gives the final feature representation $x \in \{0,1\}^d$ with dimension $d = 317 \cdot 8 = 2536$.

Table 2. Summary of datasets used for training, validation, and testing.

	Training Samples		Validation Samples	Test Samples
	Positive	Negative		
Number of Flows	21,873	999,819	1,943,980	9,696,453
Number of Bags	3,918	999,819	—	—
Number of Domains	207	38,463	46,970	45,046

5 Experiments

In this section we empirically evaluate detectors of malicious communication learned from weakly labeled data that were described in Section 4. We compare two detectors learned from the same data by different methods:

1. SVM detector learned by solving the problem (6) is used as a baseline. This method considers all instances in the positive bags to be positive and similarly for instances in the negative bags. To solve (6) we used an open-source implementation of the optimized cutting plane solver [8] available at: http://cmp.felk.cvut.cz/~xfrancv/ocas/html/index.html.
2. MIL detector learned by solving the problem (4). We solved (4) by the ASGD Algorithm 1 implemented in C++. We found empirically that setting the step-size $\gamma = \frac{1}{n}$ (where n is the number of training bags) and the number of epochs $E = 100$ works consistently well on our data. The results also depend on the initial estimate v^0. Hence, for each data we run the ASGD algorithm 10 times from randomly generated v^0 and selected the result with the smallest value of the training objective.

The source data contain only weak labels. The model selection and the final evaluation of the detectors require the ground truth labels for a subset of flows from the validation and the testing subset. The ground truth labels are obtained via submitting the flows' URL to the VirusTotal service [3]. For each submitted URL, the VirusTotal provides a report containing analysis of a set of URL scanners. The number of scanners at the time of evaluation was 62. The report is summarized by the number of positive hits, that is, the number of scanners which marked the URL as malicious. If at least three scanners mark the URL as malicious the flow is labeled as the true positive.

5.1 Main Results

Model selection. We learned the MIL detector on the training bags for different values of the cost factor α which is used to tune the ratio of the false positives and the false negatives. Each learned detector is evaluated on the validation data by computing the number of true positive flows (denoted as $\text{tp}_{\text{fp}=50}$) obtained by the detector with the decision threshold set to make the number of false

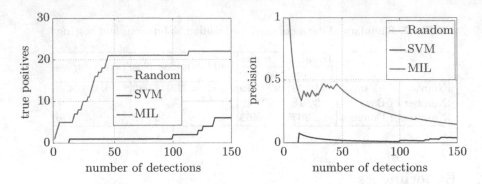

Fig. 2. The figures present results obtained on the first 150 test flows with the highest decision score computed by the MIL and the SVM detector. The left figure shows the number of true positives and the right figure the precision of the detectors as a function of the number of detected flows. We also show results for a baseline detector selecting the flows randomly.

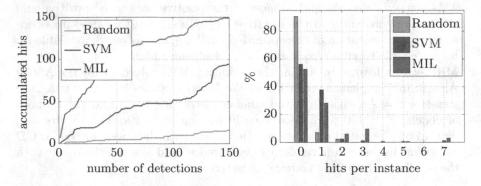

Fig. 3. The left figure shows the number of accumulated hits with respect to the number of flows selected by the MIL and the SVM detector. The right figure shows a histogram of the number of hits computed for the first 50 flows with the highest decision score. The flows with the number of hits higher than 2 are the true positives. We also show results for a baseline detector selecting the flows randomly.

positive flows on the validation data equal to 50. We also recorded the number of hits (denoted as $hits_{fp=50}$) accumulated for the flows predicted to be positive by the same detector. This evaluation procedure requires ground true labels for the first $p = 50 + tp_{fp=50}$ flows with the highest decision score $\langle w, x \rangle$. The detector with the maximal validation $tp_{fp=50}$ was selected as the final model. If more detectors have the same $tp_{fp=50}$, which happened in our case, the detector with the maximal $hits_{fp=50}$ is selected among them. The same model selection procedure was used for the SVM detector in which case, however, we also varied the regularization constant. The results of the model selection procedure are

Table 3. Summary of the errors on the training, validation and testing data for the MIL (upper table) and SVM (bottom table) detector. Each row corresponds to a detector trained with different value of the cost factor α and the regularization constant in case of the SVM. We show the number of false negatives flows fn, the number of false positive flows fp and for the MIL detector also the number of false negative bags evaluated on the training data (the value in brackets are the corresponding rates). Note that fn is at the same time the number of false negative bags because they contain a single instance each. For the validation and test data we show the number of true positives $tp_{fp=50}$, and the number of accumulated hits $hits_{fp=50}$, computed on the flows returned by the detector with the number of false positives fp set to 50. The same statistics were computed on the test data the best detector selected based on the validation results.

MIL detector

α	training			validation		testing	
	fn	fp	fn bags	$tp_{fp=50}$	$hits_{fp=50}$	$tp_{fp=50}$	$hits_{fp=50}$
0.8	4085 (0.19)	10 (1^{-5})	313 (0.08)	8	50	—	—
0.9	4326 (0.20)	3 (3^{-6})	354 (0.09)	9	65	—	—
0.95	4613 (0.21)	1 (1^{-6})	391 (0.10)	9	97	21	118
0.99	5938 (0.27)	0 (0)	530 (0.14)	9	97	—	—

SVM detector

C	α	training		validation		testing	
		fn	fp	$tp_{fp=50}$	$hits_{fp=50}$	$tp_{fp=50}$	$hits_{fp=50}$
1000	0.6	4020 (0.18)	158 (2e-4)	2	34	—	—
	0.7	5539 (0.25)	158 (2c-4)	2	29	—	—
	0.8	5602 (0.26)	158 (2e-4)	4	42	1	37
	0.9	6451 (0.29)	57 (6e-5)	3	33	—	—
10000	0.6	2068 (0.09)	158 (2e-4)	0	21	—	—
	0.7	2244 (0.10)	153 (2e-4)	0	27	—	—
	0.8	2723 (0.12)	148 (1e-4)	0	35	—	—
	0.9	3736 (0.17)	79 (8e-5)	1	31	—	—

summarized in Table 3. In the sake of space we show results only for a subset of values (α, C) selected around the best setting.

Evaluation on test data. The best MIL and SVM detectors were then evaluated on the test flows. We applied each detector on all test flows and selected top 150 instances with the highest decision score. We used the VirusTotal to obtain the ground truth labels for the selected flows. In Figure 2 we show the number of true positives and the precision as a function of increasing decision threshold. In the top 150 instances selected by the MIL detector out of 9,696,453 testing flows 22 are true positives while the baseline SVM detector found just 6 true positives. The precision of the MIL detector in the analyzed region varies from 1 to 0.14. The maximal precision attained by the SVM detector in the same region is 0.08.

We also evaluated the detectors in terms of the number of hits being the finer annotation used to define the ground true labels. Recall that the flow with the number of hits greater than 2 is marked as the true positive. The results are

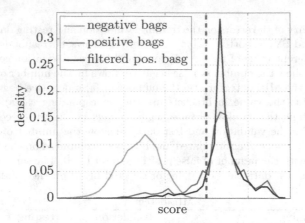

Fig. 4. A distribution of the training instances in the positive (red) and the negative (green) bags projected onto the normal vector w of the decision hyperplane of the MIL detector. The black curve corresponds to the distribution of the instances from the positive bags with the maximal distance to the decision hyperplane. The dashed blue line marks the decision hyperplane of the MIL detector.

presented in Figure 3. In particular, we show the number of accumulated hits as a function of the detected flows. We also show the histogram of the number of hits per instance measured on the first 50 flows. The highest number of flows in the top 50 instances returned by both detectors has only one hit. However, the second most frequent flows have no hit in the case of SVM but 3 hits in the case of the MIL detector.

5.2   Why Does MIL Work Better than SVM?

In this section we provide some intuition why the SVM detector ignoring the bag annotation performs worse than the MIL detector. The MIL algorithm defined by problem (4) minimizes a weighted sum of errors made by the detector on the negative and the positive bags. The errors on the positive bags are expressed by the function $\overline{\mathrm{FN}}(w, w_0)$ defined in equation (3). It is seen from (3) that the error of each positive bag is determined by a single instance which has the maximal distance from the decision hyperplane. Removal of the other non-active instances from the training set would not change the solution. Hence the MIL algorithm can be seen as two stage procedure though the stages are executed simultaneously. First, a new filtered training set is created by copying all instances from the negative bags and picking a single instance from each positive bag. Second, a supervised SVM algorithm is applied on the filtered training set which contains only bags of size one. Figure 4 visualizes the original and the filtered distribution of the training data projected on the normal vector of the decision hyperplane of the MIL detector. It is seen that the filtered distribution of the positive instances is significantly more peaky than the original one. The shape

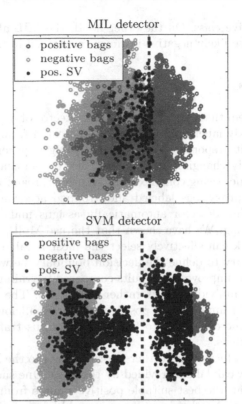

Fig. 5. A distribution of the training instances projected onto the 2D space and the support vectors selected from the positive bags by the MIL algorithm (top) and the SVM algorithm (bottom). The x-axis is a projection of the original data onto the normal vector w of the decision hyperplane of a corresponding detector and the y-axis is a projection onto the major principal component in the subspace orthogonal to w. The dashed blue line shows the decision hyperplane.

of the filtered distribution also looks smoother and better separable form the negative distribution.

The SVM and MIL algorithm use only a subset of the examples to define the decision rule. Namely, the *support vectors* which are the training examples with the signed distance to the hyperplane not higher than 1. We projected the $d = 2,536$ dimensional training data onto 2D space and displayed the support vectors selected from the positive examples by the SVM and the MIL algorithms. The 2D coordinates are obtained by projecting the original data onto the normal vector w of the decision hyperplane (x-axis) and the major principal component computed in the subspace orthogonal to w (y-axis). The visualization is shown in Figure 5. It is seen that the SVM detector uses a large set of positive support vectors heavily overlapping with the negative instances many of which probably belong to the negative class. On the other hand, the MIL algorithm ignores a large number of these likely negative instances contained in the positive bags.

6   Conclusions

We presented a system that can learn a reliable detector of malware communication using only weakly labeled data that can be created from publicly available blacklists and security reports. Relaxing the labeling requirement is important due to the constantly changing malware and due to the challenges in forensic analysis caused by increasing complexity of security threats. Our learning algorithm uses bags of proxy logs labeled and grouped according to the network domain. The resulting detector automatically assigns malicious or legitimate label to each proxy log. We have shown that the use Multiple Instance Learning (MIL) framework can effectively select the most reliable samples from the positive bags necessary to define the decision boundary between malicious and legitimate flows. This improves the results compared to training a baseline SVM detector trained on individual flows rather than bags. The result follows our intuition that not all samples in the positive bags are malicious. We have shown that the detector generalizes well to find new malicious traffic which was not marked by the blacklists.

Our ongoing work focuses on further improvements to the MIL framework to better adapt to the weak labels provided by the bags. One such improvement is to focus on more than the best suitable positive sample from the bag to define the decision boundary. Another improvement could follow from a normalization mechanism that takes into account the size of the bags. Finally, there are vast amounts of unlabeled data that can further improve the training and the final detector performance.

Acknowledgments. The research was supported by CISCO grant number 8301351C001 and the Grant Agency of the Czech Republic under Project P202/12/2071.

References

1. Cisco 2014 annual security report.
   http://www.cisco.com/web/offers/lp/2014-annual-security-report/index.html
2. List of 1 million top web sites. http://www.alexa.com
3. VirusTotal service. https://www.virustotal.com
4. Abu-Nimeh, S., Nappa, D., Wang, X., Nair, S.: A comparison of machine learning techniques for phishing detection. In: Proceedings of the Anti-phishing Working Groups 2nd Annual eCrime Researchers Summit, eCrime 2007, pp. 60–69. ACM, New York (2007)
5. Andrews, S., Tsochantaridis, I., Hofmann, T.: Support vector machines for multiple-instance learning. In: Proc. of Neural Information Processing Systems (2002)

6. Castillo, C., Donato, D., Gionis, A., Murdock, V., Silvestri, F.: Know your neighbors: Web spam detection using the web topology. In: Proceedings of SIGIR. ACM, Amsterdam, July 2007
7. Farnham, G., Leune, K.: Tools and standards for cyber threat intelligence projects. Technical report, SANS Institute InfoSec Reading Room, vol.10 (2013)
8. Franc, V., Sonnenburg, S.: Optimized cutting plane algorithm for support vector machines. In: McCallum, A., Roweis, S. (eds.) Proceedings of the 25th Annual International Conference on Machine Learning (ICML 2008), pp. 320–327. ACM, New York (2008)
9. Gu, G., Zhang, J., Lee, W.: BotSniffer: detecting botnet command and control channels in network traffic. In: Proceedings of the 15th Annual Network and Distributed System Security Symposium (NDSS 2008), February 2008
10. Ma, J., Saul, L.K., Savage, S., Voelker, G.M.: Identifying suspicious URLs: an application of large-scale online learning. In: Danyluk, A.P., Bottou, L., Littman, M.L. (eds.) Proceedings of the 26th Annual International Conference on Machine Learning, ICML 2009, Montreal, Quebec, Canada, June 14–18, 2009. ACM International Conference Proceeding Series, vol. 382, pp. 681–688. ACM (2009)
11. Perdisci, R., Lee, W., Feamster, N.: Behavioral clustering of http-based malware and signature generation using malicious network traces. In: Proceedings of the 7th USENIX Conference on Networked Systems Design and Implementation, NSDI 2010, pp. 26–26. USENIX Association, Berkeley (2010)
12. Schlesinger, M.I., Hlaváč, V.: Ten Lectures on Statistical and Structural Pattern Recognition. Kluwer Academic Publishers, Dordrecht (2002)
13. Shamir, O., Zhang, T.: Stochastic gradient descent for non-smooth optimization: Convergence results and optimal averaging schemes. In: Proc. of International Conference on Machine Learning (2012)
14. Vapnik, V.N.: The Nature of Statistical Learning Theory. Springer Verlag (1995)

Online Analysis of High-Volume Data Streams in Astroparticle Physics

Christian Bockermann[1]([⊠]), Kai Brügge[2], Jens Buss[2], Alexey Egorov[1], Katharina Morik[1], Wolfgang Rhode[2], and Tim Ruhe[2]

[1] Computer Science Department,
Technische Universität Dortmund, 44221 Dortmund, Germany
{christian.bockermann,alexey.egorov,katharina.morik}@udo.edu
[2] Astroparticle Physics Department,
Technische Universität Dortmund, 44221 Dortmund, Germany
{kai.brugge,jens.buss,wolfgang.rhode,tim.ruhe}@udo.edu

Abstract. Experiments in high-energy astroparticle physics produce large amounts of data as continuous high-volume streams. Gaining insights from the observed data poses a number of challenges to data analysis at various steps in the analysis chain of the experiments. Machine learning methods have already cleaved their way selectively at some particular stages of the overall data mangling process.

In this paper we investigate the deployment of machine learning methods at various stages of the data analysis chain in a gamma-ray astronomy experiment. Aiming at online and real-time performance, we build up on prominent software libraries and discuss the complete cycle of data processing from raw-data capturing to high-level classification using a data-flow based rapid-prototyping environment. In the context of a gamma-ray experiment, we review user requirements in this interdisciplinary setting and demonstrate the applicability of our approach in a real-world setting to provide results from high-volume data streams in real-time performance.

Keywords: Online analysis · High-volume streams · Astroparticle physics

1 Introduction

Modern astronomy studies celestial objects (stars, nebulae or active galactic nuclei) partly by observing high-energy beams emitted by these sources. By a spectral analysis of their emissions, these objects can be characterized and further insight can be derived. Plotting the energy emissions over time leads to a *light curve*, which may show pulsatile behavior and other properties that allow for a classification of the observed object. An example is the distinction of different supernova types based on the form of their light curves [12]. The creation of a spectrum of the radiated energy levels is therefore a key skill. A collection of different monitoring techniques such as satellites [22], telescopes [2,18,21] or

© Springer International Publishing Switzerland 2015
A. Bifet et al. (Eds.): ECML PKDD 2015, Part III, LNAI 9286, pp. 100–115, 2015.
DOI: 10.1007/978-3-s319-23461-8_7

water tanks [1, 4] is deployed to observe different ranges of the electromagnetic radiation produced by the sources. A central problem in all these experiments is the distinction of the crucial gamma events from the background noise that is produced by hadronic rays and is inevitably recorded. This task is widely known as the *gamma-hadron* separation problem and is an essential step in the analysis chain. The challenge in the separation step is the high imbalance between signal (gamma rays) and background noise, ranging from 1:1000 up to 1:10000 and worse, which implies large amounts of data that need to be recorded for a well-founded analysis of a source. The high sampling rate and the growing resolution of telescope cameras further require careful consideration of scalability aspects when building a data analysis chain for scientific experiments.

In this work, we investigate the *online* use of classification methods for data filtering and spectrum creation in gamma-ray astronomy. We review the data flow of prominent experiments like MAGIC or FACT, and inspect the preprocessing chain from data acquisition to the extraction of a spectrum from a machine learning point of view. With respect to the scalability requirements we discuss the use of *distributed model* application as part of the analysis chain. As recorded data is unlabeled, only simulated data can be used to train filters. These can easily be trained in a batch setting, but an application of any model is required in an online fashion. This, in turn, poses constraints on the features that can be used. Further, there exists an *interdisciplinary gap* between the domain experts (physicists) and the field of computer science that spoils a fruitful cooperation in both areas. Our contributions are as follows:

(1) We outline the data flow from observation to analysis in the real-world setting of the FACT Cherenkov Telescope.
(2) We discuss multiple spots for the use of machine learning models in the analysis and the constraints faced.
(3) Based on the streams framework [9], we provide a library of processing functions (*fact-tools*) to easily model the overall data processing and analysis chain in a rapid prototyping manner.
(4) We demonstrate the applicability of our proposed framework and the machine learning methods using real-world data of the FACT telescope.

The rest of this paper is structured as follows: In Section 2 we provide a short overview over the field of Cherenkov astronomy, subject to our study, and review the flow of data from data acquisition to the extraction of the desired information. Along this data flow, we highlight the use of machine learning models, including related works on that matter, and close with a discussion on the requirements from the view of domain experts as well as the interdisciplinary gap that we faced between the world of end-users (physics) and data engineers (computer science). In Section 3 we introduce the *fact-tools* – our high-level framework to model the data flow, which integrates state of the art tools such as WEKA [16] and MOA [7] to incorporate machine learning for various tasks. We demonstrate the use of our framework and evaluate different machine learning methods in the real-world setting of the FACT telescope in Section 4. We summarize our *lessons learnt* and future ideas in Section 5.

2   Data Analysis in Cherenkov Astronomy

The examination of sources in astronomy relies on the observation of energy emitted by these sources. Unfortunately, these energy beams cannot be observed directly, but only by an indirect measuring of the effect they are triggering in some detector medium. In the case of Cherenkov telescopes, the atmosphere is used as detector medium: particles interact with elements in the atmosphere and induce cascading air showers as they pass the atmosphere. These showers emit so-called *Cherenkov light*, which can be measured by telescopes like MAGIC or FACT. Figure 1 shows an air shower triggered by some cosmic ray beam, emitting Cherenkov light that can be captured by the telescope camera. The cone of light produced by the shower is visible in the camera for a period of about 150 nanoseconds. The camera of the telescope consists of an array of light-sensitive pixels that allow for the recording of the light impulse induced by the air shower. For a fine-grained capture of the light pulses, the camera pixels are sampled at a very high rate (e.g. 2 GHz). Figure 1 shows the layout of the FACT camera, which consists of 1440 pixels in hexagonal form. The high sampling speed requires high-performance memory for buffering the sampled data. The cameras usually continuously sample all the pixels into a ring-buffer and a hardware trigger initiates a write-out to disk storage if some pixels exceed a specified threshold (i.e. indication of shower light hitting the telescope). Upon a trigger activation, the sampled data written to disk captures a series of camera samples which amount for a time period of about 150 to 300 nanoseconds, called the *region of interest* (ROI). This fixed-length series of consecutive camera samples produced upon a trigger is called an *event* and corresponds to the light cone induced by the airshower.

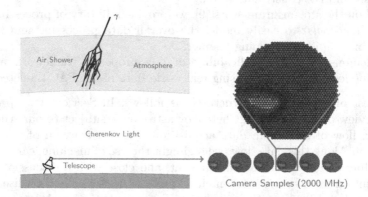

Fig. 1. An air shower produced by a particle beam hitting the atmosphere. The shower emits a cone of blue light (Cherenkov light) that will hit the telescope mirrors and is recorded in the camera. The right-hand side shows a still image of the light cone in the telescope camera (FACT telescope).

2.1   From Raw Data Acquisition to Spectral Analysis

The raw data produced by telescopes like MAGIC or FACT consists of the sampled voltages of the camera pixels for a given time period (ROI). Using these voltage levels, the following steps are required in the analysis, each of which is individually performed for each event, naturally implying a streamlined processing:

(1) *Calibration, Cleaning*: Calibrate the data, determine the pixels that are part of the light shower.
(2) *Feature Extraction*: Find features that best describe the data to solve the following steps.
(3) *Signal Separation*: Assess whether the event is induced by a gamma-ray (signal) or a hadronic shower (noise).
(4) *Energy Estimation*: Estimate the Energy of the primary gamma particle from the calculated energy correlated features.

Based on the number of signals detected and the estimated energy spectrum, properties of the observed astronomical source can be inferred. From a data analysis point of view we can map this process to the high-level data flow outlined in Figure 2. Especially the separation of signal and noise and the energy estimation are candidates for use of machine learning. The extraction of features for subsequent use of machine learning in steps (3) and (4) is a crucial step and requires back-to-back fine tuning with the learning methods. The dark arrows show additional back-to-back dependencies between the different steps. These dependencies induce a highly volatile optimization process in the development of the overall analysis chain: any changes in the calibration of cleaning may lead to slightly different feature values for the signal separation and the energy estimation step.

Fig. 2. Data processing steps from raw data acquisition to energy estimation.

The calibration and cleaning methods are highly domain specific and require careful consideration by domain experts. Given, that the electronics of such telescopes are customized prototypes, each device requires different setups that may vary with changes in the environment (temperature). These early steps in the data processing chain usually relate to hardware specifics and are fine-tuned in a manual way. In case of the FACT telescope, base voltages and gains for each pixels need to be adjusted with respect to calibration data recordings.

2.2   Feature Extraction for Machine Learning

Given domain knowledge, the gamma and hadronic particles behave different when hitting the atmosphere. As gamma particles are uncharged, they have strict directional energy and air showers created by gammas are expected to be directed straight from the source as well. Hadronic particles in contrast may be deviated by electromagnetic fields and thus will drop into the atmosphere from any direction. In addition, the atmospheric interaction of hadronic showers tends to degrade into much wilder cascades. A basic assumption is, that the structural differences in these showers are reflected in the image that the Cherenkov light emitted by these showers induces in the telescope camera. These properties are described by the so-called *Hillas parameters*, which form a set of geometric features that are widely used in gamma ray astronomy [13,19]. The features introduced by Hillas describe the orientation and size of an ellipse fitted to the area of a shower image. The ellipse is fitted to the pixels that have survived the previous *image cleaning* step, in which pixels not part of the shower are removed. The geometric orientation of the ellipse is correlated with the angular field of the telescope. Figure 3 shows a shower image after removal of non-shower pixel and the Hillas features derived. It is obvious, that the image cleaning step, in which the shower pixels are identified, has a direct impact on the ellipse, that will be fitted. Apart from the size (width, length) the orientation of the ellipse (alpha) and its offset from the origin is extracted.

In addition to these basic geometric features, other properties of shower images have been derived, such as the fluctual distribution from shower center [11] (for the Milagrito experiment) or the *surface brightness* [5]. In [14] Faleiro et al proposed the investigation of spectral statistics as discriminating features, whereas [23] evaluated an encoding of shower images using multi-fractal wavelets.

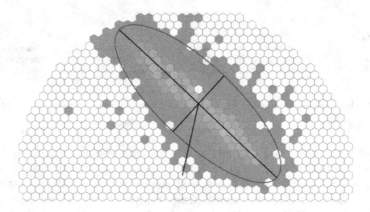

Fig. 3. Geometric *Hillas* features to support signal-noise separation. Figure shows an event in the FACT camera after image cleaning.

2.3 Signal Separation and Energy Estimation: Machine Learning

The detection of gamma induced events has been investigated as a binary classification task, widely referred to as the *gamma-hadron separation*. The features described in the previous section have all been proposed and tested in combination with a classification algorithm to achieve the best filtering of signal events from hadronic background. As the showers are distinct events, this boils down to finding some model

$$M : \mathcal{S} \to \{-1, +1\}$$

that maps a recorded shower $S \in \mathcal{S}$ represented by a set of features to one of the two possible classes. The challenge in this classification task is the highly imbalanced class distributions as a very small fraction of gamma rays needs to be separated from the large amount if showers induced by hadronic particles. The expected ratio is at the level of 1:1000 or worse, requiring a huge amount of recorded data in order to find a meaningful collection for training a classifier.

The classifiers tested with the aforementioned features range from manual threshold cuts using discriminative features [8], neural networks (combined with fractal features [23]) to support vector machines or decision trees. [8] provides a study comparing various classifiers on a fixed set of features (Hillas parameters). Random forests [10] generally provide a robust performance and have become a widely accepted method for the gamma-hadron separation in that domain.

Energy Estimation

Another field where machine learning contributes is the estimation of energy. The recorded data only reflects the image of light emitted by the air shower that was produced by the cosmic ray. Of interest to the physicists is the energy of the particles that induced the shower. The reconstruction of the energy of the primary particle can be seen as a *regression task*, finding a model

$$E : \mathcal{S} \to \mathbb{R}$$

which predicts the energy based on features obtained from the shower image. For the MAGIC telescope, Berger et al. investigated the energy reconstruction with random forests, claiming that a small set of features is suitable for a robust energy estimation, with the *size* parameter being the most important one [6].

Labeled Data by Probabilistic Simulation

A big problem when applying machine learning in astrophysics is, that particles arriving from outer space are inherently unlabeled. Using that data for supervised learning requires an additional step to obtain data for training a classifier: The solution to the labeling problem is found in data simulations using the Monte Carlo method. There exists a profound knowledge of the particle interaction in the atmosphere: Given the energy and direction of some parent particle (gamma, proton, etc.) its interaction can be described by a probabilistic model which gives a probability for particle collisions, possibly resulting in secondary particles. Each of these secondary elements may further interact with other particles of the atmosphere. This results in a cascade of levels of interactions that form the air shower. Figure 4 shows the transition of a particle and its interaction in the atmosphere. The showers previously shown in Figure 1 are examples of such simulated cascades. Unfortunately, the simulation of non-gamma showers is far more computationally extensive. Charged particles do interact with the atmosphere much more intense, resulting in more complex cascades. The simulation of atmospheric showers is performed in ray-tracing like software systems, most popular being the CORSIKA simulator [17]. The output is a simulated air shower, which needs to be run through a simulation of the telescope and camera device to produce the same raw input data as if the shower has been recorded using the real telescope.

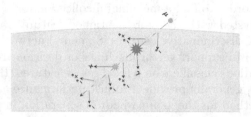

Fig. 4. Synthetic data by simulation in a stochastic process. Collision probabilities and generation of secondary particles are based on domain knowledge.

2.4   The Interdisciplinary Gap in Process Development

Looking at the big picture of the data analysis in a telescope like FACT, there is a steady development of each of the steps in progress: new features are tested for improved separation, different classifiers are investigated. The complete process from data recording to final energy estimation is continuously improved under the aspect of physics, typically resulting in diverse proprietary software solutions. The machine learning and computer science community, on the other hand, has produced a valuable collection of open-source software libraries for learning (e.g. MOA [7], WEKA [16] or RapidMiner [20]) and stream-lined process execution (e.g. Apache Storm, Samza). Unfortunately, the integration of these tools often requires specifically trained developers to adapt them to an application domain, which hinders the rapid prototyping evolution of the analytical domain software.

We generally refer to this problem as the *interdisciplinary gap* – the difficulty to apply sophisticated tools in a specific cross-disciplinary application domain. Over the collaborative research center project C3, we focused on bridging this gap by building a process design framework that provides the high-level means to define analysis chains from an end-user point of view, while keeping the power to integrate state-of-the-art software platforms such as the ones mentioned above.

3   Online Data Analysis for the FACT Telescope

After we provided a big picture of the analysis chain for the FACT telescope, we now present the *FACT-Tools*, our high-level approach to implementing an analysis chain from data acquisition to deriving the final results. The implementation focuses on an online processing of the recorded events, geared towards matching the data rate of the telescope. We build upon the streams framework, developed at the TU Dortmund University, which features a declarative XML specification of generic data flows. By providing a small set of application specific components that match the streams programming API, the domain experts gain full control of the overall process layout while retaining the possibility to integrate state-of-the-art machine learnign libraries and a possible mapping of the resulting data flows to large scale execution engines, like Apache Storm.

3.1   The streams Data Flow Framework

Modern data processing chains – like the one required for the FACT telescope – can generally be presented by their data flow. In such data flow graphs, a source emitting a sequence of records is linked to a graph of nodes, each of which provides the processing of input and the delivery of some output. Such data flow graphs are inherent to all modern data processing engines, especially in the field of data stream processing.

   With the streams framework we aim at finding an appropriate level of abstraction that allows for the design of data flows independent of a specific execution engine. The focus of streams is a light-weight middle-layer API in combination with a descriptive XML-based specification language, that can directly be executed with the streams runtime or be mapped to topologies of the Apache Storm engine. The predominant focus in the development of streams was a simplistic light-weight API and process definition that is easily applicable by domain experts and adaptable to a variety of different use cases.

   Each of the connected processes in streams consists of a pipeline of user-functions, which are applied to the processed items in order. Figure 6 shows a brick-like visualization of a process as pipeline of functions. The source nodes provides a sequence of data items that is individually processed by the process pipeline of user-functions. These user-functions are typically implemented in

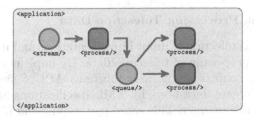

Fig. 5. The outline of an application in streams – a graph of connected processes.

single Java classes and are directly referenced by their implementation name from inside the XML element of the corresponding process. The figure shows functions for the calibration, image cleaning and extraction of Hillas parameters. With these features available, a classifier model can be applied.

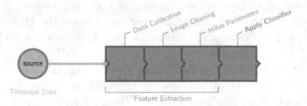

Fig. 6. A data stream *source* connected to a process that consists of four different user functions for calibration and feature extraction.

To achive the maximum level of flexibility, the data items (or tuples) that are passed from one user-function to the next, are wrapped in a simple hashmap (or dictionary), that can be accessed and enriched by each user-function. The data that can be stored in each of these can be of arbitrary serializable types, allowing for the implementation of user-functions that work on simple data types as well as frames/images (video processing) or telescope event data as we will discuss in the next section. To further ease the modelling of data flows with a set of implemented user-functions, the streams approach facilitates an automatic mapping of XML attributes to classes following the JavaBean conventions. This removes any intermediate layer between process modelling and function implementation.

```
<process input="telescope:data">
  <fact.data.DrsCalibration calibrationFile="file:/data/calib.fits" />
  <fact.image.ImageCleaning energyThreshold="2.45" />
  <fact.image.features.HillasParameters />
  <streams.weka.Apply modelUrl="http://sfb876.de/rforest.weka" />
</process>
```

Fig. 7. The XML corresponding to the pipeline of the previous figure.

3.2   FACT-Tools: Processing Telescope Data

The FACT-Tools is a collection of input implementations and user-functions that is built around the processing of telescope data. By implementing the required functionality in the context of the user-functions API of streams, this allows physicists to easily create data flows by XML specifications and test their processes in a reproducible manner. This rapid prototyping and library-style coding led to a quickly evolving setting that covers the complete data analysis chain from data acquisition to evaluation.

Table 1. The representation of a shower event as hashmap in the streams model.

Name (key)	Type	Description
EventNum	Integer	The event number in the stream
TriggerNum	Integer	The trigger number in the stream
TriggerType	Integer	The trigger type that caused recording of the event
Errors	Integer	Indicates communication errors
UnixTimeUTC	Integer	Timestamp of the recorded event in millisecond accuracy
Data	Double[432000]	The raw data array (1440 · 300 = 432000 float values)

The primal data gathered by the telescope is encoded in the FITS file format. The *Flexible Image Transport System* (FITS) is a file format proposed by NASA to store satelite images and other information in a compact, yet flexible way as it supports a variety of basic data types that can be stored. The bulk of data, recorded by the telescope for each event, is provided as a large array of values, sampled from the camera pixels. Along with those samples, additional information on the event, such as the number of the recording run, the time and high-resolution arrival times for each pixel. The FACT-Tools provides a `fact.io.FitsStream` implementation that reads this data from any input stream and emits a sequence of items (one per shower event). Table 1 shows an excerpt of the elements provided for each item.

User-Functions for Telescope Events

The core element of each event is found as key `Data` in Table 1. It holds the values sampled from each pixel upon a trigger. As a region of interest of 300 slices is written to disk for each event, this amounts to 432000 values. Based on calibration data, the sampled values need to be adjusted to match the specific voltage offsets and gains for each pixel, which may vary depending on the temperature and other environmental factors. Based on the calibrated per-pixel time series, a couple of additional user-functions can be applied, each of which reflects the computation of features such as the identification of shower- and non-shower pixels, the fitting of an ellipse to the supposed shower image and the derivation of geometric Hillas parameters or other properties of the event.

Geared towards the rapid prototyping of this data-preprocessing flow, a wide range of additional user-functions has been implemented by the domain experts, which range from additional time-calibrations to data corrections such as the removal of broken pixel data. Each of these preprocessing step is focused on improving the data quality and finding of features that may further improve the overall gamma-hadron separation task.

3.3   Integration of Machine Learning Libraries

As part of the abstract data flow design, we integrated the MOA and WEKA machine learning libraries as modules into the streams framework. Though especially MOA is geared towards online learning, the setting of the telescope data demands more for an *online application* of models: The data that is used for training the models, is synthetically generated and available as a batch data set. Typically, the size of that data is also comparably small, once the features for training have been extracted (the majority of the data volume is embodied in the raw data). A crucial aspect for the application of machine learning models, is the fact that only features, which are extracted *online* are suitable for use in such models. Any features that relate to an overall property of a data set, e.g. a normalization with respect to the sum computed over a set of instances, will not match the stream-lined setting that we aim for in the FACT data analysis.

The two user-functions `streams.weka.Train` and `streams.weka.Apply` have been implemented, which can be used to incorporate the training and application of WEKA models directly within a `streams` data flow. This ensures, that the same preprocessing setup can be used to feed the training of the model as well as its later application. The `Train` function collects a batch of user-specified instances to build its training data set. This is crucial as some features such as nominal type features require additional meta-data to be equal during training and model application. Upon building the classifier, the `Train` function outputs the serialized model in addition to the meta-data information about the attributes. In addition, its `features` parameter allows for an easy wild-card selection of features that shall be used for building the classifier. Any keys prefixed with an @ character (e.g. `@id` or `@source`) are not regarded regular features for training the classifier. Figure 8 shows the XML setting of a process for training a random forest classifier using WEKA within `streams`. Classifier options are automatically mapped to XML attributes using Java's reflection API. The approach directly supports any of the provided WEKA classifiers. The corresponding `Apply` function has previously been shown in Figure 7. It requires a `modelUrl` parameter that holds the location of the serialized model. The `streams` framework automatically handles different URL types, such as file, http or classpath resources. This eases the sharing of processes and their models as well as a distributed execution of multiple instances of the analysis with a shared separation model.

```
<process input="simulator:data">
    <!-- preprocessing left out due to out space limitations  -->
    <streams.weka.Train
            features="*,!hillas:size"
          classifier="weka.classifiers.tree.RandomForest"
            numTrees="100"
              output="/data/random-forest.weka" />
</process>
```

Fig. 8. Training a WEKA classifier specified in XML.

4   Experiments

We tested the use of WEKA within the overall processing chain of the FACT telescope as modelled with the FACT-Tools. The data we used in the experiments was generated by Monte Carlo simulations using the CORSIKA software. The dataset contains 139333 shower events, of which 100000 events stem from gamma and 39333 events from proton (hadronic) particles. The events are simulated in raw data format and passed through the standard cleaning and feature extraction chain using the FACT-Tools, resulting in 83 features suitable for separation. For the experiments, we focused on the following aspects:

1. Predictive performance of the classifier for gamma-hadron separation,
2. Improvements of separation by re-organisation of the data flow,
3. Throughput performance of the overall processing chain.

An interesting note for the performance comparisons is the optimization criterion used to assess the classification. Whereas the traditional machine learning community often uses precision, recall or accuracy for grading classifier performances, the physics field is more interested in a pure sample of gamma ray induced events. A well-accepted measure in this area is the *Q-factor* defined as

$$Q = \frac{\varepsilon_\gamma}{\sqrt{\varepsilon_p}} \quad \text{with} \quad \varepsilon_\gamma = \frac{N_\gamma^{det}}{N_\gamma} \quad \text{and} \quad \varepsilon_p = \frac{N_p^{det}}{N_p}$$

where ε_γ and ε_p represent the *gamma efficiency* (number of gammas detected divided by total number of gammas in dataset) and the *proton* or *hadron efficiency* respectively. The Q-factor aims at assessing the purity of the resulting gamma events. In addition we also provide the significance index [15].

4.1   Gamma-Hadron Separation with WEKA Classifiers

Using the basic Hillas parameters and an additional set of features build up on these, we tested different classifiers: Random Forests, an SVM implementation and a Bayesian filter. Table 2 shows the classification performance for these approaches. Each classifier was evaluated with a 10-fold cross validation and optimized parameters: The Random Forest was trained with 300 trees with 12 features and a maximum depth of 25. The SVM used an RBF kernel with $k_\gamma = 0.014$ and $C = 10$. The training set in each fold was balanced.

Table 2. Performances for gamma/hadron separation with different classifiers.

Classifier	Q Factor	Significance	Accuracy	Precision
Random Forest	4.796 ± 0.178	65.55 ± 0.358	0.969 ± 0.0021	0.959 ± 0.0029
SVM	4.013 ± 0.916	60.227 ± 1.859	0.953 ± 0.010	0.936 ± 0.025
Naive Bayes	2.267 ± 0.0609	51.65 ± 0.503	0.841 ± 0.0048	0.864 ± 0.0062

For an improved purity, the classification is weighted with the confidence provided by the classifier. Those *confidence cuts* are applied by physicists to obtain an even cleaner sample as is crucial for all subsequent analysis steps. All *gamma* predicted elements with a confidence less than some threshold are regarded as proton predictions. Though this increases the number of missed gamma events, it eliminates false positives, which may tamper with subsequent steps such as energy estimation. Figure 9 shows the impact of these confidence cuts.

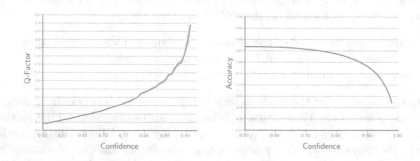

Fig. 9. Refined selection by *confidence cuts*, which improves the purity at the cost of missed signals, reflected in a decreased recall, which diminishes the accuracy.

4.2   Signal Separation with Local Models

A parameter describing the "intensity" of the shower is `size`, which incorporates the area of the ellipse and the voltage levels of the covered pixels. The `size` parameter is known to highly correlate with the energy of the original particle [6] and allows for a grouping of events based on their energies. We investigated the separation performance of Random Forests, trained on disjoint datasets defined on a partitioning using the $\log_{10}(\texttt{size})$ feature (Figure 10).

We limited this experiment to bins of $\log_{10}(\texttt{size})$, which had at least 10.000 events for testing. The right plot shows the Q-Factor for models trained and

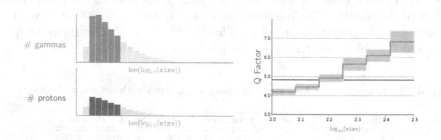

Fig. 10. Distribution of gamma and hadronic events over `size` range (left) and performance of local models per bin vs. global model (right).

evaluated on separate bins (green) and the global model trained over data from all bins (blue) without any confidence cuts applied. The figure provides the Q-Factor averaged over a 10-fold cross validation, the light green area shows the standard deviation. We only looked at the Q-Factor here, as it is the accepted criterion in that community.

4.3   Throughput Performance of the FACT-Tools

The FACT telescope records at a rate of 60 events per second, where each event amounts up to 3 MB of raw data, resulting in a rate of about 180 MB/s. Figure 11 shows the average processing time (milliseconds) of the user-functions for the complete analysis in a *log-scale*. The first two blocks of functions reflect the bulk of raw data processing. Ellipse fitting and other feature extractions which are input to the classification step are shown in bright green. The process is able to handle the full data rate of the telescope on a small scale Mac Mini.

The improved separation by the use of local models suggests a split of the data stream. Though the `size` feature is only available at a later stage in the process, it highly correlates with properties available directly after the data calibration (2nd user function) has been applied. In combination with the local models this allows for a massive parallelization by data stream grouping, when deploying the process in distributed environments such as Apache Storm. The generic abstraction provided by `streams` already allows for a direct mapping of the XML process specification to a Storm topology.

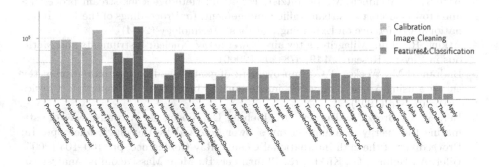

Fig. 11. Average processing times of the analysis chain functions (log scale).

5   Summary and Conclusion

In this work we reviewed the analysis chain of an Air-Cherenkov telescope and provided an online implementation of that process by the high-level abstraction framework `streams`. The resulting data flow covers the complete data processing and feature extraction. It integrates previously trained WEKA models and is able to handle the data rate of the FACT telescope in real-time. By focusing on a declarative, easy-to-use abstraction, we lowered the barrier of end-users designing their own analytical data flows, enabling direct use of state-of-the-art machine learning libraries. The generic nature of the streams abstraction and its embeddability allows for a direct integration into large scale distributed streaming engines and sets the scene to cope with the load of upcoming, more high-resolution telescopes.

Future work will further focus on improving the separation power and investigating models for energy estimation for FACT. As the scalability aspect obviously touches the data preprocessing we are looking into a direct mapping of user-functions defined using the streams API in Apache Storm and Apache Hadoop, aiming at a full code re-use without modifications.

We are also confident that this use-case can be mapped to other scenarios as we successfully tested it in steel-mill factories [24] and smart cities [3,25].

Acknowledgments. This work has been supported by the DFG, Collaborative Research Center SFB 876, project C3 (http://sfb876.tu-dortmund.de/).

References

1. Abeysekara, A.U., et al.: On the sensitivity of the HAWC observatory to gamma-ray bursts. Astroparticle Physics **35**, 641–650 (2012)
2. Anderhub, H., et al.: Fact - the first cherenkov telescope using a g-apd camera for tev gamma-ray astronomy. Nuclear Instruments and Methods in Physics Research A **639**, 58–61 (2011)
3. Artikis, A., Weidlich, M., Schnitzler, F., et al.: Heterogeneous stream processing and crowdsourcing for urban traffic management. In: Proceedings of the 17th International Conference on Extending Database Technology (2014)
4. Atkins, R., et al.: Milagrito, a tev air-shower array. Nuclear Instruments and Methods in Physics Research **449**, 478–499 (2000)
5. Badran, H.M., Weekes, T.C.: Improvement of gamma-hadron discrimination at tev energies using a new parameter, image surface brightness. Astroparticle Physics **7**(4), 307–314 (1997)
6. Berger, K., Bretz, T., Dorner, D., Hoehne, D., Riegel, B.: A robust way of estimating the energy of a gamma ray shower detected by the magic telescope. In: Proceedings of the 29th International Cosmic Ray Conference, pp. 100–104 (2005)
7. Bifet, A., Holmes, G., Kirkby, R., Pfahringer, B.: Moa: Massive online analysis. J. Mach. Learn. Res. **11**, 1601–1604 (2010)
8. Bock, R.K., Chilingarian, A., et al.: Methods for multidimensional event classification: a case study using images from a cherenkov gamma-ray telescope. Nuclear Instruments and Methods in Physics Research Section A: Accelerators, Spectrometers, Detectors and Associated Equipment **516**(2–3), 511–528 (2004)
9. Bockermann, C., Blom, H.: The streams framework. Technical Report 5, TU Dortmund University, vol. 12 (2012)
10. Breiman, L.: Random forests. Machine Learning **45**(1), 5–32 (2001)
11. Bussino, S., Mari, S.M.: Gamma-hadron discrimination in extensive air showers using a neural network. Astroparticle Physics **15**(1), 65–77 (2001)
12. Carroll, B.W., Ostlie, D.A.: An Introduction to Modern Astrophysics, 2nd edn. Pearson Addison-Wesley, San Francisco (2007)

13. De Naurois, M.: Analysis methods for atmospheric cerenkov telescopes (2006). arXiv preprint astro-ph/0607247
14. Faleiro, E., Muñoz, L., Relaño, A., Retamosa, J.: Discriminant analysis based on spectral statistics applied to TeV cosmic γ/proton separation. Astroparticle Physics **35**, 785–791 (2012)
15. Gillessen, S., Harney, H.L.: Significance in gamma-ray astronomy - the li and ma problem in bayesian statistics. Astronomy and Astrophysics **430**(1), 355–362 (2004)
16. Hall, M., Frank, E., Holmes, G., Pfahringer, B., Reutemann, P., Witten, I.H.: The weka data mining software: An update. SIGKDD Explor. Newsl. **11**(1), 10–18 (2009)
17. Heck, D., Knapp, J., Capdevielle, J.N., Schatz, G., Thouw, T.: CORSIKA: A Monte Carlo Code to Simulate Extensive Air Showers. Forschungszentrum Karlsruhe GmbH, Karlsruhe (1998)
18. Kieda, D.B.: VERITAS Collab. Status of the VERITAS ground based GeV/TeV Gamma-Ray Observatory. In: High Energy Astrophysics Division, vol. 36, p. 910. Bulletin of the American Astronomical Society, August 2004
19. Hillas, A.M.: Cerenkov light images of EAS produced by primary gamma rays and by nuclei. In: Jones, F.C. (ed.) Proceedings of the 19th International Cosmic Ray Conference, vol. 3, pp. 445–448, La Jolla, August 1985
20. Mierswa, I., Wurst, M., Klinkenberg, R., Scholz, M., Euler, T.: Yale: Rapid prototyping for complex data mining tasks. In: Ungar, L., Craven, M., Gunopulos, D., Eliassi-Rad, T. (eds.) Proceedings of the 12th ACM SIGKDD International Conference on Knowledge Discovery and Data Mining, KDD 2006, pp. 935–940. ACM, New York, August 2006
21. Petry, D., et al.: The MAGIC Telescope - prospects for GRB research. Astronomy & Astrophysics Supplement Series **138**, 601–602 (1999)
22. Pivato, G., et al.: Fermi LAT and WMAP Observations of the Supernova Remnant HB 21. The Astrophysical Journal **779**, 179 (2013)
23. Schäfer, B.M., Hofmann, W., Lampeitl, H., Hemberger, M.: Particle identification by multifractal parameters in γ-astronomy with the hegra-cherenkov-telescopes. Nuclear Instruments and Methods in Physics Research Section A: Accelerators, Spectrometers, Detectors and Associated Equipment **465**(2–3), 394–403 (2001)
24. Schlüter, J., Odenthal, H.-J., Uebber, N., Blom, H., Beckers, T., Morik, K.: Reliable bof endpoint prediction by novel data-driven modeling. In: AISTech Conference Proceedings. AISTech (2014)
25. Schnitzler, F., Artikis, A., Weidlich, M., Boutsis, I., Liebig, T., Piatkowski, N., Bockermann, C., Morik, K., Kalogeraki, V., Marecek, J., Gal, A., Mannor, S., Kinane, D., Gunopulos, D.: Heterogeneous stream processing and crowdsourcing for traffic monitoring: highlights. In: Calders, T., Esposito, F., Hüllermeier, E., Meo, R. (eds.) ECML PKDD 2014, Part III. LNCS, vol. 8726, pp. 520–523. Springer, Heidelberg (2014)

Robust Representation for Domain Adaptation in Network Security

Karel Bartos[1,2]([⊠]) and Michal Sofka[2]

[1] Faculty of Electrical Engineering,
Czech Technical University in Prague, Prague, Czech Republic
[2] Cisco Systems, Karlovo Namesti 10, 12000 Prague, Czech Republic
{kbartos,msofka}@cisco.com

Abstract. The goal of domain adaptation is to solve the problem of different joint distribution of observation and labels in the training and testing data sets. This problem happens in many practical situations such as when a malware detector is trained from labeled datasets at certain time point but later evolves to evade detection. We solve the problem by introducing a new representation which ensures that a conditional distribution of the observation given labels is the same. The representation is computed for bags of samples (network traffic logs) and is designed to be invariant under shifting and scaling of the feature values extracted from the logs and under permutation and size changes of the bags. The invariance of the representation is achieved by relying on a self-similarity matrix computed for each bag. In our experiments, we will show that the representation is effective for training detector of malicious traffic in large corporate networks. Compared to the case without domain adaptation, the recall of the detector improves from 0.81 to 0.88 and precision from 0.998 to 0.999.

Keywords: Traffic classification · Machine learning · Malware detection · HTTP traffic

1 Introduction

In supervised learning, the domain adaptation solves the problem when a joint distribution of the labels and observations differs for training (source) and testing (target) data. This can happen as a result of target evolving after the initial classifier was trained. For example, in network security, the classifier is trained from network traffic samples of malware communication which can change as a result of evolving malware. Under the assumption that the source and target distribution do not change arbitrarily, the goal of the domain adaptation is to leverage the knowledge in the source domain and transfer it to the target domain. In this work, we focus on the case where the conditional distribution of the observation given labels is different, also called a conditional shift.

The knowledge transfer can be achieved by adapting the detector using importance weighting such that training instances from the source distribution

© Springer International Publishing Switzerland 2015
A. Bifet et al. (Eds.): ECML PKDD 2015, Part III, LNAI 9286, pp. 116–132, 2015.
DOI: 10.1007/978-3-319-23461-8_8

match the target distribution [16]. Another approach is to transform the training instances to the domain of the testing data or to create a new data representation with the same joint distribution of observations and labels [1]. The challenging part is to design a meaningful transformation that transfers the knowledge from the source domain and improves the robustness of the classifier on the target domain.

In this paper, we present a new invariant representation of network traffic data suitable for domain adaptation under conditional shift. The representation is computed for bags of samples, each of which consists of features computed from network traffic logs. A bag is constructed for every user and all network communication with each domain. The representation is designed to be invariant under shifting and scaling of the feature values and under permutation and size changes of the bags. This is achieved by constructing an invariant self similarity matrix for each bag. Pairwise relevance measure is trained to reliably assign previously-unseen bags to existing categories or to create a new category.

The proposed similarity measure and the new invariant representation is applied to detect malicious HTTP traffic in network security. We will show that the classifier trained on malware communication samples from one category can successfully detect new samples from a different category. This way, the knowledge of the malware behavior is correctly transferred to the new domain which improves the classifier. Compared to the case without adaptation with 0.81 recall and 0.998 precision, the new approach has recall 0.88 and precision 0.999.

2 Problem Statement

The paper deals with the problem of supervised classification of bags of samples into categories with a lack of labeled data. The labels for positive and negative samples are often very expensive to obtain. Moreover, sample distribution typically evolves in time, so the probability distribution of training data differs from the probability distribution of test data. In contrast to the case when enough samples is available in each category and their distributions are stationary, the knowledge needs to be transferred in time within categories but also across categories using labeled samples. In the following, the problem is described in more detail.

Each sample is represented as an n-dimensional vector $\mathbf{x} \in \mathbb{R}^n$. Samples are grouped into bags, where i-th bag is a set of m_i samples $X_i = \{\mathbf{x}_1, \ldots, \mathbf{x}_{m_i}\} \in \mathcal{X}$. A single category y_i can be assigned to each bag from the set of categories $\mathcal{Y} = \{y_1, \ldots, y_N\}$. Note that not all categories are included in the training set. The probability distribution on training (labeled) and test bags for category y_j will be denoted as $P^L(X|y_j)$ and $P^T(X|y_j)$, respectively. Moreover, the probability distribution of training data differs from the probability distribution of testing data, a problem dealt with in the domain adaptation [2] (also called a conditional shift [18]):

$$P^L(X|y_j) \neq P^T(X|y_j), \ \forall y_j \in \mathcal{Y}. \tag{1}$$

The purpose of the domain adaptation is to acquire knowledge from the training (source) domain and apply it to the testing (target) domain. The relation between $P^L(X|y_i)$ and $P^T(X|y_i)$ is not arbitrary, otherwise it would not be possible to transfer any knowledge. Therefore there is a transformation τ, which transforms the feature values of the bags onto a representation, in which $P^L(\tau(X)|y_i) = P^T(\tau(X)|y_i)$. We assume that $\tau(X)$ is any representation that is invariant against shift, scale, permutation, and size changes of the bag. The goal is to find this representation, allowing to classify individual bags X_i into categories $\mathcal{Y} = \{y_1, \ldots, y_N\}$ under the conditional shift.

A number of other methods for transfer learning have been proposed, including kernel mean matching [10], kernel learning approaches [8], maximum mean discrepancy [11], or boosting [7]. These methods try to solve a general data transfer with relaxed conditions on the similarity of the distributions during the transfer. The downside of these methods is the necessity to specify the target loss function and the availability of large amount of labeled data.

Our solution to the conditional shift problem is to transform the features to a new representation. The advantage of this approach is that it is independent of the classification loss function and similarity between the probability distributions does not need to be given. The method achieves the knowledge transfer by changing the original feature values. The feature values are transformed into a new representation that is invariant against shift, scale, permutation, and size changes of the bags (number of samples within each bag). Once the data are transformed according to the proposed representation, the new feature values do not follow the original distribution and therefore they are not influenced by the shift.

To compensate for the lack of labeled data, a simple online linear transformation is applied. The transformation learns a set of weights on the new features to match the training and test distributions of the bags from the same category. At the same time, the weights are optimized to separate bags belonging to different categories. This way, bags belonging to the same category are assigned the same label during classification.

3   Invariant Representation of Bags

In this Section, an invariant representation of bags is proposed to overcome the problem of domain shift introduced in Section 2. The new representation is calculated with a transformation τ that consists of three steps to ensure that the new representation will be independent on the mean, and invariant against scaling, shifting, permutation and size of the bags. In the following, the individual steps are discussed in more detail.

3.1   Shift Invariance with Self-Similarity Matrix

As stated in Section 2, the probability distribution of bags from the training set and the testing set can be different. Therefore, in the first step, the representation of bags is transformed to be invariant against this shift. The traditional representation of i-th bag X_i that consists of a set of m samples $\{\mathbf{x}_1, \ldots, \mathbf{x}_m\}$ is typically in a form of a matrix:

$$X_i = \begin{pmatrix} \mathbf{x}_1 \\ \vdots \\ \mathbf{x}_m \end{pmatrix} = \begin{pmatrix} x_1^1 & x_1^2 & \cdots & x_1^n \\ & & \vdots & \\ x_m^1 & x_m^2 & \vdots & x_m^n \end{pmatrix},$$

where x_l^k denotes k-th feature value of l-th sample from bag X_i. This form of representation of samples and bags is widely used, as it is straightforward to compute. It is a reasonable choice in many applications with negligible difference in probability distributions. However, when the difference becomes more serious, the traditional representation often leads to unsatisfactory results. Therefore, the following transformation is proposed to overcome the difference typically caused by the dynamics of the domain, making the solution for the classification problem more effective. As a first step, the representation is transformed to be invariant against shift of the feature values.

Shift invariance guaranties that even if some original feature values of all samples in a bag are increased/decreased by a given amount, the values in the new representation remain unchanged.

Let us define a *translation invariant distance function*, which is a distance function $d : \mathbb{R} \times \mathbb{R} \to \mathbb{R}$ such that:

$$d(x_1, x_2) = d(x_1 + a, x_2 + a). \tag{2}$$

Let x_p^k, x_q^k be k-th feature values of p-th and q-th sample from bag X_i, respectively. It is possible to express the relation between the values as follows:

$$x_p^k = x_q^k - s_{pq}^k, \tag{3}$$

where s_{pq}^k is the difference between values x_p^k, x_q^k. Then it holds for each translation invariant distance function $d : \mathbb{R} \times \mathbb{R} \to \mathbb{R}$:

$$d(x_p^k, x_q^k) = d(x_p^k, x_p^k + s_{pq}^k) = d(0, s_{pq}^k) = s_{pq}^k.$$

Therefore, the new feature value $d(x_p^k, x_q^k)$ expresses the distance between the two values of k-th feature regardless of their absolute values. This value is more robust, however it could be less informative, as the information about the absolute values was removed. To compensate for the possible loss of information, the bags are represented with a matrix of these distances $d(x_p^k, x_q^k)$, which is called a self-similarity matrix S^k. Self-similarity matrix is a symmetric positive semidefinite matrix, where rows and columns represent individual samples and (i, j)-th element corresponds to the distance between i-th and j-th sample. Self-similarity matrix has been already used thanks to its properties in several applications (e.g. in object recognition [12] or music recording [14]). However, only a single self-similarity matrix for each bag has been used in these approaches.

This paper proposes to compute a set of similarity matrices, one for every feature. More specifically, a per-feature self-similarity set of matrices $\mathbf{S}_i = \{S_i^1, S_i^2, \ldots, S_i^n\}$ is computed for i-th bag X_i, where

$$
S_i^k = \begin{pmatrix} s_{11}^k & s_{12}^k & \cdots & s_{1m}^k \\ s_{21}^k & s_{22}^k & \cdots & s_{2m}^k \\ & & \vdots & \\ s_{m1}^k & s_{m2}^k & \cdots & s_{mm}^k \end{pmatrix} = \begin{pmatrix} d(x_1^k, x_1^k) & d(x_1^k, x_2^k) & \cdots & d(x_1^k, x_m^k) \\ d(x_2^k, x_1^k) & d(x_2^k, x_2^k) & \cdots & d(x_2^k, x_m^k) \\ & & \vdots & \\ d(x_m^k, x_1^k) & d(x_m^k, x_2^k) & \cdots & d(x_m^k, x_m^k) \end{pmatrix}, \quad (4)
$$

and $s_{pq}^k = d(x_p^k, x_q^k)$ is a distance between feature values x_p^k and x_q^k of k-th feature. This means that the bag X_i with m samples and n features will be represented with n self-similarity matrices of size $m \times m$.

3.2   Scale Invariance with Local Feature Normalization

As explained in the previous section, self-similarity matrix S_i of the bag X_i captures mutual distances among the samples included in X_i. Therefore, the matrix describes inner temporal dynamics of bags [12], [13]. In other words, it describes how the bag is evolving in time. In case of a bag, where all samples are the same, the matrix S_i will be composed of zeros. On the other hand, in case of a bag with many different samples, the self-similarity matrix will be composed of a wide range of values.

The next step is to transform the matrix S_i^k to be invariant against scaling. **Scale invariance** guarantees that even if some original feature values of all samples in a bag are multiplied by a common factor, the values in the new representation remain unchanged. To guarantee the scale invariance, the matrix S_i^k needs to be locally normalized onto the interval $[0, 1]$ as follows:

$$
\tilde{S}_i^k = \begin{pmatrix} \tilde{s}_{11}^k & \tilde{s}_{12}^k & \cdots & \tilde{s}_{1m}^k \\ \tilde{s}_{21}^k & \tilde{s}_{22}^k & \cdots & \tilde{s}_{2m}^k \\ & & \vdots & \\ \tilde{s}_{m1}^k & \tilde{s}_{m2}^k & \cdots & \tilde{s}_{mm}^k \end{pmatrix}, \quad \tilde{s}_{pq}^k = \frac{s_{pq}^k - \min_{i,j}(s_{ij}^k)}{\max_{i,j}(s_{ij}^k) - \min_{i,j}(s_{ij}^k)}. \quad (5)
$$

Note that the maximum and minimum value is computed only from the samples within the bag, therefore the normalization is referred to as local. After the local scaling, the matrices $\tilde{\mathbf{S}}_i = \{\tilde{S}_i^1, \tilde{S}_i^2, \ldots, \tilde{S}_i^n\}$ are invariant against shifting and scaling, focusing purely on the dynamics among the samples (matrix of differences) and not on the absolute values of the differences. An example of an input feature vector and the corresponding locally-normalized self-similarity matrix is illustrated in Figure 1 (a) and Figure 1 (b).

3.3   Permutation and Size Invariance with Histograms

Representing bags with locally-scaled self-similarity matrices $\tilde{\mathbf{S}}$ achieves the scale and shift invariance. However, as there are no restrictions on the size of the

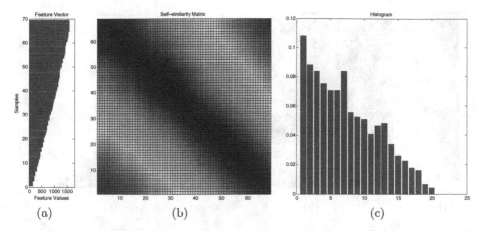

Fig. 1. Graphical illustration of the individual steps that are needed to transform the bag (set of samples) into the proposed invariant representation. First, the bag is represented with a standard feature vector (a). Then the locally normalized self-similarity matrix (b) is computed for each feature. Finally, values from the matrix will create a new histogram (c), which is invariant on the number or the ordering of the samples within the bag.

bags (i.e. how many samples are included in a bag), the corresponding self-similarity matrices can be of various sizes. The various sizes of the matrices make their comparison difficult. This is solved by introducing **size invariance** which ensures that the representation does not depend on the size of the bags. Moreover, in highly dynamic environments, the samples may occur in a variable ordering. Since the sample order does not matter for the representation of the bags, the robustness to reordering of rows and columns is guaranteed by the **permutation invariance**.

The final step of the proposed transformation is the transition from the scaled self-similarity matrices $\tilde{\mathbf{S}}_i = \{\tilde{S}_i^1, \tilde{S}_i^2, \ldots, \tilde{S}_i^n\}$ into histograms. Every matrix $\tilde{S}_i^k$ is transformed into a single histogram $\mathbf{h}_i^k$ with a predefined number of bins. Each bin of a histogram $\mathbf{h}_i^k$ represents one feature value in the proposed new representation.

Overall, i-th bag is represented as a vector $\mathbf{h}_i$ of size $n \times l$ as follows:

$$\mathbf{h}_i = (\mathbf{h}_i^1, \mathbf{h}_i^2, \ldots, \mathbf{h}_i^n), \tag{6}$$

where n is the number of features (and histograms) and l is the number of bins. The whole transformation is depicted in Figure 1. Figure 2 illustrates the invariant properties of the representation. Even though the bags from Figure 1 (a) and Figure 2 (a) have different number of samples, ordering, and range of the original feature values, the output histograms are similar.

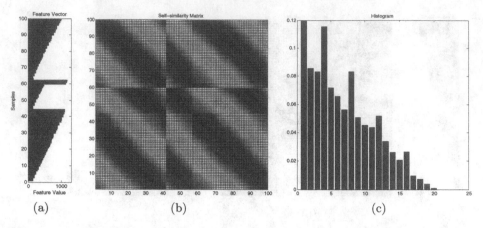

(a)   (b)   (c)

Fig. 2. Graphical illustration showing the invariant properties of the proposed representation for one feature. Even tough the bag (and thus the input feature vector (a)) has more samples than the bag from Figure 1, the histogram (c) computed from the self-similarity of the samples (b) is similar.

4   Online Similarity Learning

Representing input bags, as proposed in Section 3, ensures invariance against the conditional shift described in Section 2. Therefore, the transformed feature values can be used for learning a classifier that classifies the bags into categories.

As mentioned in Section 2, some categories may be missing in the training set. The classification method should be able to identify them in the test data and separate them from the rest of the categories. Several existing approaches have been proposed to address the classification problem with missing labels in the training set, e.g. zero-shot learning with semantic output codes [15] or through cross-modal transfer [17]. However, these approaches are typically designed for many labeled samples. When the number of labeled samples is limited, a similarity-based approach [5] can be used.

Similarity-based classifiers estimate the category label from a pairwise similarity of a testing bag and a set of labeled training bags. The comparison between two bags is performed by computing a similarity of feature vectors $\mathbf{h}_i$ and $\mathbf{h}_j$ using a similarity matrix $\mathbf{W}$. The similarity matrix is trained by using a pairwise relevance measure $r : \mathbb{R}^{n \cdot l} \times \mathbb{R}^{n \cdot l} \to \mathbb{R}$, designed to evaluate how relevant the two feature vectors are. Note that n and l denotes the number of features and the number of bins, respectively (both are defined in Section 3). The benefit of this approach lies in the fact that the algorithm requires only a limited number of labeled samples. The samples are labeled in a way to expresses relation, whether one pair of feature vectors is more relevant than the other. The relevance measure should satisfy the following conditions:

1. Let $\mathbf{h}_i, \mathbf{h}_j$ be two feature vectors from category y_m and $\mathbf{h}_k$ be from a different category y_n (or is unlabeled). Then $r(\mathbf{h}_i, \mathbf{h}_j) > r(\mathbf{h}_i, \mathbf{h}_k)$.
2. Let $\mathbf{h}_i, \mathbf{h}_j$ be two feature vectors from category y_m and $\mathbf{h}_k, \mathbf{h}_l$ be from different categories y_{n1} and y_{n2} (or are not labeled). Then $r(\mathbf{h}_i, \mathbf{h}_j) > r(\mathbf{h}_k, \mathbf{h}_l)$.

The first condition defines the basic requirement to consider two bags from the same category more relevant than two bags from different categories. The second condition ensures that two bags from the same category are more relevant to each other than two unlabeled bags. The training is done by using the passive-aggressive algorithm [6] OASIS [4] originally designed for recognizing similar images. The algorithm iteratively adjusts the weights of the similarity matrix to best fit the previous as well as the new training samples (see Algorithm 1). In [4] it has been shown that the algorithm converges fast with relatively small number of training pairs.

The algorithm finds a bilinear form $\mathbf{W}$ for which:

$$\mathbf{h}_i \, \mathbf{W} \, \mathbf{h}_j > \mathbf{h}_i \, \mathbf{W} \, \mathbf{h}_k + 1,$$

where $\mathbf{h}_i, \mathbf{h}_j$, and $\mathbf{h}_k$ are three feature vectors from the first condition mentioned earlier in this Section. In case of a hinge loss function defined as:

$$l_{\mathbf{W}}(\mathbf{h}_i \, \mathbf{h}_j, \mathbf{h}_k) = \max\{0, 1 - \mathbf{h}_i \, \mathbf{W} \, \mathbf{h}_j + \mathbf{h}_i \, \mathbf{W} \, \mathbf{h}_k\},$$

the goal is to minimize a global loss $L_{\mathbf{W}}$ over all possible triples:

$$L_{\mathbf{W}} = \sum_{i,j,k} l_{\mathbf{W}}(\mathbf{h}_i, \mathbf{h}_j, \mathbf{h}_k).$$

To minimize the global loss $L_{\mathbf{W}}$, a passive-aggressive algorithm is applied to optimize $\mathbf{W}$ over all feature vectors. The algorithm starts with the initial similarity matrix $\mathbf{W} = \mathbf{I}$ (identity matrix). In this case, the similarity is a simple dot product of the two feature vectors $\mathbf{h}_i^T \mathbf{I} \mathbf{h}_j = \mathbf{h}_i^T \cdot \mathbf{h}_j$. The algorithm then iterates over the training samples to adjust the similarity matrix $\mathbf{W}$ to satisfy the conditions (1) and (2) defined above. In each step, the algorithm randomly selects a pair of feature vectors from the same category and one feature vector from a different category (or an unlabeled bag). The purpose of each iteration is to optimize a trade-off between $\mathbf{W}$ computed so far and the current loss $l_{\mathbf{W}}$. More specifically, the algorithm solves the following convex problem with soft margin:

$$\mathbf{W}^i = \arg\min_{\mathbf{W}} \frac{1}{2} \|\mathbf{W} - \mathbf{W}^{i-1}\|_{Fro}^2 + C\xi \tag{7}$$

$$\text{s.t.} \quad l_{\mathbf{W}}(\mathbf{h}_i, \mathbf{h}_j, \mathbf{h}_k) \leq \xi \quad \text{and} \quad \xi \geq 0,$$

where $\|.\|_{Fro}$ is the Frobenius norm and the parameter C controls the trade-off. The solution of the optimization problem [4] from Equation 7 is described in Algorithm 1. The training ends after a predefined number of iterations or when the similarity between the training pairs is below a given threshold.

Algorithm 1. Training similarity matrix

---

function TRAINSIMILARITYMATRIX
  $\mathbf{W}^0 = \mathbf{I}$
  repeat
    sample three feature vectors:
      $F = \mathbf{h}_i, F_+ = \mathbf{h}_j, F_- = \mathbf{h}_k$
      such that $r(F, F_+) > r(F, F_-)$
    $\mathbf{V}^i = [F^{(1)}(F_+ - F_-), \ldots, F^{(N)}(F_+ - F_-)]^T,$
    where $F^{(i)}$ denotes i-th component of F
    $l_{W^{i-1}}(F, F_+, F_-)$
      $= \max\{0, 1 - F W^{i-1} F_+ + F W^{i-1} F_-\}$
    $\tau_i = \min\{C, \frac{l_{W^{i-1}}(F, F_+, F_-)}{\|\mathbf{V}^i\|^2}\},$
    where C is aggressiveness parameter
    $\mathbf{W}^i = \mathbf{W}^{i-1} + \tau_i \mathbf{V}^i$
  until (stopping criterion)
  return W
end function

In the testing phase, the similarity is used to create clusters of similar feature vectors, where all vectors from one cluster belong to the same category. As the last stage of the training procedure, the algorithm computes centroids $\mathbf{c}_i$ of the clusters C_i and threshold t. The threshold t is computed as an average similarity of a centroid with the rest of the vectors within a cluster. This is calculated across all clusters as follows:

$$t = \frac{\sum_{i,j} \mathbf{c}_i^T \mathbf{W} \mathbf{h}_j^{(i)}}{\text{number of all feature vectors } \mathbf{h}_j^{(i)}}, \tag{8}$$

where $\mathbf{h}_j^{(i)}$ denotes that j-th feature vector from i-th cluster. In case of a vector not similar to any of the existing centroids (the similarity is below the threshold t), this vector will create a new centroid and thus a new category.

5   Application in Network Security

We applied the combination of the proposed representation with the similarity learning to classify unseen malware bags in network security domain. The next section provides specification of the datasets, followed by the results from the experimental evaluation.

5.1   Specification of the Datasets

This section provides detailed description of the datasets and features used in the experimental evaluation. The datasets are divided into two disjoint parts: training, and testing. Both datasets were obtained from 1 month of real network traffic of 80 international companies (more than 500,000 users) in form of proxy

Fig. 3. URL decomposition into seven parts.

Table 1. List of features extracted from proxy logs. Features from the right column are applied on all URL parts.

Features	Features applied on all URL parts + referer
duration	length
HTTP status	digit ratio
is URL encrypted	lower case ratio
is protocol HTTPS	upper case ratio
number of bytes up	vowel changes ratio
number of bytes down	has repetition of '&' and '='
is URL in ASCII	starts with number
client port number	number of non-base64 characters
server port number	has a special character
user agent length	max length of consonant stream
MIME-Type length	max length of vowel stream
number of '/' in path	max length of lower case stream
number of '/' in query	max length of upper case stream
number of '/' in referer	max length of digit stream
is second-level domain rawIP	ratio of a character with max occurrence

logs. These logs contain HTTP/HTTPS flows, where one flow represents one communication between a user and a server. More specifically, one flow is a group of packets with the same source and destination IP address, source and destination port, and protocol. As flows from the proxy logs are bidirectional, both directions of a communication are included in each flow.

A flow consists of the following fields: user name, source IP address, destination IP address, source port, destination port, protocol, number of bytes transferred from client to server and from server to client, flow duration, timestamp, user agent, URL, referer, MIME-Type, and HTTP status. The most informative field is URL, which can be decomposed further into 7 parts as illustrated in Figure 3. We extracted 317 features from the flow fields (see the list in Table 1). Features from the right column are applied on all URL parts, including the URL itself and the referer.

Flows are grouped into bags, where each bag contains flows with the same user (or source IP) and the same second-level domain. Thus, each bag represents communication of a user with a particular domain. The size of a bag is at least 5 flows to be able to compute a representative histogram from feature values. As the datasets were originally unlabeled, we used available blacklists and other malware feeds from Collective Intelligence Framework (CIF) [9] to add positive

Table 2. Number of flows and bags of malware categories and background traffic.

Malware Category	Samples	
	Flows	Bags
C&C malware	30,105	532
DGA malware	3,772	105
DGA exfiltration	1,233	70
Click fraud	9,434	304
Trojans	1,230	12
Background	867,438	15,000

Table 3. Summary of the SVM results from the baseline and the proposed representation. Both classifiers have the same results on the training set, however SVM classifier where the bags were represented with the proposed self-similarity approach achieved better performance on the test data.

Representation	Training Data					Test Data				
	TP	FP	TN	precision	recall	TP	FP	TN	precision	recall
baseline	304	0	6976	1.0	1.0	584	13	7987	0.998	0.81
self-similarity	304	0	6976	1.0	1.0	**633**	**6**	**7994**	**0.999**	**0.88**

labels to the training dataset. All bags with domains marked as malicious by CIF (or by other external tools) were labeled as positive.

There are 5 malware categories: malware with command & control channels (marked as C&C), malware with domain generation algorithm (marked as DGA), DGA exfiltration, click fraud, and trojans. The summary of malicious categories is shown in Table 2. The rest of the background traffic is considered as legitimate.

5.2   Experimental Evaluation

This section shows the benefits of the proposed representation for a two-class and a multi-class classification problem in network security. The feature vectors described in Section 5.1 correspond to input feature vectors $\{x_1, \ldots, x_m\}$ defined in Section 2. These vectors were transformed to the proposed histogram representation $\{h^1, \ldots, h^n\}$, as described in Section 3. Each histogram h^i had 32 bins ($l = 32$). The proposed approach was compared with a baseline representation, where each bag is represented as a joint histogram of the input feature values $\{x_1, \ldots, x_m\}$. This means that one histogram was computed from values of every feature and bag, and the histograms were then concatenated to one final feature vector for each bag. Note that the baseline representation differs from the proposed representation in the fact that the baseline does not compute histograms from self-similarity matrices, but directly from the input feature values. Comparing these two approaches will show the importance of the self-similarity matrix, when dealing with domain adaptation problems.

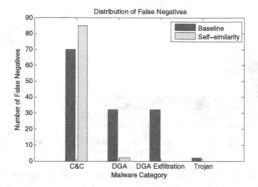

Fig. 4. Analysis of false negatives for both approaches. Thanks to the proposed self-similarity representation, SVM classifier was able to correctly classify all DGA exfiltration, trojan, and most of DGA malware bags, with a slight increase of false negatives for C&C.

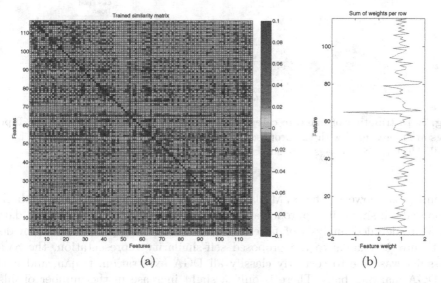

Fig. 5. Graphical illustration of a similarity submatrix W trained according to Algorithm 1 (a) and the corresponding sum of weights for each row (b). The matrix can also serve for feature selection, as some features have a negligible weight.

First, a two-class SVM classifier was evaluated on both representations. To demonstrate the conditional shift of positive bags, only click fraud bags were used in the training set as positive bags. A total of 6976 negative bags were included in the training set. The SVM classifier was evaluated on bags from C&C and DGA malware, DGA exfiltration, trojans, and 8000 negative background bags. The results are shown in Table 3. Both classifiers have the same results on the

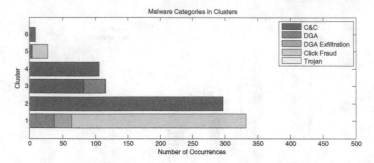

Fig. 6. Distribution of malware categories in clusters with more than 5 bags. Input bags are represented with the baseline approach. C&C bags are scatted across more clusters, and trojan malware bags were not clustered at all.

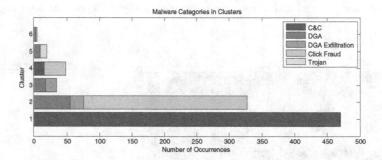

Fig. 7. Distribution of malware categories in clusters with more than 5 bags. Input bags are represented with the proposed approach. Most C&C bags were placed into a single cluster. Trojan bags were successfully found in cluster 5.

training set, however the SVM classifier using the data represented with the proposed self-similarity approach achieved better performance on the test data.

More detailed analysis of false negatives for both approaches in provided in Figure 4. Thanks to the proposed self-similarity representation, the SVM classifier was able to correctly classify all DGA exfiltration, trojan, and most of DGA malware bags. There is only a slight increase in the number of false negatives for C&C. Overall, the proposed self-similarity representation shows better robustness than the baseline approach.

Next, the performance on a multi-class problem with missing labels is evaluated with the similarity learning algorithm described in Section 4. Two malware categories were included in the training set (click fraud and C&C) together with 5000 negative bags. Similarity matrix W, trained according to the Algorithm 1, is depicted in Figure 5.

In the next experiment, similarity matrix W was used to create an adjacency matrix of all bags in the test set, where i, j-th component of this matrix is computed as $\mathbf{h}_i^T \mathbf{W} \mathbf{h}_j$. This means that i, j-th component expresses the distance between i-th and j-th bag in a metric space defined by the learned similarity

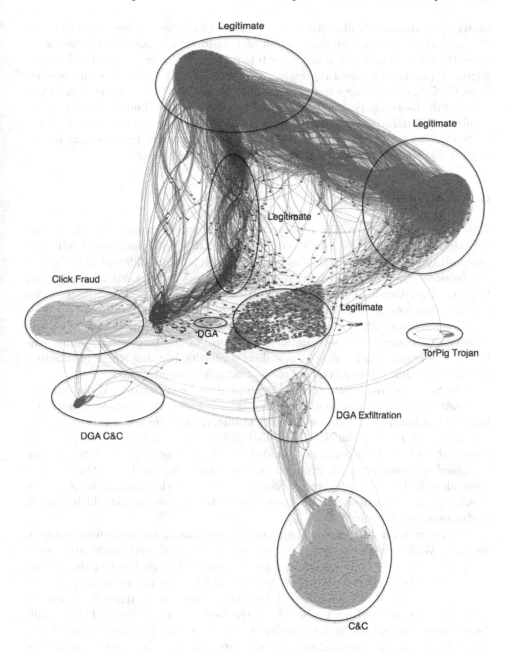

Fig. 8. Graphical illustration of the clustering results, where the input bags were represented with the proposed representation. Legitimate bags are concentrated in three large clusters on the top and in a group of non-clustered bags located in the center. Malicious bags were clustered into six clusters.

matrix W. Modularity clustering [3] was used to cluster the bags according to the adjacency matrix. The distribution of categories in malicious clusters with more than 5 bags is depicted in Figure 6 (for the baseline representation) and in Figure 7 (for the proposed representation). In contrast to the baseline results, most C&C bags are concentrated in a single cluster. Moreover, trojan bags were successfully found (in cluster 5) as opposed to the baseline. The overall clustering results are illustrated in Figure 8. The legitimate bags are concentrated in three large clusters on the top and in a group of non-clustered bags located in the center, while the malicious bags were clustered to six clusters.

6   Conclusion

This paper proposed a robust representation of bags of samples suitable for the domain adaptation problems with conditional shift. Under conditional shift, the probability distributions of the observations given labels is different in the training (source) and testing (target) data which complicates standard supervised learning algorithms. The new representation is designed to be invariant under common changes between the source and target data, namely shifting and scaling of the feature values and permutation and size changes of the bags. This is achieved by computing a self-similarity measure of the bags using sample features. The representation is used in online similarity learning which results in a robust algorithm for multi-class classification with missing labels.

The proposed representation was evaluated and compared with the baseline representation without adaptation in two network security use cases. First, in a binary classification of malicious network traffic, the new invariant representation improved the recall of an SVM classifier from 0.81 to 0.88 and the precision from 0.998 to 0.999. Second, in a modularity clustering of network traffic, the proposed approach correctly grouped malware according to their categories and even identified a new category, previously unseen in the training data. These results demonstrate the invariant properties of the representation which make it useful in network security.

There are several remaining challenges in the domain adaptation for network security. With constantly evolving malware, conditional shift might still occur even when the new malware families are represented as outlined in this paper. There are other types of malware, some of which have not been identified or fully understood, that have different behavioral patterns making it impossible to transfer knowledge from the source to the target domain. Some of these challenges might be solved by introducing nonlinearity to the malware similarity measure. As in the presented online similarity learning, the measure could use the known samples to learn the differences between malicious and legitimate traffic. This is the direction of our future research.

References

1. Ben-David, S., Blitzer, J., Crammer, K., Pereira, F., et al.: Analysis of representations for domain adaptation. Advances in neural information processing systems **19**, 137 (2007)
2. Blitzer, J., McDonald, R., Pereira, F.: Domain adaptation with structural correspondence learning. In: Proceedings of the 2006 Conference on Empirical Methods in Natural Language Processing, pp. 120–128. Association for Computational Linguistics (2006)
3. Brandes, U., Delling, D., Gaertler, M., Gorke, R., Hoefer, M., Nikoloski, Z., Wagner, D.: On modularity clustering. IEEE Transactions on Knowledge and Data Engineering **20**(2), 172–188 (2008)
4. Chechik, G., Sharma, V., Shalit, U., Bengio, S.: Large scale online learning of image similarity through ranking. The Journal of Machine Learning Research **11**, 1109–1135 (2010)
5. Chen, Y., Garcia, E.K., Gupta, M.R., Rahimi, A., Cazzanti, L.: Similarity-based classification: Concepts and algorithms. The Journal of Machine Learning Research **10**, 747–776 (2009)
6. Crammer, K., Dekel, O., Keshet, J., Shalev-Shwartz, S., Singer, Y.: Online passive-aggressive algorithms. The Journal of Machine Learning Research **7**, 551–585 (2006)
7. Dai, W., Yang, Q., Xue, G.-R., Yu, Y.: Boosting for transfer learning. In: Proceedings of the 24th International Conference on Machine learning, pp. 193–200. ACM (2007)
8. Duan, L., Tsang, I.W., Xu, D.: Domain transfer multiple kernel learning. IEEE Transactions on Pattern Analysis and Machine Intelligence **34**(3) (2012)
9. Farnham, G., Leune, K.: Tools and standards for cyber threat intelligence projects. Technical report, SANS Institute InfoSec Reading Room, p. 10 (2013)
10. Gretton, A., Smola, A., Huang, J., Schmittfull, M., Borgwardt, K., Schölkopf, B.: Covariate shift by kernel mean matching. Dataset shift in machine learning (2009)
11. Iyer, A., Nath, S., Sarawagi, S.: Maximum mean discrepancy for class ratio estimation: convergence bounds and kernel selection. In: Proceedings of the 31st International Conference on Machine Learning (ICML 2014), pp. 530–538 (2014)
12. Junejo, I.N., Dexter, E., Laptev, I., Perez, P.: View-independent action recognition from temporal self-similarities. IEEE Transactions on Pattern Analysis and Machine Intelligence **33**(1), 172–185 (2011)
13. Körner, M., Denzler, J.: Temporal self-similarity for appearance-based action recognition in multi-view setups. In: Wilson, R., Hancock, E., Bors, A., Smith, W. (eds.) CAIP 2013, Part I. LNCS, vol. 8047, pp. 163–171. Springer, Heidelberg (2013)
14. Müller, M., Clausen, C.: Transposition-invariant self-similarity matrices. In: Proceedings of the 8th International Conference on Music Information Retrieval (ISMIR), pp. 47–50 (2007)
15. Palatucci, M., Pomerleau, D., Hinton, G.E., Mitchell, T.M.: Zero-shot learning with semantic output codes. In: Advances in Neural Information Processing Systems (NIPS), pp. 1410–1418 (2009)

16. Shimodaira, H.: Improving predictive inference under covariate shift by weighting the log-likelihood function. Journal of statistical planning and inference **90**(2), 227–244 (2000)
17. Socher, R., Ganjoo, M., Manning, C.D., Ng, A.: Zero-shot learning through cross-modal transfer. In: Advances in Neural Information Processing Systems, pp. 935–943 (2013)
18. Zhang, K., Schölkopf, B., Muandet, K., Wang, Z.: Domain adaptation under target and conditional shift. In: Dasgupta, S., Mcallester, D. (eds.) Proceedings of the 30th International Conference on Machine Learning (ICML 2013), JMLR Workshop and Conference Proceedings, vol. 28, pp. 819–827 (2013)

Safe Exploration for Active Learning
with Gaussian Processes

Jens Schreiter[1]([✉]), Duy Nguyen-Tuong[1], Mona Eberts[1], Bastian Bischoff[1],
Heiner Markert[1], and Marc Toussaint[2]

[1] Robert Bosch GmbH, 70442 Stuttgart, Germany
jens.schreiter@de.bosch.com
[2] University of Stuttgart, MLR Laboratory, 70569 Stuttgart, Germany

Abstract. In this paper, the problem of safe exploration in the active learning context is considered. Safe exploration is especially important for data sampling from technical and industrial systems, e.g. combustion engines and gas turbines, where critical and unsafe measurements need to be avoided. The objective is to learn data-based regression models from such technical systems using a limited budget of measured, i.e. labelled, points while ensuring that critical regions of the considered systems are avoided during measurements. We propose an approach for learning such models and exploring new data regions based on Gaussian processes (GP's). In particular, we employ a problem specific GP classifier to identify safe and unsafe regions, while using a differential entropy criterion for exploring relevant data regions. A theoretical analysis is shown for the proposed algorithm, where we provide an upper bound for the probability of failure. To demonstrate the efficiency and robustness of our safe exploration scheme in the active learning setting, we test the approach on a policy exploration task for the inverse pendulum hold up problem.

1 Introduction

Active learning (AL) deals with the problem of selective and guided generation of labeled data. In the AL setting, an agent guides the data generation process by choosing new informative samples to be labeled based on the knowledge obtained so far. Providing labels for new data points, e.g. image labels as by Lang and Baum [1992] or measurements of the system output in case of physical systems, like by Hans et al. [2008], can be very costly and tedious. The overall goal of AL is to create a data-based model, without having to supply more data than necessary and, thus, reducing the agent annotation effort or the measurements on machines. For regression tasks, the AL concept is sometimes also referred to optimal experimental design, see Fedorov [1972].

In this paper, we consider the problem of safe data selection while jointly learning a data-based regression model on the explored input space. Given failure conditions, the goal is to actively select a budget of measurement points for approximating the model, and keeping the probability of measurement failures

© Springer International Publishing Switzerland 2015
A. Bifet et al. (Eds.): ECML PKDD 2015, Part III, LNAI 9286, pp. 133–149, 2015.
DOI: 10.1007/978-3-319-23461-8_9

to a minimum at the same time. In practice, safe data selection is highly relevant, especially, when measurements are performed on technical systems, e.g. combustion engines and test benches. For such technical systems, it is important to avoid critical points, where the measurements can damage the system. Thus, the main objective is (i) to approximate the system model from sampled data, (ii) using a limited budget of measured points, and (iii) ensuring that critical regions of the considered system are avoided during measurements.

We consider active data selection problems from systems with compact input spaces $\mathbb{X} \subset \mathbb{R}^d$. Within this constrained area $\mathbb{X}$ we have regions, where sampling and measuring is undesirable and can damage the system. For technical systems, for example, operation of the system in specific regions can result in exceeding the allowed physical limits such as temperatures and pressures. If the agent chooses a sample in such a region, it is considered as failure. To anticipate a failure, we assume that the agent observes some feedback from the system for each data point. This feedback indicates the health status of the system. In case of the combustion engine, this feedback is given by the engine temperature, for example.

The safety of an actively exploring algorithm is defined over the probability of failure, i.e. we call an exploration scheme safe at the level of $1 - \delta$, if the failure probability when querying an instance is lower than a certain threshold δ. The user can define δ sufficiently small to achieve an acceptable risk of failure. However, reducing this probability of failure comes at the cost of decreased sample efficiency, i.e. more samples will be required, as the agent will take smaller steps and explore more carefully. Throughout the paper, we use the notations $\mathbb{X}_+$ for safe and $\mathbb{X}_-$ for unsafe regions of the confined input space $\mathbb{X}$ to distinguish between safe and hazardous input areas of the system.

In this work, we employ Gaussian processes (GP's) to learn the regression model from a limited budget of incrementally sampled data points. Our exploration strategy for determining the next query points is based on the differential entropy criterion, cf. Krause and Guestrin [2007]. Furthermore, we employ a problem specific GP classifier to identify safe and unsafe regions in $\mathbb{X}$. The basic idea of this discriminative GP is learning a decision boundary between two classes, preferably without sampling a point in the unsafe region $\mathbb{X}_-$. To the best of our knowledge, such a safe exploring scheme has never been considered before in the AL context. We further show a theoretical analysis of our proposed safe exploration scheme with respect to the AL framework.

The remainder of the paper is organized as follows. In Section 2, a brief overview on some existing work on safe exploration approaches is given. In Section 3, we introduce our setting first and, subsequently, describe the proposed algorithm while providing details about our exploration strategy and the employed safety constraint. Section 4 provides some theoretical results on our proposed safe exploration technique. Experiments on a toy example and on learning a control policy for the inverse pendulum are shown in Section 5. A discussion in Section 6 concludes the paper.

2   Related Work

Most existing work for safe exploration in unknown environments arose in the reinforcement learning setting. The strategy of Moldovan and Abbeel [2012] for safe exploration in finite Markov decision processes (MDP's) relies on the restriction of suitable policies which ensure ergodicity at a user-defined safety level, i.e. there exists a policy with high probability to get back to the initial state. In the risk-sensitive reinforcement learning approach by Geibel [2001], the ergodic assumption for MDP's is dropped by introducing fatal absorbing states. The risk of a policy is thereby defined over the probability for ending in a fatal state. Therefore, the authors present a model-free reinforcement learning algorithm which is able to find near-optimal policies under bounded risk. The work by Gillula and Tomlin [2011] provides safety guarantees via a reachability analysis, when an autonomous robot explores its environment online. Here, the robot is observed by an aerial vehicle which avoid it from taking unsafe actions in the state space. This hybrid control system assumes bounded actions and disturbances to ensure a safe behavior for all current observable situations. Instead of bounding the action space, Polo and Rebollo [2011] introduce learning from demonstrations for dynamic control tasks. To safely explore the continuous state space in their reinforcement learning setting, a previously defined safe policy is iteratively adapted from the demonstration with small additive Gaussian noise. This approach ensures a baseline policy behavior which is used for safe exploration of high-risk tasks, e.g. hovering control of a helicopter. Galichet et al. [2013] consider a multi-armed risk-aware bandit setting to prevent hazards when exploring different tasks, e.g. energy management problems. They introduce a reward-based framework to limit the exploration of dangerous arms, i.e. with a negative exploration bonus for risky actions. However, their approach is highly dependent on the designed reward function which has significant impact on the probability for damaging the considered system. Similarly, Hans et al. [2008] define a reward-based safety function to assess each state of the MDP and assume that there exists a safe return policy to leave critical states with non-fatal actions. Although this assumption may not hold generally, it allows the usage of dynamic programming to solve an adapted Bellman optimality equation to get a return policy.

Strategies for exploring unknown environments without considering safety issues has also been reflected in the framework of global optimization with GP's. For example, Guestrin et al. [2005] propose an efficient submodular exploration criterion for near-optimal sensor placements, i.e. for discrete input spaces. In Auer [2002], a framework is presented which yields a compromise between exploration and exploitation through confidence bounds. Srinivas et al. [2012] show, that under reasonable assumptions strong exploration guaranties can be given for Bayesian optimization with GP's. Due to the fact that the exploration tasks may lead to NP-hard problems, cf. Guestrin et al. [2005], the additional introduction of safety will increase the complexity which must be handled by the AL scheme on a higher level.

3   Safe Exploration for Active Learning

In this section, we introduce our safe exploration approach for active learning in detail. The basic idea is to jointly learn a discriminative function during exploration of the input space. Using the sampled data, we incrementally learn a model of the system. We employ Gaussian processes to learn the model, as well as to build the discriminative function. The overall goal is to learn an accurate model of the system, while avoiding data sampling from unsafe regions indicated by the discriminative function as much as possible. We assume that additional information is available, when our exploration scheme is getting close to the decision boundary, cf. Hans et al. [2008]. For real-world applications in combustion engine measurement, for example, the engine temperature is an indicator for the proximity to the decision boundary. Thus, it is possible to design a discriminative function for recognizing whether our active learner is getting close to unsafe input locations. For the safe region of the input space, we assume that it is compact and connected. In Figure 1, we illustrate the described setting for the input space $\mathbb{X} \subset \mathbb{R}^2$. Next, we give an introduction to Gaussian processes and proceed by describing our exploration scheme and defining the safety constraint. The derived algorithm will be summarized in Section 3.4.

3.1   Gaussian Processes

In this paper, we use GP's in the regression and classification context to design a safe active learning scheme. A GP is a collection of random variables, where any finite subset of them has a joint multivariate normal distribution. In the following, we adopt the notations by Rasmussen and Williams [2006]. For both, the regression and the classification case, we use a centered Gaussian prior given by $p\left(\,\cdot\mid \boldsymbol{X}\right)=\mathcal{N}(\,\cdot\mid \boldsymbol{0}, \boldsymbol{K})$, where the matrix $\boldsymbol{X}$ contains row-wise all associated input points $\boldsymbol{x}_i \in \mathbb{R}^d$ which induce the covariance matrix $\boldsymbol{K}$. The dot $\cdot$ acts as

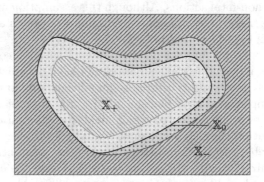

Fig. 1. Partition of the input space $\mathbb{X}$ into a safe explorable area $\mathbb{X}_+$ and an unsafe region $\mathbb{X}_-$ separated by the unknown decision boundary $\mathbb{X}_0$. Over the dotted area, a discriminative function is learned for recognizing whether the exploration becomes risky.

placeholder to distinguish between the latent target values f and the discriminative function g for the safety constraint. Throughout the paper, we use the stationary squared exponential covariance function

$$k(\boldsymbol{x}, \boldsymbol{z}, \boldsymbol{\theta}) = \sigma^2 \exp\big(-\tfrac{1}{2}(\boldsymbol{x} - \boldsymbol{z})^T \boldsymbol{\Lambda}^{-2}(\boldsymbol{x} - \boldsymbol{z})\big), \tag{1}$$

where the signal magnitude σ^2 and the diagonal matrix $\boldsymbol{\Lambda} \in \mathbb{R}^{d \times d}$ of length-scales are summarized in the set $\boldsymbol{\theta}$ of hyperparameters. We assume for both GP's that the hyperparameters are previously given. Here, recall from Krause and Guestrin [2007] that determining the hyperparameters in advance is similar to defining a grid over the whole input space $\mathbb{X} \subset \mathbb{R}^d$.

In the following, the cases for regression and classification are considered separately. For regression, the m data points are row-wise composed in $\boldsymbol{X} \in \mathbb{R}^{m \times d}$. The likelihood $p(\boldsymbol{y} \mid f, \boldsymbol{X})$ is given through the model $y_i = f(\boldsymbol{x}_i) + \varepsilon_i$ with centered Gaussian noise $\varepsilon_i \sim \mathcal{N}(0, \sigma^2)$. The m inputs $\boldsymbol{X}$ and their associated targets $\boldsymbol{y} \in \mathbb{R}^m$ are summarized in $\mathcal{D}_m = (\boldsymbol{y}, \boldsymbol{X})$. Due to the fact that the likelihood in the regression case is Gaussian, exact inference is possible which yields a Gaussian posterior. The marginal likelihood of the regression model given the data then satisfies

$$p(\boldsymbol{y} \mid \boldsymbol{X}) = \mathcal{N}(\boldsymbol{y} \mid \boldsymbol{0}, \boldsymbol{K} + \sigma^2 \boldsymbol{I}), \tag{2}$$

cf. Rasmussen and Williams [2006]. For the predictive distribution of test points $\boldsymbol{x}_*$, we obtain

$$p(y_* \mid \boldsymbol{x}_*, \mathcal{D}_m) = \mathcal{N}\Big(y_* \mid \boldsymbol{k}_*^T \boldsymbol{\alpha}, k_{**} - \boldsymbol{k}_*^T (\boldsymbol{K} + \sigma^2 \boldsymbol{I})^{-1} \boldsymbol{k}_* + \sigma^2\Big), \tag{3}$$

where $k_{**} = k(\boldsymbol{x}_*, \boldsymbol{x}_*, \boldsymbol{\theta}_f) = \sigma_f^2$ is the covariance value, $\boldsymbol{k}_* \in \mathbb{R}^m$ is the corresponding covariance vector induced by the test point and the training points, and with the prediction vector $\boldsymbol{\alpha} = (\boldsymbol{K} + \sigma^2 \boldsymbol{I})^{-1} \boldsymbol{y} \in \mathbb{R}^m$.

To distinguish between safe and unsafe regions in our active learning framework, we use GP classification. Here, the latent discriminative function $g \colon \mathbb{X} \to \mathbb{R}$ is learned and, subsequently, mapped to the unit interval to describe the class likelihood for each point. This mapping is realized by means of a point symmetric sigmoid function, where we use the cumulative Gaussian $\Phi(\cdot)$ throughout the paper. For the class affiliation probability, we thus obtain $\Pr[c = +1 \mid \boldsymbol{x}] = \Phi(g(\boldsymbol{x}))$. Given the discriminative function g, the class labels $c_i \in \{-1, +1\}$, $i \in \{1, \ldots, k\}$, of the data points $\boldsymbol{x}_1, \ldots, \boldsymbol{x}_k$ are independent random variables, so that the likelihood factorizes over the data points. This results in the non-Gaussian likelihood

$$p(\boldsymbol{c} \mid \boldsymbol{g}, \boldsymbol{X}) = \prod_{i=1}^{k} \Phi(c_i \, g(\boldsymbol{x}_i)), \tag{4}$$

where the matrix $\boldsymbol{X} \in \mathbb{R}^{k \times d}$ is row-wise composed of the input points. Due to the non-Gaussian likelihood (4), the posterior is not analytically tractable for this probit model. To solve this problem, we use the efficient and accurate Laplace approximation to calculate an approximate Gaussian posterior $q(g \mid c, X)$ similar to Nickisch and Rasmussen [2008]. The approximate posterior distribution forms the basis for the subsequent prediction of class affiliations for test points. Note that our classification model is an extended variant of the general classification task and will be explained in more detail in Section 3.3. For more details regarding GP classification, we refer to Rasmussen and Williams [2006].

3.2   Exploration Strategy

The exploration strategy applied in this paper is a selective sampling approach based on the posterior entropy of the GP model for the functional relationship $f(x)$. This entropy criterion has been frequently used in the active learning literature, e.g. in Seo et al. [2000] or Guestrin et al. [2005]. To the best of our knowledge, safe exploration in combination with this criterion has not been considered before. The main goal for this uncertainty sampling task, cf. Settles [2010], is to find an optimal set of feasible data points X_{opt} of given size m and determined hyperparameters θ_f such that the differential entropy of the model evidence (2) is maximal, i.e.

$$X_{\mathrm{opt}} = \underset{X \subset \mathbb{X},\ |X|=m}{\arg\max}\ \mathrm{H}\big[y \mid X\big]. \tag{5}$$

Here, the differential entropy of the evidence (2) depends on the determinant of the associated covariance matrix, i.e.

$$\mathrm{H}\big[y \mid X\big] = \frac{1}{2}\log\big|2\pi e\left(K + \sigma^2 I\right)\big|, \tag{6}$$

see e.g. Cover and Thomas [2006]. Note that Ko et al. [1995] showed that the problem of finding a finite set X_{opt} is NP-hard. Nevertheless, the following lemma summarizes some nice results about the differential entropy in our GP setting.

Lemma 1. *Let $\mathcal{D}_m$ be a non-empty data set and $\sigma^2 \geq (2\pi e)^{-1}$. Then, for some stationary covariance function with magnitude σ_f^2, the differential entropy (6) of the GP evidence (2) is a non-negative, monotonically increasing, and submodular function.*

The detailed proof of the latter lemma uses known results from Nemhauser et al. [1978] and Cover and Thomas [2006] and is given in the supplemental material. In Lemma 1 the lower bound for the noise σ^2 can be easily achieved by additionally scaling the target values y and the magnitude σ_f^2 with some suitable constant, cf. the proof of Lemma 1. The properties of the entropy induced by Lemma 1 guarantee that the greedy selection scheme

$$x_{i+1} = \underset{x_* \in \mathbb{X}}{\arg\max}\ \mathrm{H}\big[y_* \mid x_*, \mathcal{D}_i\big] \tag{7}$$

for our agent yields a nearly optimal subset of input points, where the distribution of y_* results from (3). More precisely, as shown by Nemhauser et al. [1978], it holds true that the greedy selection scheme (7) yields a model entropy which is greater than 63% of the optimal entropy value induced by X_{opt}. This guarantee induces an efficient greedy algorithm and, with some foresight to the introduction of safety in Section 3.3, that it will generally not be possible to design an optimal and fast safe active learning algorithm, cf. Moldovan and Abbeel [2012].

As empirically shown in Ramakrishnan et al. [2005], the described entropy based sampling strategy tends to select input locations close to the border and induces a point set X which is nearly uniformly distributed in the confined input space $\mathbb{X}$. This is a favorable behavior for modeling with GP's, which additionally enables us to slightly extrapolate the borders of the technical system with our GP model.

For the visualization of the exploration process, we use theoretical results by Cover and Thomas [2006] and the bounds

$$\tfrac{m}{2}\log\!\left(2\pi e\sigma^2\right) \le \mathrm{H}\!\left[\,y\mid X\,\right] \le \tfrac{m}{2}\log\!\left(2\pi e(\sigma_f^2+\sigma^2)\right),$$

derived in the proof of Lemma 1, to normalize the gain in the differential entropy (6). After sampling the m-th query point, we thus define the normalized entropy ratio *NER* by

$$NER(X) = \frac{\tfrac{2}{m}\,\mathrm{H}\!\left[\,y\mid X\,\right] - \log\!\left(2\pi e\sigma^2\right)}{\log\!\left(1+\tfrac{\sigma_f^2}{\sigma^2}\right)}. \tag{8}$$

If this ratio is close to one when X enlarges during the exploration process, we gain nearly maximal entropy for the selected queries. Otherwise, if $NER \approx 0$, the current query x_m does not explore the input space very much.

3.3 Safety Constraint

Our approach to introduce safety is based on additional information to describe the discriminative function g, when getting close to the decision boundary. In practice, without any feedback from the physical system or some user-defined knowledge, it is not possible to explore the environment safely, cf. Valiant [1984]. We encode this additional feedback in a possibly noisy function $h\colon \mathbb{X}\to(-1,1)$ to train the discriminative function g around the decision boundary. To define a likelihood for this heterogeneous model, we assume that sampling from the system to get values $h_j = h(x_j)$ or labels $c_i = c(x_i)$ leads to consistent data. That is, depending on the location of the data point (cf. Figure 1), we obtain either labels or discriminative function values. For $c \in \mathbb{R}^k$, $h \in \mathbb{R}^l$, $g \in \mathbb{R}^n$, and $k+l=n$, the model likelihood results in

$$p(c,h\mid g,X) = \prod_{j=1}^{l}\mathcal{N}\!\left(h_j\,\middle|\,g_j,\tau^2\right)\prod_{i=1}^{k}\Phi\!\left(c_i\,g_i\right), \tag{9}$$

where we used (4) and a Gaussian regression model for h with noise variance τ^2. Employing the Gaussian prior for all $g_i = g(x_i)$ and the Laplace approximation, we get the Gaussian posterior approximation $q(g \mid c, h, X)$ with respect to the exact posterior

$$p(g \mid c, h, X) \approx q(g \mid c, h, X) = \mathcal{N}(g \mid \mu, \Sigma), \tag{10}$$

where $\mu = \arg\max_{g}(p(g \mid c, h, X))$ and $\Sigma = (W + K^{-1})^{-1}$. Here, the diagonal Matrix W is given by

$$W = -\frac{\partial^2}{\partial g \partial g^T} \log\left(p(c, h \mid g, X)\right)\Big|_{g=\mu}.$$

For the predictive distribution of test cases, we then obtain

$$q(g_* \mid x_*, c, h, X) = \mathcal{N}(g_* \mid \mu_{g_*}, \sigma_{g_*}^2) \tag{11}$$

through the efficient Laplace approximation by Nickisch and Rasmussen [2008], where $\mu_{g_*} = k_*^T K^{-1} \mu$, $\sigma_{g_*}^2 = k_{**} - k_*^T W^{\frac{1}{2}} B^{-1} W^{\frac{1}{2}} k_*$ and $B = I + W^{\frac{1}{2}} K W^{\frac{1}{2}}$. The safety constraint in our AL approach should ensure that the probability of making a failure is small, e.g. less than $1 - p$ for some $p \in (0, 1)$. Formally, our safety constraint is thus given by

$$\Pr[g_* \geq 0 \mid x_*, c, h, X] \geq p. \tag{12}$$

For a successful safe exploration, the function h must fulfill some conditions, which will be presented in Section 4.

3.4   Safe Active Learning: The Algorithm

In this subsection, we present our entropy based active learning framework extended by a constraint inducing safety, cf. Algorithm 1. Since the entropy from the greedy selection rule (7) is a monotonic function in the posterior variance, we can reduce the query strategy and search for points where our regression model is maximally uncertain. We can also simplify the safety constraint (12) by defining the confidence parameter $\nu = \Phi^{-1}(p) \in \mathbb{R}$. Thus, we obtain for our agent the optimization problem

$$x_{i+1} = \arg\max_{x_* \in \mathbb{X}} \operatorname{Var}[y_* \mid x_*, \mathcal{D}_i]$$
$$\text{s.t.:} \quad \mu_{g_*} - \nu \sigma_{g_*} \geq 0, \tag{13}$$

where the moments μ_{g_*} and $\sigma_{g_*}^2$ are defined as in (11). This optimization task is solved via second order optimization techniques. For notational simplicity, let $\mathcal{D}_i$ contain all data for both GP's. In addition to the hyperparameters given previously, we assume that at least $m_0 \geq 1$ safe starting points, i.e. points with positive label c_i, are given at the beginning of the exploration. As stopping criterion for the above problem, we used the maximal number of queries n. It is also possible to replace the latter by a bound for the current model accuracy or the number of feasible points m.

Algorithm 1 Safe Active Learning with GP's
Require: $\mathcal{D}_{m_0}$, ν, $\boldsymbol{\theta}_f$, σ^2, $\boldsymbol{\theta}_g$, τ^2
1: $i = m_0$
2: train model and discriminative GP function on $\mathcal{D}_{m_0}$
3: **while** $i < n$ **do**
4:     $i = i + 1$
5:     get $\boldsymbol{x}_i$ from solving (13)
6:     sample y_i, c_i or h_i and add them to $\mathcal{D}_i$ with query $\boldsymbol{x}_i$
7:     train model and discriminative GP function on $\mathcal{D}_i$
8: **end while**

4 Theoretical Analysis

The goal of this section is to investigate our proposed safe AL algorithm. It should be noted that the main objective of our exploration strategy is to avoid samples from unsafe regions $\mathbb{X}_-$ as much as possible. However, a desirable property of an exploration scheme is that it induces a nearly space-filling, i.e. uniform, distribution of the queries in the whole input space $\mathbb{X}$. More precisely, an exploration strategy is called space-filling, if it yields a low-discrepancy input design $\boldsymbol{X}$ of size n such that the discrepancy

$$D(\boldsymbol{X}) = \sup_{B \in \mathcal{B}(\mathbb{X})} \left| \frac{1}{n} \left| \{ \boldsymbol{x}_i \mid \boldsymbol{x}_i \in B \} \right| - \mu_d(B) \right| \leq \gamma \frac{\log^d(n)}{n}$$

for some positive constant γ independent of n and the normalized Lebesgue measure $\mu_d(B)$ for any B which is contained in the Borel algebra $\mathcal{B}(\mathbb{X})$. An example strategy which yields a low-discrepancy design is the Sobol sequence, see Sobol [1976]. Intuitively, it is clear that a space-filling exploration scheme will cover the input space $\mathbb{X}$ and, thus, query in dangerous regions $\mathbb{X}_-$. This supposition is confirmed by the following theorem, where the proof by contradiction is given in the appendix.

Theorem 1. *For every space-filling exploration strategy on $\mathbb{X}$ with a Lebesgue measurable subset $B \subset \mathbb{X}_- \subset \mathbb{X}$ such that $\mu_d(B) > \epsilon$ for some $\epsilon > 0$, there exists a query $\boldsymbol{x}_n \in B$ for an adequate n possibly depending on ϵ.*

For the initialization of our scheme, we assumed that we have at least one safe starting point with positive label. However, depending on the hyperparameter $\boldsymbol{\theta}_g$ and especially the confidence parameter ν, the optimization problem (13) may be empty, i.e. there may not exist any $\boldsymbol{x}_* \in \mathbb{X}$ fulfilling the safety constraint. This behavior is due to the classification part for learning the discriminative function g. Namely, this problem occurs if the user wants a very safe exploration which yields a high confidence parameter ν, but the small GP classification model, i.e. consisting only of a few data points, is too uncertain to enable exploration under such a strong safety constraint. To solve this problem and to obtain a more confident discrimination model, the user has to define an initialization point set

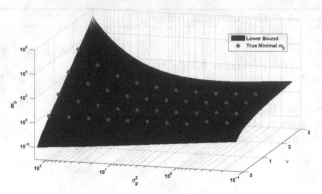

Fig. 2. Lower bound for the number of necessary initialization points m_0 given by Theorem 2 to ensure a non-empty optimization problem (13). The red stars indicate some true necessary set sizes, calculated according to the explanations in the proof.

where potentially all points lie close to each other. The next theorem provides a lower bound for the size m_0 of the initial point set. This result is especially relevant, when employing the proposed safe AL in practice. The proof is moved to the supplemental material.

Theorem 2. *To ensure a non-empty safety constraint in the optimization problem* (13) *for our GP setting, we need at least an initial point set of size*

$$m_0 \geq \left(2 \mathcal{N} \big(\tfrac{1}{2} (\sqrt{1 + 4\nu\sigma_g} - 1) \big) \right)^{-1} \min \left(\frac{\nu}{\sqrt{3}\sigma_g}, \sqrt{\frac{\nu}{\sqrt{3}\sigma_g^3}} \right).$$

In Figure 2, the bound from Theorem 2 is illustrated and compared to some true necessary set sizes m_0. As it is hard to restrict the explicit expressions for μ_{g_*} and $\sigma_{g_*}^2$ depending on m_0, the bound of the theorem is adequately tight. Nevertheless, the magnitude and the asymptotic behavior are captured by the lower bound of the theorem. Note that the safety constraint of the optimization problem (13) is always satisfied, if ν is non-positive, i.e. $p \leq 0.5$, which follows from the centered GP prior.

Finally, we will bound the probability of failure for our active exploration scheme to ensure a high level of safety. Having already queried $i-1$ points and if our prior GP assumptions are correct, the probability of failure when sampling $\boldsymbol{x}_* \in \mathbb{X}$ without considering the safety constraint (12) is given by

$$\Pr\big[g_* \leq 0 \mid \boldsymbol{x}_*, \mathcal{D}_{i-1}\big] = 1 - \Phi\Big(\frac{\mu_{g_*}}{\sigma_{g_*}}\Big), \tag{14}$$

where the moments of g_* are explained by the exact posterior distribution followed from (9). In other words, the discriminative function is not positive in the case of failure. We are interested in an upper bound for the probability of making at least one failure, when querying n data points with our safe active learning scheme summarized in Algorithm 1. Our result is stated in the following theorem, where the sketch of the proof is subsequently given.

Theorem 3. *Let* $\mathbb{X} \subset \mathbb{R}^d$ *be compact and non-empty, pick* $\delta \in (0,1)$ *and define* $\nu = \Phi^{-1}\left(1 - \frac{\delta}{n-m_0}\right)$. *Ensure that the discriminative prior and posterior GP is correct. Suppose also that at least an initialization point set of size* $m_0 < n$ *with respect to Theorem 2 is given. Selecting* n *possible queries satisfying the safety constraint, our active learning scheme presented in Algorithm 1 is unsafe with probability* δ, *i.e.*

$$\Pr\left[\bigvee_{i=m_0+1}^{n}\left(g_i \leq 0 \mid \boldsymbol{x}_i : \mu_{g_i} - \nu\sigma_{g_i} \geq 0, \mathcal{D}_{i-1}\right)\right] \leq \delta.$$

Proof (Theorem 3). Firstly, we need to bound the probability of failure (14) for every possible query $\boldsymbol{x}_*$ fulfilling the safety constraint. For an arbitrary but firmly selected input point $\boldsymbol{x}_i \in \mathbb{X}$

$$\Pr\left[g_i \leq 0 \mid \boldsymbol{x}_i : \mu_{g_i} - \nu\sigma_{g_i} \geq 0, \mathcal{D}_{i-1}\right]$$
$$\leq \Pr\left[g_i \leq \mu_{g_i} - \nu\sigma_{g_i} \mid \boldsymbol{x}_i : \mu_{g_i} - \nu\sigma_{g_i} \geq 0, \mathcal{D}_{i-1}\right]$$
$$= 1 - \Phi(\nu),$$

holds true under our condition. That is, the safety constraint (12) induces a probability of failure which is less than $1 - p$ in each iteration, remembering the relationship $p = \Phi(\nu)$. Furthermore, we use the union bound to obtain

$$\Pr\left[\bigvee_{i=m_0+1}^{n}\left(g_i \leq 0 \mid \boldsymbol{x}_i : \mu_{g_i} - \nu\sigma_{g_i} \geq 0, \mathcal{D}_{i-1}\right)\right]$$
$$\leq \sum_{i=m_0+1}^{n}\Pr\left[\left(g_i \leq 0 \mid \boldsymbol{x}_i : \mu_{g_i} - \nu\sigma_{g_i} \geq 0, \mathcal{D}_{i-1}\right)\right]$$
$$= (n - m_0)\left(1 - \Phi(\nu)\right) = \delta.$$

Note that the m_0 initialization points are feasible with probability 1 under the assumptions of the theorem.   □

The lower bound of Theorem 3 provides a safety level of Algorithm 1 greater than or equal to $1 - \delta$. The user is then able to choose a sufficiently small δ, when carrying out our algorithm. After determining δ, we calculate ν and, if necessary, p as explained in the theorem. It is clear that, for fixed safety level, ν has to increase, if n increases. In this case, the number of necessary initialization points also increases, see Theorem 2.

   To get a more detailed illustration of the bound in Theorem 3, we assume that each query is selected independently of all others. In contrast to the proof of the safety bound, where we did not assume independence, we then have $1 - p^{n-m_0} \leq \delta$. Intuitively, we anticipate an upper bound for the expected number of failures when sampling n points. Examples for this bound are shown in the next section.

   Moreover, it is necessary for our AL scheme that the function h is a lower bound of the true discriminative function of the system. In this case, we may loose information when exploring the system. However, the validity of the safety level compared to Theorem 3 is the main requirement for us. The function h must also satisfy the conditions for the specified discriminative GP prior, e.g. continuity and mean square differentiability.

5   Evaluations

In this section, we verify the presented Algorithm 1 on a 1-dimensional toy example and, subsequently, evaluate our safe exploration scheme on an inverse pendulum policy search problem.

Firstly, we learn to approximate the cardinal sine function $f(x) = 10\,\mathrm{sinc}(x - 10)$ with additive Gaussian noise $\varepsilon \sim \mathcal{N}(0, 0.0625)$ on the interval $\mathbb{X} = [-10, 15]$. We define a safe region $\mathbb{X}_+ = [-5, 11]$, and, consequently, unsafe regions $\mathbb{X} \setminus \mathbb{X}_+$ for which we always sample negative labels. To recognize when exploring gets dangerous we define $h(x) = \frac{1}{2}(x + 5)^2$ for $-5 \le x < -4$ and $h(x) = \frac{1}{2}(x - 11)^2$ for $10 < x \le 11$. The observation noise for the function h is set to $\mathcal{N}(0, 0.01)$. Otherwise, i.e. within $[-4, 10]$, we sample positive class labels. Hyperparameters of the target and discriminative GP are set to $\sigma_f^2 = 4$, $\lambda_f = 3$, and $\sigma_g^2 = 1$, $\lambda_g = 10$, respectively. The m_0 starting points with respect to Theorem 2 are uniformly sampled in the range $[-0.5, 0.5]$. Finally, we wish to select $n = 40$ input points for different probability levels δ. Table 1 shows the selected safety levels with corresponding confidence parameter ν and the necessary number of starting points m_0 derived in the proposed theorems. The results in the presented table, i.e. the final differential entropy (6) and the number (#) of failures are averaged over ten runs. The maximum possible differential entropy value with 40 points is 19.50, which is almost reached in all cases. The expected number of failures provides only an upper bound for the true bound, since independence is assumed in its calculation over $(n - m_0)\,p$. They are additionally compared in Figure 3. Due to this strong assumption, the upper bound for the expected number of failures will not be very tight. The normalized entropy ratios (8) for all cases and averaged over ten runs are illustrated in Figure 4. Note that for the calculation of the differential entropy H only the safe queries are taken into account, analogously to Table 1. Therefore, the curves for the normalized

Table 1. Averaged results over ten runs of the 1-dimensional toy example. The entropy and the number of failures is slightly decreasing for smaller δ. Otherwise, the parameter ν and m_0 are growing with decreasing δ as explained in the text.

δ	ν	m_0	H	# failures	# expected failures
0.05	2.99	4	17.39	0.0	1.8
0.10	2.77	4	18.54	0.2	3.4
0.20	2.54	4	18.73	0.6	6.5
0.30	2.40	3	18.69	1.1	9.6
0.40	2.30	3	18.88	1.6	12.3
0.50	2.21	3	18.87	2.3	14.6

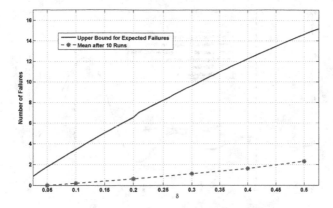

Fig. 3. Expected number of failures calculated under the independence assumption (upper bound) for the toy example compared with the failures obtained by averaging over ten runs of the safe learning algorithm.

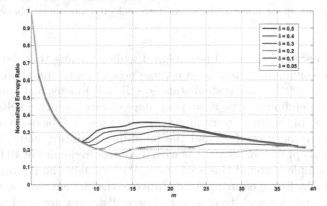

Fig. 4. Normalized entropy ratios for different safety levels and averaged over ten runs. In the beginning phase the ratios fall until the safety constraint is satisfied, since we sample only around zero. After that the gain in entropy increases until the safe region is explored. This behavior and the value of the gain depends on δ and the confidence parameter ν, respectively.

entropy ratios end before the final number of queries is reached. The plot also shows the effects of the decreasing failure level δ for the number of initialization points and the gain in entropy, cf. the theorems in Section 4, where we explore faster and a greater input region with higher δ. In Figure 5 we present the final result for one run of the safe active learning Algorithm 1 after selecting 40 queries. The cardinal sine function is learned accurately over the safe input region. Furthermore, the plot shows that the small definition regions for the function h around the decision boundary are sufficient for safe exploration.

As a second test scenario, we consider exploration of control parameters for the inverse pendulum hold up problem. Here, we wish to learn the mapping between parameters of a linear controller and performance on keeping the pole

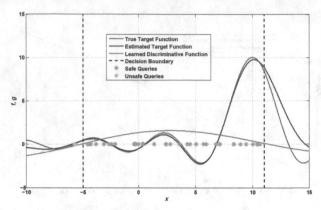

Fig. 5. Final result for safe active learning of a generalized cardinal sine function in the secure interval $[-5, 11]$ with 40 queries and $\delta = 0.30$. Only one selected query fails, i.e. falls below the lower decision boundary. All other chosen data points cover the safe input space $\mathbb{X}_+$ well. The final discriminative GP separates the safe and unsafe regions of the whole input space adequately, even if we selected no query above the upper border.

upwards. The parameters should be explored while avoiding that the pendulum falls over. The simulated system is an inverse pendulum mounted on a cart, see Deisenroth et al. [2015]. The system state $s_t \in \mathbb{S}$ is 4-dimensional (cart position z_t and velocity, pendulum angle ϑ_t and angular velocity), which results in 5 open parameters x, x_0 of the linear controller $\pi(s_t) = x^T s_t + x_0$. The desired goal state is defined by cart position $0\,cm$ and the pendulum pointing upwards with $0°$. The controller is applied for 10 seconds with a control frequency of $10\,\mathrm{Hz}$ starting from a state sampled around this goal state. We evaluate the performance of a given controller by first measuring the distance of the current state s_t to the goal state, i.e. $\|s_t - \mathbf{0}\|$, for each time step t. These distances are used to compute a saturating cost $r_t = 1 - \exp(-\|s_t\|^2)$. Additionally, we average over all time steps and ten different starting states to yield a meaningful interpretation of the cost. To that we add Gaussian noise $\varepsilon \sim \mathcal{N}(0, 0.001)$. The hyperparameters of the target and discriminative GP are previously learned over a uniformly distributed point set of size 100 over the entire input space, i.e. policy parameter space

Table 2. Median with respect to the number of failures over ten runs for the policy exploration task from the cart pole. Here the entropy decreases as δ increases, except for the lower values of δ, since the number of unsafe queries increases strongly.

δ	ν	m_0	H	# failures	# expected failures
0.01	4.26	8	393.3	0	9.9
0.05	3.89	7	397.7	6	48.4
0.10	3.72	6	412.8	29	94.6
0.20	3.54	6	402.7	57	180.2
0.30	3.43	5	395.6	80	257.9
0.40	3.35	5	389.3	101	328.1
0.50	3.29	5	380.7	127	391.6

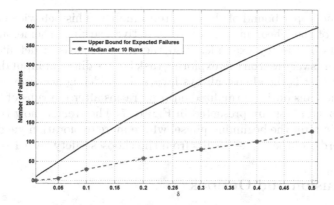

Fig. 6. Expected number of failures calculated under the independence assumption (upper bound) for the policy search task with the failures obtained by the median of ten runs of the safe learning algorithm.

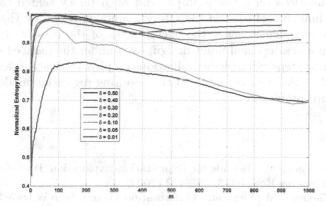

Fig. 7. Median of normalized entropy ratios for different safety levels over ten runs of the policy search task. In the beginning phase the ratios are nearly maximal, i.e. close to one, since we obtain many very safe queries with positive class labels which yield to a strong exploration behavior. After the beginning phase, the safe region is roughly explored until a slight valley is reached. Then the most queries are sampled from the inner region.

$X \subset \mathbb{R}^5$. The function h is defined over the deviation from the unstable goal state for each time step, i.e. $2 - |z_t| \, [\, |z_t| > 1 \, cm\,] - |\vartheta_t| \, [\, |\vartheta_t| > 1°\,]$, where $[\cdot]$ is the indicator function. These local errors are averaged over all time steps of the roll-out and all starting states to get the value of h. If $h \geq 0.95$ we get a positive class label and a negative one if $h \leq -1$ or the pendulum falls down. In this case we set $n = 1000$ for various values of δ. Table 2 summarizes the resulting values for ν and m_0 given by the theorems in Section 4. The value of the differential entropy H increases until $\delta = 0.10$, where we yield a good tradeoff between a low number of failures and a fast exploration. In Figure 6 we show that the median of the number of failures after ten runs of our safe exploration scheme is much lower than the upper bound of the expected number. This behavior clarifies the effectiveness of the Theorem 3 for our safe active learning scheme, analogously to the results of the toy example. Also the nearly heuristic definition of the function h works very well on this exploring task for five different input dimensions. A reason for that is the almost always favorable modeling accuracy when using GP's. The medians of the normalized entropy ratios after ten runs of the inverse pendulum control task are presented in Figure 7. The trends show the effect of the different δ after the beginning phase, where more exploration yields a higher curve. The order of the curves results from the chooses safety level $1 - \delta$ too.

6   Conclusion and Outlook

In this paper, we propose a novel and provable safe exploration strategy in the active learning (AL) context. This approach is especially relevant for real-world AL applications, where critical and unsafe measurements need to be avoided. Empirical evaluations on a toy example and on a policy search task for the inverse pendulum control confirm the safety bounds provided by the approach. Moreover, the experiments show the effectiveness of the presented algorithm, when selecting a near optimal input design, even under the induced safety constraint. The next steps will include evaluations on physical systems and – on the theoretical side – the error bounding of the resulting regression model, which is generally a hard problem.

References

Auer, P.: Using Confidence Bounds for Exploitation-Exploration Trade-Offs. Journal of Machine Learning Research **3**, 397–422 (2002)

Cover, T.M., Thomas, J.A.: Elements of Information Theory. John Wiley & Sons (2006)

Deisenroth, M.P., Fox, D., Rasmussen, C.E.: Gaussian Processes for Data-Efficient Learning in Robotics and Control. Transactions on Pattern Analysis and Machine Intelligence **37**, 408–423 (2015)

Fedorov, V.V.: Theory of Optimal Experiments. Academic Press (1972)

Galichet, N., Sebag, M., Teytaud, O.: Exploration vs exploitation vs safety: risk-aware multi-armed bandits. In: Ong, C.S., Ho, T.B. (eds.) Proceedings of the 5th Asian Conference on Machine Learning, JMLR: W&CP, vol. 29, pp. 245–260 (2013)

Geibel, P.: Reinforcement learning with bounded risk. In: Brodley, C.E., Danyluk, A.P. (eds.) Proceedings of the 18th International Conference on Machine Learning, pp. 162–169 (2001)

Gillula, J.H., Tomlin, C.J.: Guaranteed safe online learning of a bounded system. In: Amato, N.M. (ed.) Proceedings of the International Conference on Intelligent Robots and Systems, pp. 2979–2984 (2011)

Guestrin, C., Krause, A., Singh, A.: Near-Optimal sensor placements in gaussian processes. In: De Raedt, L., Wrobel, S. (eds.) Proceedings of the 22nd International Conference on Machine Learning, pp. 265–275 (2005)

Hans, A., Schneegaß, D., Schäfer, AM., Udluft, S.: Safe Exploration for reinforcement learning. In: Verleysen, M. (ed.) Proceedings of the European Symposium on Artificial Neural Networks, pp. 143–148 (2008)

Ko, C., Lee, J., Queyranne, M.: An Exact Algorithm for Maximum Entropy Sampling. Operations Research **43**, 684–691 (1995)

Krause, A., Guestrin, C.: Nonmyopic active learning of gaussian processes: an exploration–exploitation approach. In: Ghahramani, Z. (ed.) Proceedings of the 24th International Conference on Machine Learning, pp. 449–456 (2007)

Lang, K.J., Baum, E.B.: Query learning can work poorly when a human oracle is used. In: Proceedings of the International Joint Conference on Neural Networks, pp. 335–340 (1992)

Moldovan, T.M., Abbeel, P.: Safe exploration in markov decision processes. In: Langford, J., Pineau, J. (eds.) Proceedings of the 29th International Conference on Machine Learning, pp. 1711–1718 (2012)

Nemhauser, G.L., Wolsey, L.A., Fisher, M.L.: An Analysis of the Approximations for Maximizing Submodular Set Functions. Mathematical Programming **14**, 265–294 (1978)

Nickisch, H., Rasmussen, C.E.: Approximations for Binary Gaussian Process Classification. Journal of Machine Learning Research **9**, 2035–2078 (2008)

Polo, F.J.G., Rebollo, F.F.: Safe reinforcement learning in high-risk tasks through policy improvement. In: Proceedings of the Symposium on Adaptive Dynamic Programming and Reinforcement Learning, pp. 76–83 (2011)

Ramakrishnan, N., Bailey-Kellogg, C., Tadepalli, S., Pandey, V.N.: Gaussian processes for active data mining of spatial aggregates. In: Kargupta, H., Kamath, C., Srivastava, J., Goodman, A. (eds.) Proceedings of the 5th SIAM International Conference on Data Mining, pp. 427–438 (2005)

Rasmussen, C.E., Williams, C.K.I.: Gaussian Processes for Machine Learning. The MIT Press (2006)

Seo, S., Wallat, M., Graepel, T., Obermayer, K.: Gaussian process regression: active data selection and test point rejection. In: Proceedings of the International Joint Conference on Neural Networks vol. 3, pp. 241–246 (2000)

Settles, B.: Active Learning Literature Survey. In: Computer Sciences Technical Report University of Wisconsin, Madison (2010)

Sobol, I.M.: Uniformly Distributed Sequences with an Additional Uniform Property. USSR Computational Mathematics and Mathematical Physics **16**, 236–242 (1976)

Srinivas, N., Krause, A., Kakade, S.M., Seeger, M.W.: Information-Theoretic Regret Bounds for Gaussian Process Optimization in the Bandit Setting. Transactions on Information Theory **58**, 3250–3265 (2012)

Valiant, L.G.: A Theory of the Learnable. Communications of the ACM **27**, 1134–1142 (1984)

Semi-Supervised Consensus Clustering for ECG Pathology Classification

Helena Aidos[1]($\boxtimes$), André Lourenço, Diana Batista[1],
Samuel Rota Bulò[3], and Ana Fred[1]

[1] Instituto de Telecomunicações, Instituto Superior Técnico,
Universidade de Lisboa, Lisbon, Portugal
haidos@lx.it.pt
[2] Instituto Superior de Engenharia de Lisboa, Lisbon, Portugal
[3] FBK-irst, Trento, Italy
[4] CardioID Technologies, Lisbon, Portugal

Abstract. Pervasive technology is changing the paradigm of healthcare, by empowering users and families with the means for self-care and general health management. However, this requires accurate algorithms for information processing and pathology detection. Accordingly, this paper presents a system for electrocardiography (ECG) pathology classification, relying on a novel semi-supervised consensus clustering algorithm, which finds a consensus partition among a set of baseline clusterings that have been collected for the data under consideration. In contrast to typical unsupervised scenarios, our solution allows exploiting partial prior knowledge of a subset of data points. Our method is built upon the evidence accumulation framework to efficaciously sidestep the cluster correspondence problem. Computationally, the consensus partition is sought by exploiting a result known as Baum-Eagon inequality in the probability domain, which allows for a step-size-free optimization. Experiments on standard benchmark datasets show the validity of our method over the state-of-the-art. In the real world problem of ECG pathology classification, the proposed method achieves comparable performance to supervised learning methods using as few as 20% labeled data points.

Keywords: Electrocardiography · ECG · Semi-supervised learning · Consensus clustering · Evidence accumulation clustering

1 Introduction

Heart disease, or more formally cardiovascular disease (CVD), is the first cause of death worldwide. An estimated 17.3 million people died from CVD in 2008, representing 30% of all global deaths. Among these deaths, an estimated 7.3 million were due to coronary heart disease and 6.2 million were due to stroke. In the US, about 0.6 million people die from heart disease every year (25% of the deaths).

© Springer International Publishing Switzerland 2015
A. Bifet et al. (Eds.): ECML PKDD 2015, Part III, LNAI 9286, pp. 150–164, 2015.
DOI: 10.1007/978-3-319-23461-8_10

These statistics trigger our work, which provides a semi-supervised Electrocardiography (ECG) pathology classification system that tries to mitigate the aforementioned serious threats. The system builds upon the pervasive healthcare framework, where devices are becoming more handy, user-friendly and comfortable for the user, focusing on usability and allowing continuous (or quasi-continuous) monitoring of biosignals. The aim is to automatically classify ECG data streams acquired by monitoring devices, giving alerts of abnormal situations. The use of the semi-supervised learning paradigm is motivated by the existence of prior knowledge about classes in this domain, namely pathologies, which can be gathered from annotated records of some patients, but a larger number of records has no annotation, being this an expensive and time consuming process. This large amount of unsupervised data carries important information that a supervised learning approach would neglect, while being exploited by a semi-supervised learning approach.

Several clustering algorithms have been proposed exploiting side-information (e.g., [2,3,8,14,16]), using typically must-link and cannot-link constraints. Basu et al. [2] proposed a method for actively picking must-link and cannot-link constraints by selecting the most informative examples from the training set. On the other hand, Li et al. [16] presents a framework integrating consensus clustering and semi-supervised learning from a nonnegative matrix factorization perspective, allowing the must-link and cannot-link constraints to be enforced within the clustering algorithm. Gao et al. [14] proposed a framework that incorporates the predictive power of multiple supervised and unsupervised models, deriving a consensus label partition for a set of objects.

In this paper we propose a semi-supervised learning algorithm based on consensus clustering, i.e. the problem of finding a consensus partition among a set (or ensemble) of baseline clusterings that have been collected for some data under consideration. Our method follows the Evidence Accumulation Clustering (EAC) paradigm [12], which summarizes the information of the clustering ensemble into a pairwise co-association matrix, where each entry corresponds to the number of times a given pair of objects is placed in the same cluster. The advantage of the pairwise voting mechanism is that it subsumes the problem of cluster correspondence among partitions. Several algorithms for consensus clustering have been proposed based on EAC [1,13,17,19,23]. In [23] the problem of extracting a consensus partition is formulated as a matrix factorization problem, in a similar fashion as [16]. In [18,19] the consensus partition is estimated through a probabilistic model for the co-association matrix, while in [17] a generalization of [23] is introduced to cope with partial observations of the co-association matrix. In contrast to the typical unsupervised scenario, which is addressed by the aforementioned works, our method allows to exploit partial knowledge of the cluster assignment of a subset of data points to constrain the solution space of the consensus partition. Computationally, the consensus partition is sought by exploiting a result known as Baum-Eagon inequality [5] in the probability domain, which allows for a step-size-free optimization.

The rest of the paper is organized as follows: in Section 2, we describe the application context of our algorithm, positioning it on the future workflow of pervasive healthcare. In Section 3, the proposed algorithm is described. Section 4 is devoted to the experimental validation on standard benchmark datasets and presents results for the real world problem of classifying pathologies in ECG. Finally, in Section 5, we draw conclusions and outline future works.

2   ECG Pathology Classification

The use of Electrocardiography (ECG) as a diagnostic technique is a well established medical practice rooted in the pioneering work by Einthonven in the end of the 19th century. Clinical practice relies mainly on the widespread short-term ($<$ 1 minute) 12-lead ECG for diagnosis and, in selected cases, on Holter monitors ($\sim$ 24 hour assessment), providing information for the diagnosis and prevention of a wide array of cardiovascular disorders [7,9]. Nevertheless, the outreach of ECG data acquisition and processing can still be significantly improved in the context of a pervasive healthcare framework with the off-the-person ECG paradigm [25]. The goal of off-the-person approaches is not to replace existing data acquisition procedures, but to enhance and complement current practices with a simplified sensor setup that can be transparently brought to the subject, in multiple aspects of his everyday life. This enables a more comprehensive assessment of cardiovascular function, contributing to the development of preventive behaviors and methodologies. Also, it opens the door to many potential applications, such as continuous monitoring, cardiac dysrhythmia detection, and ECG biometrics [20,22], among others.

The commercial exploration of such concepts has already began, and one of the most successful products is AliveCor[1], a Heart monitor for mobile devices. It consists on a 1-lead ECG acquisition system that can be installed on a mobile device, which records the ECG using the hands of the user. The system enables the detection of Atrial Fibrillation (AF), and upload of the recorded information for expert revision.

(a) AliveCor Heart Monitor.   (b) CardioID Keyboard.

Fig. 1. Examples of pervasive healthcare devices.

[1] http://www.alivecor.com/

This type of devices fits in the category of pervasive healthcare, where sensors do not need to be with the person, but instead are embedded into everyday use objects. A major advantage of this approach is the fact that the sensor placement does not require a voluntary action from the user, unlike, for example, wearable on-the-person devices, aligned with future medical trends [26]. Figure 1 illustrates two examples, the AliveCor Heart Monitor, and a keyboard with integrated electrodes developed by CardioID[2] that enables continuous ECG monitoring. These approaches produce enormous amounts of data. As an example, in a typical acquisition setup, where the signal is sampled at a frequency of 1 KHz, and with a resolution of 12 bits/sample, a total amount 5 MB of information is acquired per hour, corresponding to 123 MB/day/person or 44 GB/year/person, leading to a scenario where cloud computing is the most desirable approach.

Fig. 2. Global Architecture.

The global architecture of such a system is illustrated in Figure 2, where pervasive healthcare devices stream data to the cloud, and automatic classification algorithms process the data. In this paper we focus on the classification algorithm and propose a novel semi-supervised consensus clustering (SSCC) algorithm to categorize the data.

3   Semi-Supervised Consensus Clustering (SSCC)

Consensus clustering is the problem of organizing a set of n data points $\mathcal{X} = \{x_1, \ldots, x_n\}$ into groups, starting from the output of different clustering algorithms[3] that have been run on $\mathcal{X}$, or on sub-sampled versions thereof. This set (*a.k.a. ensemble*) of clusterings is denoted by $\mathcal{E} = \{\phi_1, \ldots, \phi_m\}$, where each $\phi_u \in \mathcal{J}_u \to \{1, \ldots, k_u\}$ is a function encoding a partition of a subset of data points indexed by $\mathcal{J}_u \subseteq \mathcal{I} = \{1, \ldots, n\}$ into k_u clusters. Partitions not comprising all data points, *i.e.* such that $\mathcal{J}_u \subsetneq \mathcal{I}$, indicate clusterings of sub-sampled versions of $\mathcal{X}$. The use of sub-sampling is motivated, *e.g.* in the presence of large-scale datasets, or to promote diversity in the ensemble [10].

---

[2] http://www.cardio-id.pt/
[3] Or different parametrizations of the same algorithm.

As the name suggests, consensus clustering tries to find a good representative for all clusterings in the ensemble $\mathcal{E}$. Formally, we call *consensus partition* a partition having minimum divergence from the other partitions in the ensemble:

$$\phi^* \in \operatorname*{arg\,min}_{\hat{\phi} \in \mathcal{I} \to \{1,\ldots,k\}} \sum_{u=1}^{m} d(\hat{\phi}, \phi_u), \qquad (1)$$

where $d(\cdot, \cdot)$ is a divergence measure between partitions.

In this paper, we depart from a purely unsupervised approach in favor of a *semi-supervised* perspective, by assuming partial knowledge of the cluster assignments (*a.k.a.* labels) of a subset of data points. Accordingly, we denote by $\mathcal{L} \subset \mathcal{I}$ the indices of data points that are labeled, and by $\ell_i \in \{1, \ldots, k\}$ the label given to the ith data point, $i \in \mathcal{L}$. We can then use this a prior knowledge to constrain the solution space of (1) to obtain a semi-supervised consensus clustering formulation, *i.e.*

$$\phi^* \in \operatorname*{arg\,min}_{\hat{\phi} \in \mathcal{I} \to \{1,\ldots,k\}} \sum_{u=1}^{m} d(\hat{\phi}, \phi_u) \qquad (2)$$

$$\text{s.t. } \hat{\phi}(i) = \ell_i \text{ for all } i \in \mathcal{L}.$$

The same knowledge could in principle be exploited at the ensemble construction phase. However, constraining the clusterings will lead to a drop of the ensemble's diversity, thus loosing one of the most desirable properties of an ensemble [15]. Moreover, there is a vast number of unsupervised clustering algorithms available for the ensemble construction, and only a limited number of algorithms that have been extended to include constraints.

In order to compare two clusterings, we face the so-called cluster correspondence problem, *i.e.* two partitions are the same if we can turn one into the other by a proper re-labeling of the clusters, and if two partitions are different, we would like to measure their divergence under the best possible re-labeling. There is however a way to sidestep the cluster correspondence problem, by adopting a pairwise divergence measure like the following one:

$$d(\hat{\phi}, \phi_u) = \sum_{i,j \in \mathcal{J}_u} \left[\mathbb{1}_{\hat{\phi}(i)=\hat{\phi}(j)} - \mathbb{1}_{\phi_u(i)=\phi_u(j)} \right]^2, \qquad (3)$$

which counts the number of times two data points are clustered together in $\hat{\phi}$, but not in ϕ_u, and vice versa. In (3), $\mathbb{1}_P$ denotes the indicator function for the truth value of proposition P.

The objective function in (2) is related to the evidence accumulation framework [12]. Indeed, it can be re-written in terms of the so-called *co-association* matrix, which is defined as

$$c_{ij} = \begin{cases} \frac{1}{N_{ij}} \sum_{u \in \mathcal{U}_{ij}} \mathbb{1}_{\phi_u(i)=\phi_u(j)} & N_{ij} > 0, \\ 0 & \text{otherwise}, \end{cases} \qquad (4)$$

where $\mathcal{U}_{ij} \subseteq \{1, \ldots, m\}$ denotes the indices of those clusterings, where both data points x_i and x_j have been clustered, *i.e.* $\mathcal{U}_{ij} = \{u \in \{1, \ldots, m\} : i, j \in \mathcal{J}_u\}$, and $\mathbb{N}$ is a matrix with entries $\mathbb{N}_{ij} = |\mathcal{U}_{ij}|$ if $i \neq j$, and 0 otherwise. The relation with the co-association matrix is established as follows:

$$\sum_{u=1}^{m} d(\hat{\phi}, \phi_u) = \sum_{u=1}^{m} \sum_{i,j \in \mathcal{J}_u} \left[\mathbb{1}_{\hat{\phi}(i)=\hat{\phi}(j)} - \mathbb{1}_{\phi_u(i)=\phi_u(j)} \right]^2$$

$$= \sum_{i,j \in \mathcal{I}} \sum_{u \in \mathcal{U}_{ij}} \left[\mathbb{1}_{\hat{\phi}(i)=\hat{\phi}(j)} + \mathbb{1}_{\phi_u(i)=\phi_u(j)} - 2\mathbb{1}_{\hat{\phi}(i)=\hat{\phi}(j)}\mathbb{1}_{\phi_u(i)=\phi_u(j)} \right]$$

$$= \sum_{i,j \in \mathcal{I}} \left[\mathbb{N}_{ij}\mathbb{1}_{\hat{\phi}(i)=\hat{\phi}(j)} + \sum_{u \in \mathcal{U}_{ij}} \mathbb{1}_{\phi_u(i)=\phi_u(j)} - 2\mathbb{1}_{\hat{\phi}(i)=\hat{\phi}(j)} \sum_{u \in \mathcal{U}_{ij}} \mathbb{1}_{\phi_u(i)=\phi_u(j)} \right]$$

$$= \sum_{i,j \in \mathcal{I}} \mathbb{N}_{ij} \left[\mathbb{1}_{\hat{\phi}(i)=\hat{\phi}(j)} + \mathbb{C}_{ij} - 2\mathbb{1}_{\hat{\phi}(i)=\hat{\phi}(j)}\mathbb{C}_{ij} \right]$$

$$= \sum_{i,j \in \mathcal{I}} \mathbb{N}_{ij} \left[\mathbb{1}_{\hat{\phi}(i)=\hat{\phi}(j)} - \mathbb{C}_{ij} \right]^2 + \sum_{i,j \in \mathcal{I}} \mathbb{N}_{ij}\mathbb{C}_{ij}(1 - \mathbb{C}_{ij}). \tag{5}$$

Note that the right-most term in (5) is regarded as a constant for the optimization in (2), thus not affecting the minimizers.

With the objective of re-writing (2) in matrix form, we introduce a different, but equivalent, representation of the consensus partition in terms of a matrix $\mathbb{Z} = [\boldsymbol{z}_1, \dots, \boldsymbol{z}_n] \in \mathcal{S}_{01}^{k \times n}$, where $\mathcal{S}_{01}^{k \times n}$ denotes the set of binary, left-stochastic matrices. The equivalence follows from the fact that any $\phi \in \mathcal{I} \rightarrow \{1, \dots, m\}$ has a one-to-one corresponding matrix $\mathbb{Z} \in \mathcal{S}_{01}^{k \times n}$ with $(\mathbb{Z}_{ki} = 1) \iff (\phi(i) = k)$. Under this variable change, the term $\mathbb{1}_{\phi(i)=\phi(j)}$ becomes $\boldsymbol{z}_i^\top \boldsymbol{z}_j$.

By exploiting the matrix representation and the relation in (5), the semi-supervised consensus clustering formulation in (2) can be cast into the following equivalent one (with omitted constant terms):

$$\mathbb{Z}^\star \in \arg \min_{\mathbb{Z} \in \mathcal{S}_{01}^{k \times n}} \|\mathbb{C} - \mathbb{Z}^\top \mathbb{Z}\|_{\mathbb{N}}^2$$
$$\text{s.t. } \mathbb{Z}_{\ell_i i} = 1 \text{ for all } i \in \mathcal{L}, \tag{6}$$

where $\| \cdot \|_{\mathbb{N}}$ is the Frobenious matrix norm weighted by $\mathbb{N}$, *i.e.* $\|\mathbb{A}\|_{\mathbb{N}} = \sqrt{\sum_{ij} \mathbb{N}_{ij}\mathbb{A}_{ij}^2}$. The optimization problem in (6) is non-convex and finding a global solution is hard. For this reason we opt for a relaxed version of it with the binary-valued matrix variable $\mathbb{Z} \in \mathcal{S}_{01}^{k \times n}$ being replaced with real-valued one $\mathbb{Y} \in \mathcal{S}^{k \times n}$, where $\mathcal{S}^{k \times n}$ denotes the set of real, left-stochastic, matrices, *i.e.* nonnegative matrices with columns summing up to 1. The relaxed optimization problem becomes

$$\mathbb{Y}^\star \in \arg \min_{\mathbb{Y} \in \mathcal{S}^{k \times n}} \|\mathbb{C} - \mathbb{Y}^\top \mathbb{Y}\|_{\mathbb{N}}^2$$
$$\text{s.t. } \mathbb{Y}_{ki} = \mathbb{1}_{k=\ell_i} \text{ for all } (k, i) \in \{1, \dots, k\} \times \mathcal{L}. \tag{7}$$

Given a solution $\mathbb{Y}^\star$ we recover a putative solution $\phi^\star$ to (2) by taking $\phi^\star(i) \in \arg \max_{k \in \{1, \dots, k\}} \mathbb{Y}_{ki}^\star$.

In order to optimize (7) we follow an approach similar to [23,24] by making use of a result known as *Baum-Eagon inequality*:

Theorem 1 (Baum-Eagon [5]). *Let* $Y \in \mathcal{S}^{k \times n}$ *and let* $f(Y)$ *be a homogeneous* [4] *polynomial in the variables* Y_{ki} *with nonnegative coefficients. Define the mapping* $\hat{Y} = M(Y)$ *as follows:*

$$\hat{Y}_{ki} = Y_{ki} \frac{\partial}{\partial Y_{ki}} f(Y) \bigg/ \sum_{h=1}^{k} Y_{hi} \frac{\partial}{\partial Y_{hi}} f(Y) \tag{8}$$

for all $i \in \{1, \ldots, n\}$ *and* $k \in \{1, \ldots, k\}$. *Then* $f(M(Y)) > f(Y)$ *unless* $M(Y) = Y$. *In other words, M is a growth transformation for the polynomial* f.

The Baum-Eagon inequality is an effective tool for the maximization of polynomial functions in probability domain. The idea is to rewrite (7) into a maximization of a polynomial with nonnegative coefficients in a way to preserve $Y^\star$ as the optimal solution. By doing so, we can use the Baum-Eagon inequality to obtain a step-size-free optimizer by re-iterating the update rule $Y^{t+1} = M(Y^t)$ starting from a matrix $Y^0 \in \mathcal{S}^{k \times n}$ with positive entries. The theorem indeed guarantees a strict increase of the objective at each step until a fixed-point is reached.

We can turn (7) into a maximization problem by changing the sign of the objective function. However, the resulting function in Y will not be a polynomial with nonnegative coefficients. Nevertheless, there is a trick that can be exploited to transform the problem in the desired form. We will use the fact that $E_n = Y^\top E_k Y$ for any $Y \in \mathcal{S}^{k \times n}$, where E_n denotes a $n \times n$ matrix of ones:

$$-\|C - Y^\top Y\|_N^2 = -\|Y^\top Y\|_N^2 + 2\langle C, Y^\top Y \rangle_N + \text{const}$$

$$= \underbrace{\|Y^\top E_k Y\|_N^2 - \|E_n\|_N^2}_{=0} - \|Y^\top Y\|_N^2 + 2\langle C, Y^\top Y \rangle_N + \text{const}$$

$$= \|Y^\top (E_k - I) Y\|_N^2 + 2\langle C, Y^\top Y \rangle_N + \text{const}, \tag{9}$$

where $\langle A, B \rangle_N = \sum_{ij} N_{ij} A_{ij} B_{ij}$ is a weighted matrix dot product, and I is a properly-sized identity matrix. Note that in the derivation "const" represents additive terms not depending on the variable Y, thus not affecting the optimization results. We can now take the quantity in (9) (with constant terms omitted) as the polynomial f with nonnegative coefficients to be maximized, *i.e.*

$$f(Y) = \|Y^\top (E_k - I) Y\|_N^2 + 2\langle C, Y^\top Y \rangle_N.$$

and maximizers of $f(Y)$ on the feasible set of (7) will correspond to minimizers of (7) as required.

The last thing to care about is that Theorem 1 assumes $Y \in \mathcal{S}^{k \times n}$, but our feasible domain is a convex subset thereof, due to the integration of the supervisions. Hence, it is not clear whether the theorem applies also to the constrained setting we have. To show that the theorem actually does apply,

---

[4] The same result was proven to hold also in the case of non-homogeneous polynomials [6].

assume without loss of generality that the labeled data points are the last ones, such that we can write $\mathtt{Y} = [\mathtt{Y}^u, \mathtt{Y}^l]$, where $\mathtt{Y}^l$ is entirely specified with the label information as $\mathtt{Y}^l_{ki} = \mathbb{1}_{k=\ell_i}$ for all $i \in \mathcal{L}$, while $\mathtt{Y}^u \in \mathcal{S}^{k \times (n - |\mathcal{L}|)}$ are variables to be optimized. Since $g(\mathtt{Y}^u) = f(\mathtt{Y})$ is a polynomial in $\mathtt{Y}^u$ with nonnegative coefficients, we can apply Theorem 1 to obtain a growth transformation for g, and thus find a solution also to the constrained optimization problem in (7). In practice, it is not necessary to compute g explicitly for the optimization, because it is sufficient to avoid updating the labeled entries of $\mathtt{Y}$ during the computation of the update rule based on f.

4   Experiments

In this section we validate the proposed algorithm (SSCC) both on standard benchmark datasets and on a real world problem, namely the pathology classi-fication of ECG data. In the architecture described in Figure 2, the raw ECG is acquired and preprocessed to extract relevant features, which are then fed to the algorithms to classify the pathologies. We adopt the feature extraction process proposed in [4], which is described in Section 4.2.

4.1   Data Description

We evaluated the proposed algorithm in two different scenarios: using some benchmark datasets from the UCI Machine Learning repository[5], and using the MIT-BIH (Massachusetts Institute of Technology - Beth Israel Hospital) arrhythmia database [21].

In the first scenario, we used four benchmark datasets to validate our algo-rithm, namely breast cancer, iris, wine and std yeast cell. The *Breast cancer* dataset consists of 683 patterns having nine features belonging to two classes. The *Iris* dataset consists of three species of Iris plants, characterized by four features and 50 samples in each class. The *Wine* dataset consists of the results of a chemical analysis of wines grown from the same region in Italy divided into three classes with 178 patterns, and described by 13 features. The *Std yeast cell* is composed by 384 genes over two cell cycles of yeast cell data, and is characterized by 17 features and it has five classes.

The second scenario consists in classifying pathologies in ECGs from the MIT-BIH arrhythmia database. Each record is approximately 30 minutes long and has in total 48 two-channel Holter records. The upper signal is usually a modified limb lead II and the lower signal is most often a modified lead V1. All signals were digitized at a sample rate of 360 Hz. The database includes different sets of annotations verified by more than one cardiologist: beats are identified and labeled, and the beginning of all rhythms is indicated.

---

[5] http://archive.ics.uci.edu/ml/

4.2   Feature Extraction

For the classification of pathologies in ECGs, we will focus on the discrimination between normal sinus and the most common arrhythmia, atrial fibrillation (AF). Only modified limb II records are used and each record is split according to rhythm annotations (note that lead II and lead I, introduced in section 2, contain information about the frontal plane, and can be used to detect atrial fibrillation). Each record is segmented in windows of 60 seconds, leading to 98 AF segments and 911 normal sinus rhythm segments. These segments are the objects that will be classified. Two types of features are obtained: spectral features extracted using the wavelet transform, and time domain features used to provide information about heart rate characteristics.

The spectral features were obtained by the power spectral density (PSD) of the wavelet decomposition of the signals. The decomposition of the signals are performed up to the sixth level using the redundant discrete wavelet transform [11], obtaining six detailed and one approximated set of coefficients. Afterwards, the PSD of each set of wavelets coefficients was estimated using Welch's method [27], and the integral over the range $[0, 55]$ Hz was computed, leading in total to seven features per pattern.

Besides the spectral features we considered two additional time-domain features: average of RR interval [7] and standard deviation of RR intervals.

4.3   Setup

We constructed the ensemble for the proposed methodology by performing 200 runs of k-means, with k uniformly chosen between $\sqrt{N}/2$ and $\sqrt{N}$ (N is the number of samples of the dataset), for the benchmark datasets, and between 2 and 20, for the MIT-BIH database. Since the labels of the points are only used to extract the consensus partition, we present also the results of an unsupervised consensus clustering method (PPC) [23].

We compared SSCC also against a semi-supervised consensus clustering approach called Bipartite Graph-based Consensus Maximization (BGCM) [14], which can be seen as a consensus method, where the ensemble is constructed with supervised and unsupervised methods. In this paper we constructed in total six partitions in the ensemble: three from supervised methods, namely k-nearest neighbor, with $k \in \{1, 3, 5\}$, and three from an unsupervised method, the k-means, with k equal to the true number of classes. Moreover, this algorithm has two parameters, α and β, corresponding to the price paid from deviating from the estimated labels of groups and observed labels of objects. We set those parameters to 2 and 8, respectively.

In the MIT-BIH database, the classes are unbalanced, so we randomly selected 98 out of the 911 normal sinus rhythm segments. Also, in order to test the influence of the percentage of labeled points in each algorithm, we randomly selected 5% of labeled points in each class and the remaining are unlabeled points to be classified by the algorithms. This procedure was repeated 50 times, and was also run with minimum 10% and maximum 60% labeled data points.

Thus, we assessed the performance of each algorithm as an average error rate of 50 runs. The same scheme was applied to the benchmark datasets from the UCI Machine Learning repository, creating datasets with balanced classes.

4.4 Validation of the Algorithm: Benchmark Datasets

Figure 3 presents the average error rates for the four benchmark datasets considered. For SSCC and BGCM, we only present the results for 10% and 20% of labeled points, since as we increase the percentage of labeled points, there is a decrease in error rates. PPC is an unsupervised approach, which means it does not require any labeled points, thus the error remains constant when we increased the percentage of labeled points.

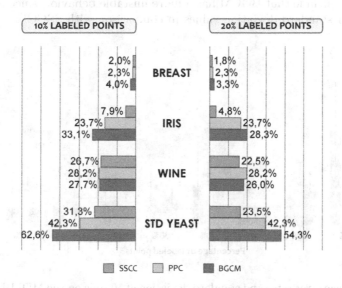

Fig. 3. Average error rates of 50 runs.

Notice that, for 10% of labeled points, SSCC achieves approximately 8% of error rate for the Iris dataset, performing three or four times better than the unsupervised version (which does not use any labeled points) and BGCM. That difference is even more accentuated when we have 20% of labeled points, where SSCC achieves less than 5% of error rate. The higher difference in error rates between SSCC and the other two algorithms is also visible in the Std yeast dataset, although a little less accentuated. In the remaining datasets, SSCC is still the best algorithm, however it is only better 1% or 2% than PPC and BGCM. Overall, the proposed methodology performs better than the other two algorithms considered in these experiments.

4.5   Application on Real Datasets: ECG MIT-BIH Database

Figure 4 presents the results of applying each algorithm to the MIT-BIH arrhythmia dataset, when the percentage of labeled points are varying. As can be observed by analyzing the figure, when the percentage of labeled points is of only 5%, SSCC and PPC have similar performances (SSCC shows slightly better results). On the other hand, BGCM has an error rate of approximately 17%, which is almost twice the error rate of SSCC. Moreover, as it could be expected, when the percentage of labeled points increases, both semi-supervised approaches show a significant improvement in the error rates, achieving around 3% of misclassified points when 60% of the dataset is labeled. In particular, when 30% or more points are labeled, the two semi-supervised approaches are quite similar, on average. However, by carefully analyzing the standard deviation, it is possible to conclude that BGCM has a more unstable behavior, since it presents much higher standard deviation values in comparison with SSCC.

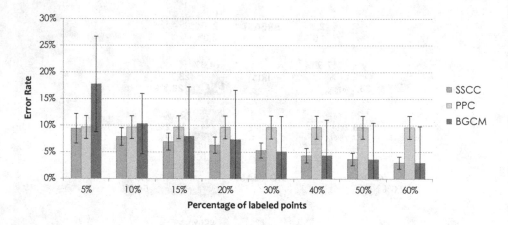

Fig. 4. Average error rates and standard deviation of 50 runs on the MIT-BIH arrhythmia database.

In [4], a supervised study was conducted using this database, and using 75% of data to train the classifier and 25% to test, the 1-nearest neighbor has an error rate of $2.6\% \pm 1.1\%$, and an artificial neural network with 14 hidden neurons has $3.0\% \pm 1.2\%$. The results presented in this study are quite comparable with the supervised study, since with only 20% of labeled patterns, SSCC has an error rate of $6.3\% \pm 1.5\%$, and with 60% of labeled patterns, SSCC has achieved $3.0\% \pm 1.1\%$.

Unlike normal sinus rhythm, atrial fibrillation has an inherent irregular RR interval, allowing for an easy visualization of the dataset. Accordingly, figure 5 presents the two time domain features considered in this study. The labeling produced by each algorithm corresponds to one run out of the 50 runs, and for the BGCM and SSCC, we are fixing the amount of labeled patterns at 10%. Notice that PPC incorrectly labeled the patterns in the frontier of both classes and the normal patterns that are mixed in the atrial fibrillation class. In fact, that couple of patterns that are mixed with the atrial fibrillation are incorrectly classified by SSCC and, some of them, by BGCM. On the other hand, SSCC correctly classified almost all the patterns in the frontier of both classes. The BGCM algorithm incorrectly classified a large amount of normal sinus rhythm in the lower left corner, due to k-means initialization. Notice that, for this algorithm, the number of components k in k-means must be set to the true number of classes, so that cloud of green points was assigned to atrial fibrillation. The blue dot in the middle corresponds to a pattern with the true labeled known.

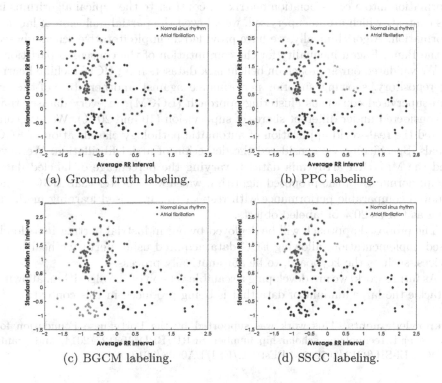

(a) Ground truth labeling.   (b) PPC labeling.

(c) BGCM labeling.   (d) SSCC labeling.

Fig. 5. Scatterplot of two features (average and standard deviation of RR intervals) from the ECG MIT-BIH dataset, when we have 10% of labeled patterns. The blue dots correspond to the normal sinus, the red ones to the atrial fibrillation, and the green dots are the points incorrectly classified by each algorithm.

Moreover, the SSCC algorithm incorrectly classified a few patterns that are closer to the atrial fibrillation class, the same patterns were identified by PPC. Since those patterns are so close to the atrial fibrillation, it may be worth to re-analyze those segments to ensure that they are in fact normal rhythms, and do not correspond to abnormal patterns closely resembling normal ones.

5   Conclusions

In real world problems it is unfeasible to ask experts to annotate all the available data. This is particularly true for the real-world scenario addressed in this paper, namely the automatic pathology classification of electrocardiographic (ECG) signals, where the amount of annotated data is very small compared to the amount of available data.

In this paper we proposed a semi-supervised consensus clustering algorithm, which automatically allows to label the unknown objects using only a small subset of known information. Our approach is based on the evidence accumulation clustering, a consensus clustering paradigm that summarizes the ensemble information into a co-association matrix. In contrast to the typical algorithms in this context, which are unsupervised, we allowed the partial inclusion of labeled information. Algorithmically, we have provided a simple iterative scheme based on the Baum-Eagon inequality for the computation of the consensus partition.

We validated our approach on benchmark datasets from UCI Machine Learning repository, showing superior performance against another state-of-the-art semi-supervised consensus clustering approach (BGCM), in every dataset that we considered under different shares of supervision (10 and 20 %). We also considered the real-world application of automatic pathology classification of ECG signals. Specifically, we considered the detection of atrial fibrillation. We have used the MIT-BIH arrhytmia dataset, varying the percentage of labeled data. The performance of the proposed algorithm was always better than BGCM and achieved comparable performance with respect to supervised learning methods using as few as 20% of labeled objects.

The proposed approach will be deployed by our industrial partner in a cloud-based implementation, allowing ECG data acquired using pervasive healthcare devices, such as the keyboard, to be automatically processed.

As future work, we are developing a scalable version of the algorithm, allowing to tackle the large amount of data that is being produced in this context.

Acknowledgments. This work was supported by the Portuguese Foundation for Science and Technology, scholarship number SFRH/BPD/103127/2014, and grants PTDC/EEI-SII/2312/2012 and PEst-OE/EEI/LA0008/2013.

References

1. Aidos, H., Fred, A.: Consensus of clusterings based on high-order dissimilarities. In: Celebi, M.E. (ed.) Partitional Clustering Algorithms, pp. 313–351. Springer International Publishing (2015)
2. Basu, S., Banerjee, A., Mooney, R.J.: Active semi-supervision for pairwise constrained clustering. SDM **4**, 333–344 (2004)
3. Basu, S., Davidson, I., Wagstaff, K.: Constrained Clustering: Advances in Algorithms, Theory, and Applications. Chapman and Hall/CRC (2008)
4. Batista, D., Fred, A.: Spectral and time domain parameters for the classification of atrial fibrillation. In: International Conference on Bio-inspired Systems and Signal Processing (BIOSIGNALS 2015), pp. 329–337 (2015)

5. Baum, L.E., Eagon, J.A.: An inequality with applications to statistical estimation for probabilistic functions of Markov processes and to a model for ecology. Bull. American Math. Society **73**, 360–363 (1967)
6. Baum, L.E., Sell, G.R.: Growth transformations for functions on manifolds. Pacific J. Math. **27**, 221–227 (1968)
7. Chung, E.K.: Pocket guide to ECG diagnosis. Blackwell Science (1996)
8. Covoes, T.F., Hruschka, E., Ghosh, J.: A study of k-means-based algorithms for constrained clustering. Intelligent Data Analysis **17**, 485–505 (2013)
9. Drew, B.J., Califf, R.M., Funk, M., Kaufman, E.S., Krucoff, M.W., Laks, M.M., Macfarlane, P.W., Sommargren, C., Swiryn, S., Van Hare, G.F.: Practice standards for electrocardiographic monitoring in hospital settings an american heart association scientific statement from the councils on cardiovascular nursing, clinical cardiology, and cardiovascular disease in the young. Circulation **110**(17), 2721–2746 (2004)
10. Fern, X.Z., Brodley, C.E.: Solving cluster ensemble problems by bipartite graph partitioning. In: Proceedings of the Twenty-first International Conference on Machine Learning, ICML 2004, p. 36. ACM, New York (2004). http://doi.acm.org/10.1145/1015330.1015414
11. Fowler, J.E.: The redundant discrete wavelet transform and additive noise. Signal Processing Letters, IEEE **12**(9), 629–632 (2005)
12. Fred, A., Jain, A.: Combining multiple clusterings using evidence accumulation. IEEE Transactions on Pattern Analysis and Machine Intelligence **27**(6), 835–850 (2005)
13. Fred, A.L., Lourenço, A., Aidos, H., Rota Bulò, S., Rebagliati, N., Figueiredo, M., Pelillo, M.: Learning similarities from examples under the evidence accumulation clustering paradigm. In: Pelillo, M. (ed.) Similarity-Based Pattern Analysis and Recognition. Advances in Computer Vision and Pattern Recognition, pp. 85–117. Springer London (2013)
14. Gao, J., Liang, F., Fan, W., Sun, Y., Han, J.: Graph-based consensus maximization among multiple supervised and unsupervised models. In: Advances in Neural Information Processing Systems, pp. 585–593 (2009)
15. Kuncheva, L., Vetrov, D.: Evaluation of stability of k-means cluster ensembles with respect to random initialization. IEEE Transactions on Pattern Analysis and Machine Intelligence **28**(11), 1798–1808 (2006)
16. Li, T., Ding, C., Jordan, M.: Solving consensus and semi-supervised clustering problems using nonnegative matrix factorization. In: dSeventh IEEE International Conference on Data Mining, ICDM 2007, pp. 577–582, October 2007
17. Lourenço, André, Bulò, Samuel Rota, Rebagliati, Nicola, Fred, Ana, Figueiredo, Mário, Pelillo, Marcello: Consensus clustering using partial evidence accumulation. In: Sanches, João M., Micó, Luisa, Cardoso, Jaime S. (eds.) IbPRIA 2013. LNCS, vol. 7887, pp. 69–78. Springer, Heidelberg (2013)
18. Lourenço, A., Rota Bulò, S., Rebagliati, N., Fred, A.L.N., Figueiredo, M.A.T., Pelillo, M.: Probabilistic evidence accumulation for clustering ensembles. In: Marsico, M.D., Fred, A.L.N. (eds.) ICPRAM 2013 - Proceedings of the 2nd International Conference on Pattern Recognition Applications and Methods, pp. 58–67. SciTePress (2013)
19. Lourenço, A., Rota Bulò, S., Rebagliati, N., Fred, A.L., Figueiredo, M.A., Pelillo, M.: Probabilistic consensus clustering using evidence accumulation. Machine Learning **98**(1–2), 331–357 (2015)
20. Lourenço, A., Silva, H., Fred, A.: Unveiling the biometric potential of Finger-Based ECG signals. Computational Intelligence and Neuroscience 2011 (2011)

21. Moody, G.B., Mark, R.G.: The impact of the MIT-BIH arrhythmia database. Engineering in Medicine and Biology Magazine, IEEE **20**(3), 45–50 (2001)
22. Odinaka, I., Lai, P.H., Kaplan, A., O'Sullivan, J., Sirevaag, E., Rohrbaugh, J.: ECG biometric recognition: A comparative analysis. IEEE Trans. Inf. Forensics Security **7**(6), 1812–1824 (2012)
23. Rota Bulò, S., Lourenço, A., Fred, A., Pelillo, M.: Pairwise probabilistic clustering using evidence accumulation. In: Hancock, E., Wilson, R., Windeatt, T., Ulusoy, I., Escolano, F. (eds.) SPR and SPR 2010. LNCS, vol. 6218, pp. 395–404. Springer, Heidelberg (2010)
24. Rota Bulò, S., Pelillo, M.: Probabilistic clustering using the baum-eagon inequality. In: ICPR, pp. 1429–1432 (2010)
25. da Silva, H.P., Carreiras, C., Lourenço, A., Fred, A., das Neves, R.C., Ferreira, R.: Off-the-person electrocardiography: performance assessment and clinical correlation. Health and Technology, 1–10 (2015)
26. Topol, E.: The Creative Destruction of Medicine. Basic Books (2012)
27. Welch, P.: The use of fast Fourier transform for the estimation of power spectra: a method based on time averaging over short, modified periodograms. IEEE Transactions on audio and electroacoustics, 70–73 (1967)

Two Step Graph-Based Semi-supervised Learning for Online Auction Fraud Detection

Phiradet Bangcharoensap[1]([⊠]), Hayato Kobayashi[2], Nobuyuki Shimizu[2],
Satoshi Yamauchi[2], and Tsuyoshi Murata[1]

[1] Tokyo Institute of Technology, Meguro, Tokyo 152-8552, Japan
phiradet.b@ai.cs.titech.ac.jp, murata@cs.titech.ac.jp
[2] Yahoo Japan Corporation, Minato, Tokyo 107-6211, Japan
{hakobaya,nobushim,satyamau}@yahoo-corp.jp

Abstract. We analyze a social graph of online auction users and propose an online auction fraud detection approach. In this paper, fraudsters are those who participate in their own auction in order to drive up the final price. They tend to frequently bid in auctions hosted by fraudulent sellers, who work in the same collusion group. Our graph-based semi-supervised learning approach for online auction fraud detection is based on this social interaction of fraudsters. Auction users and their transactions are represented as a social interaction graph. Given a small set of known fraudsters, our aim was to detect more fraudsters based on the hypothesis that strong edges between fraudsters frequently exist in online auction social graphs. Detecting fraudsters who work in collusion with known fraudsters was our primary goal. We also found that *weighted degree centrality* is a distinct feature that separates fraudsters and legitimate users. We actively used this fact to detect fraud. To this end, we extended the *modified adsorption* model by incorporating the weighted degree centrality of nodes. The results, from real world data, show that by integrating the weighted degree centrality to the model can significantly improve accuracy.

Keywords: Online auction fraud detection · Graph-based semi-supervised learning · Weighted degree centrality

1 Introduction

Over the last decade, online auctions have quickly become popular e-commerce services. The extensive profits attract many users to commit fraud in online auction websites. Online auction fraud is increasingly recognized as one of serious global concerns.

Generally, online auction fraud can be categorized into three types, according to the time when the fraudulent activity is committed: pre-auction, in-auction, and post-auction [6]. Pre-auction fraud occurs prior to an auction, for example selling of low quality product. Post-auction frauds are committed afterwards, such as non-delivery of products. Both pre-auction and post-auction frauds can

A. Bifet et al. (Eds.): ECML PKDD 2015, Part III, LNAI 9286, pp. 165–179, 2015.
DOI: 10.1007/978-3-319-23461-8_11

be directly verified with physical evidence. The remaining type of fraud is in-auction, which is the main target of this research. There are many kinds of in-auction fraud, as shown in Figure 1. The main focus of this research was *competitive shilling* in which fraudsters participate in their own auction as bidders with another user ID in order to drive up the final price. When such a fraud takes place, a legitimate winner has to pay more than a reasonable final price. Hereafter, the term *fraud* refers to this definition.

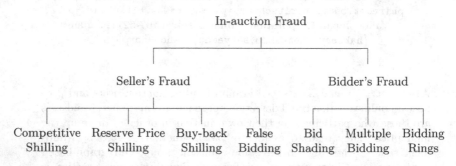

Fig. 1. Categorization of in-auction fraud [6]

Fraudulent bidders tend to frequently bid in auctions hosted by a particular seller(s) working in the same collusion group. Therefore, there is a very high tendency that a connections between fraudsters exist. In network science, we call this phenomenon *Homophily*. If we represent auction users and their activities as a social graph, groups of fraudulent users working in collusion should have strong links. It is analogous to one of the key assumptions of graph-based semi-supervised learning models (graph-based SSL). In general, graph-based SSL models try to assign a similar label to adjacent nodes. In other words, the models try to maintain the *smoothness* between adjacent nodes. This analogy, between *homophily* and *smoothness*, motivates us to investigate the potential of graph-based SSL models in online auction fraud detection.

There are two main contributions of this study. First, to the best of our knowledge, no study has been conducted to apply a graph-based SSL model for solving this serious Internet crime. We discuss the application of the state-of-the-art graph-based SSL model called modified adsorption (MAD) [14] to detect auction fraud. Furthermore, we found that the sum of the interactions between nodes with their neighbors can be used to distinguish between legitimate users and fraudsters. This sum is called *weighted degree centrality*. We argue that the weighted centrality of fraudsters is considerably higher than that of typical users. This fact alone sheds lights on the behaviors and social interactions of auction users, contributing to our understanding of the Web and its users. Even though MAD involves edge weights as one type of information for propagating labels, a higher total weight of edges does not imply a higher likelihood of there being a fraud in the context of MAD. Therefore, we extended the model by incorporating the weighted degree centrality in the sense that it can be used to detect fraud. Our extended model, called the 2-STEP model emphasizes that a higher weighted degree centrality implies a higher chance of being fraudsters. This is our second contribution which involves the social behavior to detect auction frauds. According to experiments on real world data, our 2-STEP model significantly increases result accuracy.

The remainder of this paper is divided into six sections. We begin by giving details about the data we obtained from an online auction site in Section 2. In Section 3, we describe our proposed approach. We explain the performance evaluation of our approach in Section 4. Next, we discuss the results from the evaluation and discuss a possible extension for future work in Section 5. We describe related work in Section 6. Finally, we conclude our paper in Section 7.

2 Data Description

We first discuss the details of the data used in this research. The data are auction transactions, a set of known fraudsters, and a set of trustworthy users. The transactions were transformed into a social interaction graph. Section 2.2 gives the formal definition of the graph.

2.1 Resources

The following three sub-sections give more detail about the data used in this research.

Online Auction Transaction. We obtained online auction transactions from YAHUOKU![1], which is one of the largest online auction sites in Japan and operated by Yahoo Japan Corporation. The dataset contains comprehensive bidding and selling activities on the website. Each record is a five-element tuple — (selling time, product ID, seller ID, bidding time, bidder ID). All user IDs are anonymized for preserving privacy. It is possible that the seller ID or bidder ID of two different transactions are identical. One user ID can be either of a seller or bidder. There are around 16 million transactions with around 3 million products and 2 million users.

Set of Known Fraudsters. This set consists of IDs of users suspected to be fraudsters according to the definition of competitive shilling in Section 1. It is important to note that this set includes only a partial set of possible fraudsters because it is almost impossible to extract all fraudsters in this dataset due to its size. The detection of fraudsters incurs high costs because it is performed manually. Approximately, 550 users are listed as fraudsters. The detailed description about ground-truth labeling cannot be disclosed since it is confidential business information. The set is used for training models and measuring performance. The *Set of Known Fraudsters* is referred to as $\mathcal{F}$.

---

[1] http://auctions.yahoo.co.jp/

Set of Store Users. Some online auction user IDs are registered as official stores. These users can be placed on a whitelist, which contains trustworthy users who are unlikely to commit fraud. In our auction transactions, around 10,000 accounts are registered as stores. The *Set of Store Users* is referred to as S.

2.2 Graph Construction

In this research, we represent online auction transactions as a weighted undirected graph $G(V, E, \mathbf{W})$, where V is the set of n nodes, E is the set of edges, and $\mathbf{W} \in \mathbb{R}^{n \times n}$ is the edge weight matrix ($\mathbb{R}$ is the set of real numbers). A node $v \in V$ represents an auction user ID. The set $E \subset V \times V$ represents interaction between nodes. An edge $e = (u, v) \in E$ indicates that u has a bid on an auction hosted by v, or vice versa. Each edge weight $\mathbf{W}_{uv} \in \mathbb{R}_+$ reflects the total number of u's products that are bidded by v. To remove noise, users participating in less than five transactions were removed. Finally, the graph contains around 0.8 million nodes and 3 million edges.

The largest group of nodes contains users who have never hosted any auction, but have only bid. This group of nodes occupies around 70% of the entire graph. Let us define this group of users as *bidder*. Another group contains users who have never bid on any auction, but only hosted. Approximately 15% of nodes fit into this category. We call this category *seller*. There is no link between nodes within the *bidder* and within *seller* groups. The last group contains users who both host and bid — *mixed*. Nodes in *mixed* can link with the previous two groups and within *mixed* themselves. Suppose we represent the graph as a directed graph whose edges originating from a *bidder* to a *seller*. It is obvious that the *seller* should not have any outgoing edges. Our model is based on an information propagation model. The propagation process cannot flow information to the entire graph when the graph contains many sink nodes. Therefore, we represent the transactions as an undirected graph. Figure 2 shows the degree distribution of our graph.

3   Proposed Approach

In this section, we now formally define the problem we want to solve and propose solutions.

3.1   Problem Definition

The definition of the problem that we try to solve in this paper is summarized as below. We construct a weighted undirected graph $G(V, E, \mathbf{W})$ from a real world online auction dataset as described in Section 2.2. We assign a score, indicating likelihood of committing fraud activity, to all nodes. The score assignment is based on label propagation approach as described in Section 3.3. Nodes are ranked according to the score in descending order.

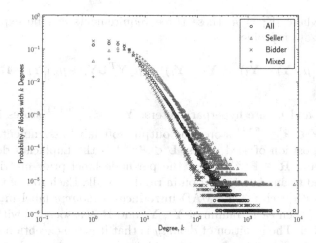

Fig. 2. Degree (k) distribution of the YAHUOKU! network

Input: A weighted undirected graph $G(V, E, \mathbf{W})$, a set of known fraudsters $\mathcal{F}$, and a set of store users $\mathcal{S}$.

Output: An ordered list of $\{v | v \in V \wedge v \notin \mathcal{F} \cup \mathcal{S}\}$.

Goal: Actual fraudulent nodes are expected to be ranked at the top of the output ordered list.

Approach: Propagating label information from seed known fraudsters and legitimate users to unknown users.

3.2 Modified Adsorption (MAD)

MAD is a graph-based semi-supervised learning model proposed by Talukdar and Crammer [14]. This research adopted the modified adsorption (MAD) to propagate information from known fraudsters to the whole graph. We used the Junto Label Propagation Toolkit[2] as an implementation of MAD.

The MAD takes as the input a weighted undirected graph. The weight of edges represents the degree of similarity or correlation between nodes. A few nodes, or instances, are labeled — called seed nodes. MAD propagates labels from the few seed nodes to all nodes. Finally, all nodes are assigned a score indicating the likelihood of being each label. To deal with noisy initial labels, MAD allows the initial labels to change.

The model trade offs between three requirements: *accuracy* — the initial labels of seed nodes should be retained, *smoothness* — similar labels should be assigned to neighbor nodes, and *regularity* — output labels should be as uninformative as possible. We denote $\mathcal{L}$ as the set of m possible labels, and $\mathbf{M}_l$ as the lth column of any matrix $\mathbf{M}$. Given a weighted undirected graph

---

[2] https://github.com/parthatalukdar/junto/

$G(V, E, \mathbf{W})$, where $|V| = n$, these three requirements can be expressed as a convex optimization problem as

$$\min_{\hat{\mathbf{Y}}} \sum_{l \in \mathcal{L}} \Big[\mu_1 (\mathbf{Y}_l - \hat{\mathbf{Y}}_l)^T \mathbf{S} (\mathbf{Y}_l - \hat{\mathbf{Y}}_l)) + \mu_2 \hat{\mathbf{Y}}_l^T \mathbf{L} \hat{\mathbf{Y}}_l + \mu_3 \left\| \hat{\mathbf{Y}}_l - \mathbf{R}_l \right\|^2 \Big], \quad (1)$$

where μ_1, μ_2, and μ_3 are hyperparameters, $\mathbf{Y} \in \mathbb{R}_+^{n \times (m+1)}$ stores initial label information, $\hat{\mathbf{Y}} \in \mathbb{R}_+^{n \times (m+1)}$ stores the output soft label assignment, $\mathbf{S} \in \mathbb{R}^{n \times n}$ indicates the position of seed nodes, $\mathbf{L} \in \mathbb{R}^{n \times n}$ is the Laplacian derived from the given G, and $\mathbf{R} \in \mathbb{R}^{n \times (m+1)}$ is the per-node label prior matrix, which is strongly related to an abandon action in random-walk. Each row of the matrices is associated with each $v \in V$. MAD introduces a *dummy* label in addition to the labels in $\mathcal{L}$, then each column of $\mathbf{Y}$, $\hat{\mathbf{Y}}$, and $\mathbf{R}$ is associated with $l \in \mathcal{L}$ and the *dummy* label. The intuition of *dummy* is that it is the exceptional case of all possible labels $\mathcal{L}$. The score of *dummy* is high when weights of edges originating from the node tends to be uniformly distributed, in other words the entropy of the weights is high. In terms of scalability, it has been proven that MAD is parallelizable in MapReduce [15].

3.3   MAD for Online Auction Fraud Detection

In this section, we discuss how to apply MAD to auction fraud detection. The key idea is to use the information from the set of known fraudsters and set of store sellers to assign initial labels. We denote the set of possible labels $\mathcal{L}$ as {*fraud, legitimate*}. We denote the 1st, 2nd, and 3rd column of $\mathbf{Y}$, $\hat{\mathbf{Y}}$, and $\mathbf{R}$ as associated with labels *fraud, legitimate*, and *dummy*, respectively.

MAD allows embedding world knowledge to a model by putting weights over initial labels. Precisely, it assigns a non-negative number to an element $\mathbf{Y}_{vl}$ in matrix $\mathbf{Y}$, where $\mathbf{Y}_{vl}$ is the vth row and l^{th} column of $\mathbf{Y}$. In other words, $\mathbf{Y}_{vl}$ is the weight of node v over the label l. We create $\mathbf{Y}$ with conditions as

$$\mathbf{Y}_{vl} = \begin{cases} \alpha & \text{if } l = 1, v \in \mathcal{F}; \\ \beta & \text{if } l = 2, v \in \mathcal{S}; \\ 0 & \text{otherwise}, \end{cases} \quad (2)$$

where α and β are parameters, $\mathcal{F}$ is the set of known fraudsters, and $\mathcal{S}$ is the set of store users. The parameters α and β reflect the degree of association between v and the labels *fraud* and *legitimate*, respectively. Thus, seed nodes are $V_l = \{v | v \in V \wedge v \in \mathcal{F} \cup \mathcal{S}\}$. Let V_u denotes unlabeled nodes such that $V_u = V - V_l$, typically $|V_l| \ll |V_u|$. As previously mentioned, MAD outputs a soft label matrix $\hat{\mathbf{Y}}$ as a result of the label propagation process. Now, we assign each v a score reflecting the likelihood of being fraudsters, called *fraud score*, as

$$\varphi(v, \hat{\mathbf{Y}}) = \frac{\hat{\mathbf{Y}}_{v1}}{\sum_{l=1}^{m+1} \hat{\mathbf{Y}}_{vl}}. \quad (3)$$

Finally, we sort all $v \in V$ according to the *fraud score* in descending order. Then, users working in collusion with the known fraudsters are expected to be ranked at the top of the output ordered list. In Eq. 3, the score associated with the *dummy* label is a part of the denominator. Another alternative of the *fraud score* can be derived without the contribution of the *dummy* label. It is defined as

$$\bar{\varphi}(v, \hat{\mathbf{Y}}) = \frac{\hat{\mathbf{Y}}_{v1}}{\sum\limits_{l=1}^{m} \hat{\mathbf{Y}}_{vl}}. \tag{4}$$

In Section 4.3, we compare the performance of $\varphi(.)$ and $\bar{\varphi}(.)$. We discuss the results of this comparison in Section 5.

3.4   2-STEP Model

We now present an extension of the approach described in the previous section. We give another alternative definition of *fraud score*. We found that fraudsters tend to have many heavy links to their neighbors. Let us define the *weighted degree centrality* of v as the sum of edge weights originating from v, as

$$k_w(v) = \sum_{u \in N(v)} \mathbf{W}_{uv}, \tag{5}$$

where $\mathbf{W}_{uv}$ is weight of the edge (u, v), and $N(v)$ is the set of neighbors of v. The $k_w(v)$ is high if v has a heavy link(s). Figure 3 shows the fraction of nodes having weighted degree centrality k_w. It can be observed that fraudsters have a higher probability $p(k_w)$ than legitimate users when weighted degree centrality is high ($k_w > 20$). In contrast, fraudsters have lower probability when the centrality is low. Fraudulent users in the same group always interact together in order to inflate the auction or reputation. Many heavy links appear between them. In general, few popular legitimate users gain a great deal of attraction. Therefore, there is a higher tendency for fraudsters to have high weighted degree centrality, as shown in Figure 3.

MAD uses information from edge weight $\mathbf{W}_{uv}$ in the second term of Eq. 1. However, it does not use the weighted degree centrality. Suppose there is a fraudulent bidder, b, who has one heavy link to a fraudulent seller, s. Some legitimate users connect with the fraudulent seller s. In this case, the confidence of the *fraud* label of s, $\hat{\mathbf{Y}}_{s1}$, decreases because the legitimate information propagates to s. Because s, who is the only neighbor of b, is interfered by legitimate users, then b's score of the *fraud* label, $\hat{\mathbf{Y}}_{b1}$, is low as well. Therefore, this fraudulent bidder will not be ranked at the top of the ranking results. If we incorporate the observed behavior that users who have heavy edge(s) tends to be fraudsters, the tendency of being fraudsters of b increases.

Therefore, we provide another definition of *fraud score*. This definition combines the result from MAD and the *weighted degree centrality*. We modify the aforementioned definition of the *weighted degree centrality* by penalizing each

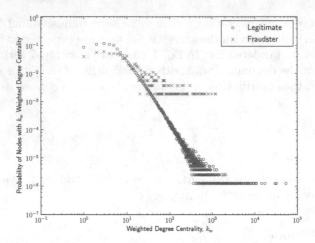

Fig. 3. Weighted degree centrality (k_w) distribution of legitimate users and known fraudsters

edge weighted with the $\varphi(.)$ of the neighbor. This definition of *fraud score* can be mathematically defined as

$$\rho(v, \hat{\mathbf{Y}}) = \varphi(v, \hat{\mathbf{Y}}) + \frac{\gamma}{|N(v)|} \sum_{u \in N(v)} \mathbf{W}_{uv}\, \varphi(u, \hat{\mathbf{Y}}), \qquad (6)$$

where γ is a parameter, $\varphi(.)$ is defined in Eq. 3, W_{uv} is the weight of the edge (u, v), and $N(v)$ is the set of neighbors of v. The $\varphi(.)$ can be replaced with $\bar{\varphi}(.)$. We call this extension *2-STEP* model.

4   Experiments

We evaluated the performance of the models described in the previous sections. This section describes the evaluation methodology and reports the results.

4.1   Evaluation Metric

The output of the proposed approach is an ordered list of nodes. Therefore, the normalized discounted cumulative gain (NDCG) [7] — a well-known evaluation metric for the information retrieval task — is used. It is defined as

$$\text{NDCG} = \frac{\text{DCG}}{\text{IDCG}}, \qquad (7)$$

$$\text{DCG} = \sum_{i=1}^{p} \frac{2^{r(i)} - 1}{\log_2(i+1)},$$

$$\text{IDCG} = \sum_{i=1}^{\min(p, |Q|)} \frac{1}{\log_2(i+1)},$$

where p is the maximum number of nodes that are considered, $|Q|$ is the number of actual fraudsters in testing data, and r(i) is the relevance value of the ith node. In this work, the relevance value is binary, r(i) $\in \{0, 1\}$. r(i) is 1 if and only if the ith node of the output list is fraudulent. NDCG ranges from 0.0 to 1.0. NDCG assumes that it is less useful for users when a relevant instance is ranked at a lower position of the result. Thus, NDCG penalizes relevant instances logarithmically proportional to the position of the instance. A higher NDCG indicates much better performance.

4.2 Methodology

Our experiments were conducted on real world data acquired from YAHUOKU!, as mentioned in Section 2. We performed 5-fold cross validation in all experiments. Known fraudsters, sellers, bidders, mixed are distributed among 5 partitions, so that the distribution of user types in each fold resembles the whole dataset. The total number of fraudsters is far less than the total number of legitimate users. In each iteration, one chunk of the known fraudsters list was treated as the testing set, Q, and the remaining were used as training set. All stores were treated as a training set in every iteration.

The straightforward manner to evaluate performance is calculating NDCG on all test nodes. As previously mentioned in Section 2.2, auction users can be categorized into three groups — *bidder*, *seller*, and *mixed*. The degree distribution in Figure 2 shows that they tend to have different behaviors. We would like to gain more insight into the performance of detecting different types of fraudsters. After a model has assigned a *fraud score* to nodes, we rank the nodes in a specific category only. The remaining categories are ignored. It should be noted that the training set, or seed nodes, still contains every type of nodes. We annotate the caption *all* to the results obtained from evaluating all types of nodes. We annotate the caption *bidder*, *seller*, or *mixed*, if the results were measured on a specific user type.

4.3 Results

We set $\alpha = \beta$ for MAD and the 2-STEP model. We investigated the effect of different α. We tried $\alpha \in \{0.2, 0.4, 0.6, 0.8, 1.0\}$ and found that there was no statistically significant difference in NDCG. We then used $\alpha = 0.4$ which gave the lowest variance in the parameter tuning experiment. We set $\mu_1 = 1$, $\mu_2 = 0.01$, and $\mu_3 = 0.01$. For the 2-STEP model, we set $\gamma = 1$.

In Section 3.3, we gave two alternative definitions of *fraud score*. One definition is based on information from the *dummy* label and the other is not. We conducted a two-tailed paired t-test to compare the average NDCG obtained from $\varphi(.)$ and $\bar{\varphi}(.)$. Table 1 summarizes the comparison of $\varphi(.)$ and $\bar{\varphi}(.)$ on all and individual types of fraudulent users. The result indicates that the *dummy* label has a significant advantage. From now on, we used $\varphi(.)$ as the main *fraud score* for MAD and 2-STEP.

Table 1. Comparison of $\varphi(.)$ and $\bar{\varphi}(.)$ on all and separate node types ($\langle\mathrm{NDCG}\rangle$ = mean of NDCG and SD = standard deviation)

Node Type	$\varphi(.)$		$\bar{\varphi}(.)$		p-value
	$\langle\mathrm{NDCG}\rangle$	SD	$\langle\mathrm{NDCG}\rangle$	SD	
All	0.431	0.015	0.406	0.019	0.002
Bidder	0.423	0.026	0.397	0.035	0.008
Seller	0.336	0.049	0.284	0.029	0.007
Mixed	0.374	0.044	0.319	0.024	0.006

We now compare our 2-STEP model (Section 3.4) with MAD (Section 3.3), weighted degree centrality (Eq. 5), and eigenvector centrality. In this experiment, we measured NDCG on the whole output ordered list, $p = |V_u|$. The eigenvector centrality is a well-known centrality measure defined as the principal eigenvector of a graph's adjacency matrix. PageRank is one of its variants [5]. The two centrality-based model are unsupervised. The results are summarized in Figure 4(a). The centrality-based methods, weighted degree centrality and eigenvector centrality, could not precisely spot fraudsters. Our 2-STEP model outperformed MAD obviously, with 0.490 over 0.437 NDCG on average. Furthermore, Figure 4(b), 4(c), and 4(d) show the NDCG in ranking fraudulent *bidder*, *sellers*, and *mixed* separately. The results follow the same pattern as the previous result. These results imply that our extension of MAD outperformed the other models for every kind of user.

Figure 5 compares the NDCG of our 2-STEP model, MAD, and a semi-supervised model (CD). The results from the two centrality-based methods are not included in the figure because it is obvious that they cannot perform well in this online auction fraud detection. We used Viswanath et al. [17] as the baseline in this experiment. The baseline system uses a local community detection schema, Mislove's algorithm [10], to detect Sybil as mentioned in Section 6. In this experiment, we set $p \in \{100, 500\}$. We did not calculate the NDCG on the whole output since the software implementation of the baseline system we used did not provide sorted results of the whole dataset. The result conforms the previous experiments that the 2-STEP model clearly outperformed MAD and the baseline.

5   Discussion

It is apparent from Table 1 that the *dummy* label has a significant advantage. As mentioned in Section 3.2, the score of *dummy* label is directly proportional to the entropy of weights of edges originating from the node. This implies that a user who uniformly interacts with others tends to be legitimate. The experimental results confirm that incorporating the entropy of edge weights can significantly improve the system. This result is consistent with the previous computational linguistics study [15]. Our result verifies that the effect determined in computational linguistics even appears in online auction fraud detection.

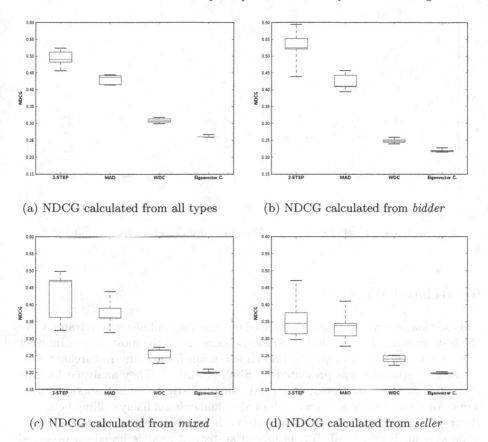

(a) NDCG calculated from all types   (b) NDCG calculated from *bidder*

(c) NDCG calculated from *mixed*   (d) NDCG calculated from *seller*

Fig. 4. NDCG of 2-STEP, MAD, weighted degree centrality (WDC), and eigenvector centrality where we set p equals the whole length of output list (box = 25th and 75th percentiles; central red line = median; and bar = min and max values)

According to Figure 4, the 2-STEP and MAD models exhibit low NDCG in ranking *seller*. The propagation method is used to satisfy the three requirements discussed in Section 3.2. One of them is *smoothness* — two adjacent nodes should be assigned similar labels. Suppose there is a fraudulent seller v who can successfully fool many legitimate users, the information from the legitimate users propagate to the fraudulent seller. In this case, the seller's *fraud* score, $\hat{\mathbf{Y}}_{v1}$, decreases relative to the *legitimate* score, $\hat{\mathbf{Y}}_{v2}$. We hypothesize that this is the primary reason fraudulent sellers are more difficult to detect. An effective method to solve this problem should be designed for future work.

We are also sure that the method is easily combined with the approach using user's information because it only use basic information of transaction. In future work, we will implement this hybrid approach in real service. The hybrid approach is expected to improve the effectiveness of the detection system in terms of not only on precision and resistance against the wrong user information.

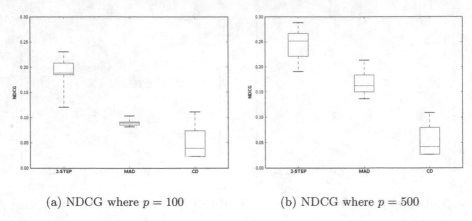

(a) NDCG where $p = 100$ (b) NDCG where $p = 500$

Fig. 5. NDCG of 2-STEP, MAD, and community detection-based model [10]

6   Related Work

This section surveys related work focused on detecting online auction fraud. Over the last decade, online auction fraud has become one of the most serious Internet crimes; therefore, it has attracted much attention from many researchers. One of the first attempts was presented by Shah et al. [13]. They analyzed bidding strategies on eBay and revealed normal characteristics of online auction fraudsters. An association analysis was adopted to find cases of likely shilling behavior. However, they did not propose any systematic schema that can be used in large scale system. Rubin et al. [12] proposed statistical models based on observed fraudulent behaviors. Their statistical bidder profiles are based on suspicious patterns in which shilling bidders are strongly associated with sellers, and shills rarely win auctions. Recently, supervised machine learning techniques have been used. Tsang et al. [16] used the C4.5 algorithm in WEKA to detect fraudulent bidders based on their bidding history. In general, these two research tried to propose a set of rules to detect fraudsters. However, sophisticated fraudsters usually have very flexible, adaptive, and various strategies. Therefore, it would be difficult to detect fraudsters via generalized rules — as the Vapnik's principle that when trying to solve some problem, one should not solve a more difficult problem as an intermediate step [3]. Yoshida and Ohwada [19] used a one-class support vector machine (SVM) and a decision tree to learn bidding attributes based on bidding history and user's evaluation results. In real world-wide-web situations, fraudsters can easily control and lie in their profile and rating. If a fraud detection system deeply rely on the user's inputs, these miss-leading information would easily defeat the precision of the system.

Many recent attention has focused on graph-based approach since objects in graph have long-range correlations [2]. Markov random field modeling was used to solve this problem [4,11] in which belief propagation was used to detect near bipartite cores in an undirected graph, which was expected to be an abnormal pattern. However, fraudsters in our dataset rarely form the near bipartite core structure, then this schema could not be effectively used in our context. In 2012, Shi-Jen et al. [8] adapted PageRank and k-core clustering algorithm to detect collusive groups in online auction. As shown in Section 4.3, the eigenvector centrality, which PageRank is derived from, exhibited unpleasant results in our dataset.

Online auction fraud detection can be recognized as a member of anomaly detection problem. A large body of literature has investigated the potential of graph-based anomaly detection algorithm. Akoglu et al. discovered several rules in graph-based features from 1-step neighborhood around a node [1]. Therefore, fraud or anomaly score of a node depends on only 1-step neighbors. In contrast, our information propagation-based models can better use long-range correlations of objects in a graph. In 2010, Viswanath et al. demonstrated that community detection algorithm can be utilized to avoid multiple identity, or Sybil, attacks [17]. Sybil are malicious attackers who create multiple identities and influence the working of systems that rely upon open membership such as collaborative content rating and recommendation system. It is noticeable that Sybil and our focused fraudsters share a common behavior that they are groups of identities aiming for committing unacceptable activities. The scheme works by detecting local communities around trusted nodes because Sybil nodes tend to poorly connected to the rest of network. In another word, they assume that the *homophily* behavior tends to happen. In addition, the *homophily* behavior was employed in the area of social security [18] and accounting fraud detection [9]. Please refer to Akoglu et al.'s survey [2] for more extensive review about graph-based anomaly detection.

As described in Section 1, *homophily* behavior tends to occur in online auction fraud networks. There is a very high tendency that online auction fraudsters have connections together. Even if, the behavior has been widely used to solve many fraud detection problems, to the best of our knowledge, there is no publication that focused on such behavior for online auction fraud detection. The homophily behavior is analogous with one of the main principle concepts of graph-based SSL models. Therefore, this work proposed a graph-based SSL model for online auction fraud detection.

7   Conclusion

We proposed an online auction fraud detection approach involving the extension of a graph-based semi-supervised learning model. The development of our approach was motivated by the *homophily* behavior of fraudsters. We extended the *modified adsorption* model to propagate information from a small set of known fraudsters to the entire graph. We found that fraudsters tend to have many heavy interactions with neighbors. We integrated this suspicious social behavior into this extended model, which we call the 2-STEP model. The experiments, on real-world data, suggest that our approach significantly improves accuracy.

Acknowledgments. We would like to thank the Yahoo! JAPAN patrol team for their assistance and support, especially Hiroyuki Kobayashi and Yuichi Nakatsu for regarding their data preparation.

References

1. Akoglu, L., McGlohon, M., Faloutsos, C.: Oddball: spotting anomalies in weighted graphs. In: Zaki, M.J., Yu, J.X., Ravindran, B., Pudi, V. (eds.) PAKDD 2010. LNCS, vol. 6119, pp. 410–421. Springer, Heidelberg (2010)
2. Akoglu, L., Tong, H., Koutra, D.: Graph based anomaly detection and description: a survey. Data Mining and Knowledge Discovery, 1–63 (2014)
3. Chapelle, O., Schölkopf, B., Zien, A.: Semi-supervised learning. MIT, Cambridge (2010)
4. Chau, D.H., Pandit, S., Faloutsos, C.: Detecting fraudulent personalities in networks of online auctioneers. In: Fürnkranz, J., Scheffer, T., Spiliopoulou, M. (eds.) PKDD 2006. LNCS (LNAI), vol. 4213, pp. 103–114. Springer, Heidelberg (2006)
5. Chau, D.H., Nachenberg, C., Wilhelm, J., Wright, A., Faloutsos, C.: Polonium: tera-scale graph mining and inference for malware detection. In: SIAM International Conference on Data Mining (SDM), pp. 131–142 (2011)
6. Dong, F., Shatz, S.M., Xu, H.: Combating online in-auction fraud: Clues, techniques and challenges. Computer Science Review **3**(4), 245–258 (2009)
7. Järvelin, K., Kekäläinen, J.: IR evaluation methods for retrieving highly relevant documents. In: Proceedings of the 23rd Annual International ACM SIGIR Conference on Research and Development in Information Retrieval, SIGIR 2000, pp. 41–48. ACM, New York (2000)
8. Lin, S.J., Jheng, Y.Y., Yu, C.H.: Combining ranking concept and social network analysis to detect collusive groups in online auctions. Expert Systems with Applications **39**(10), 9079–9086 (2012)
9. McGlohon, M., Bay, S., Anderle, M.G., Steier, D.M., Faloutsos, C.: Snare: a link analytic system for graph labeling and risk detection. In: Proceedings of the 15th ACM SIGKDD International Conference on Knowledge Discovery and Data Mining, KDD 2009, pp. 1265–1274. ACM, New York (2009)
10. Mislove, A., Viswanath, B., Gummadi, K.P., Druschel, P.: You are who you know: inferring user profiles in online social networks. In: Proceedings of the 3rd ACM International Conference on Web Search and Data Mining, WSDM 2010, pp. 251–260. ACM, New York (2010)
11. Pandit, S., Chau, D.H., Wang, S., Faloutsos, C.: Netprobe: a fast and scalable system for fraud detection in online auction networks. In: Proceedings of the 16th International Conference on World Wide Web, WWW 2007, pp. 201–210. ACM, New York (2007)
12. Rubin, S., Christodorescu, M., Ganapathy, V., Giffin, J.T., Kruger, L., Wang, H., Kidd, N.: An auctioning reputation system based on anomaly. In: Proceedings of the 12th ACM Conference on Computer and Communications Security, CCS 2005, pp. 270–279. ACM, New York (2005)
13. Shah, H.S., Joshi, N.R., Sureka, A., Wurman, P.R.: Mining ebay: bidding strategies and shill detection. In: Zaïane, O.R., Srivastava, J., Spiliopoulou, M., Masand, B. (eds.) WebKDD 2003. LNCS (LNAI), vol. 2703, pp. 17–34. Springer, Heidelberg (2003)

14. Talukdar, P.P., Crammer, K.: New regularized algorithms for transductive learning. In: Buntine, W., Grobelnik, M., Mladenić, D., Shawe-Taylor, J. (eds.) ECML PKDD 2009, Part II. LNCS, vol. 5782, pp. 442–457. Springer, Heidelberg (2009)
15. Talukdar, P.P., Pereira, F.: Experiments in graph-based semi-supervised learning methods for class-instance acquisition. In: Proceedings of the 48th Annual Meeting of the Association for Computational Linguistics, ACL 2010, pp. 1473–1481. Association for Computational Linguistics (2010)
16. Tsang, S., Koh, Y.S., Dobbie, G., Alam, S.: Detecting online auction shilling frauds using supervised learning. Expert Systems with Applications 41(6), 3027–3040 (2014)
17. Viswanath, B., Post, A., Gummadi, K.P., Mislove, A.: An analysis of social network-based sybil defenses. SIGCOMM Comput. Commun. Rev. 40(4), 363–374 (2010)
18. Vlasselaer, V.V., Akoglu, L., Eliassi-Rad, T., Snoeck, M., Baesens, B.: Guilt-by-constellation: fraud detection by suspicious clique memberships. In: 2015 48th Hawaii International Conference on System Sciences (HICSS), pp. 918–927 (2015)
19. Yoshida, T., Ohwada, H.: Shill bidder detection for online auctions. In: Zhang, B.-T., Orgun, M.A. (eds.) PRICAI 2010. LNCS, vol. 6230, pp. 351–358. Springer, Heidelberg (2010)

Watch-It-Next: A Contextual TV Recommendation System

Michal Aharon[1], Eshcar Hillel[1], Amit Kagian[1], Ronny Lempel[2],
Hayim Makabee[3], and Raz Nissim[1]([⊠])

[1] Yahoo Labs, Haifa, Israel
{michala,eshcar,akagian,raz}@yahoo-inc.com
[2] Outbrain, Netanya, Israel
rlempel@gmail.com
[3] Pontis, Ra'anana, Israel
hayim.makabee@pontis.com

Abstract. As consumers of television are presented with a plethora of available programming, improving recommender systems in this domain is becoming increasingly important. Television sets, though, are often shared by multiple users whose tastes may greatly vary. Recommendation systems are challenged by this setting, since viewing data is typically collected and modeled per *device*, aggregating over its users and obscuring their individual tastes.

This paper tackles the challenge of TV recommendation, specifically aiming to provide recommendations for the next program to watch following the currently watched program the device. We present an empirical evaluation of several recommendation methods over large-scale, real-life TV viewership data. Our extentions of common state-of-the-art recommendation methods, exploiting the current watching context, demonstrate a significant improvement in recommendation quality.

1 Introduction

In recent years, online experiences have made increasing use of recommendation technology and user modeling techniques. Media sites recommend what to read, music streaming sites recommend what to listen to, VOD subscription services recommend what to watch, and more. In many of these instances, recommendation effectively transforms an online service to be personalized - tailored to its single or primary user. This occurs when recommendations are exposed in an experience consumed through a personal device such as a smartphone or a laptop, or when they are exposed on personal accounts such as a social network account. However, in other cases recommendation technology is applied in settings where multiple users share an application or device. Examples include game consoles or high end smart TVs in household living rooms, family accounts in VOD subscription services or eCommerce sites, and shared desktops or tablets in homes. These multi-user cases represent a challenge to recommender systems, as recommendations given to one user may actually be more suitable for another

© Springer International Publishing Switzerland 2015
A. Bifet et al. (Eds.): ECML PKDD 2015, Part III, LNAI 9286, pp. 180–195, 2015.
DOI: 10.1007/978-3-s319-23461-8_12

user. In general, the utility of recommendations in multi-user settings is lower, sometimes to the point of frustrating users when the recommendations they receive are dominated by items suitable for others.

This paper tackles the TV recommendation problem by applying context to implicitly disambiguate the user, or users, that are watching it. Specifically, we address the household smart TV situation, where a television set runs software that identifies what is watched on it, and taps that knowledge to recommend to its owners what they should watch at any given time. Obviously, individual household members may have very different viewing habits and tastes. Furthermore, when certain combinations of household members watch TV together, they may watch programs that differ from what any of them would have watched alone. Thus, every combination of viewers at any time may require different recommendations.

A key observation we tap is that often viewers watch multiple programs in succession, in what we call a TV *viewing session*. Thus, a given watched show influences our prior belief over the show that might be watched next. The data at our disposal does not identify which household members were watching television at any point in time, hence we measure the accuracy of predicting the next watched show – as is common in offline evaluation of recommendation algorithms, we associate improved prediction with better recommendation.

Our methods decouple the recommendation from the learning algorithm that produces the model. Essentially, we *exploit models produced by context-less learning* in order to produce contextual recommendations. This allows our recommender to easily work with any learning component that maps users and items to low-dimensional vectors, and to benefit from any optimization improving this component. Specifically, trained models leverage the currently watched show as context for recommending the next show to watch, *without changing how modeling is performed*. To determine the affinity of the individual(s) watching the device and a show, we perform a simple *3-way product* of the vectors representing the device, the currently watched show (context), and the shows to potentially watch next (items). The context enables the algorithm to recommend significantly more relevant watching options than those output by state-of-the-art context-less recommenders.

We implemented a recommender system, *WatchItNext*, that demonstrates the usefulness of our methods on top of two learning models: Latent Factor Modeling and Latent Dirichlet Allocation. We provide an empirical analysis of *WatchItNext* on two recommendation scenarios: (1) *exploratory setting* focusing on recommending items which have yet to be consumed by a user (e.g., for a recommended TV series, no episode has been previously watched[1]), and (2) *habitual setting* where recommendations also include items that were previously watched on the device (such as previous episodes of a series). Using large-scale real-life offline viewing data, which processed 4-months long viewing histories of 340,000 devices, we demonstrate improvements over these two TV recommendation scenarios. Our personalized and contextual recommendation methods significantly

---

[1] We model viewing patterns at the series level, not at the episode level.

outperform personalized non-contextual as well as contextual non-personalized baselines.

The rest of this paper is organized as follows. Section 2 surveys background and related work. Section 3 presents the algorithms used to tackle the problem. Section 4 details the experimental setup and Section 5 reports our experimental results. We conclude in Section 6.

2   Background

This section describes two methods that serve as building blocks in our algorithms section, followed by a review of previous work related to the TV and in general, the multi-user device recommendation problem.

2.1   Building Blocks

Latent Factor Model (LFM). Collaborative Filtering methods analyze relationships and interactions between users and items according to consumption patterns or user ratings. Latent Factor Modeling is a common approach in this field [19]. Following the assumption that the underlying behavioral pattern is low-rank, LFM approximates a user-item relationship matrix as a multiplication of two low rank matrices, one representing the users and the other representing the items. Therefore, a (latent) vector $\bar{u} \in \mathbb{R}^n$ is assigned for each user, and another vector $\bar{i} \in \mathbb{R}^n$ for each item. The recommendation score for user u and item i is computed by the inner product of the corresponding vectors, $\langle \bar{u}, \bar{i} \rangle$. Often, the latent vectors are modeled using an iterative optimization process, such as gradient descent.

Latent Dirichlet Allocation (LDA). *Latent Dirichlet Allocation* is a generative probabilistic model of a corpus, originally proposed for modeling text documents [9]. The intuition behind this model is that documents exhibit multiple topics, which are distributions over a fixed vocabulary W. The LDA algorithm performs data analysis to compute an approximation of the conditional distributions of the hidden variables (*topics*), given the observed variables (words of the documents). In its standard application, LDA gets as input a set of documents, along with their bag-of-words representation, often represented as a sparse matrix in which the (i, d) entry is the number of times the word i appeared in the document d. It produces a set of n topics, where each topic is a multinomial distribution vector in $R^{|W|}$ representing the probability of each word to appear in a document discussing that topic. The documents' latent topics can later be inferred according to the words they contain.

2.2   Related Work

Standard recommender systems log viewing history per device. A possible approach to enhance personalization in multi-user devices is using *context-aware* and in particular *time-aware* recommender systems. In the smart TV use case, this is backed by the assumption that each member of the household has regular time slots in which she watches TV. For example, children are more likely to watch TV in the afternoon than late at night.

We review the general context-aware approach, existing video and TV content recommendation systems, and some papers that handle the multi-user problem in other domains, and compare these solutions with ours.

Context-Aware Recommendations. Context-Aware Recommender Systems (CARS) exploit information on the situation in which the user is interacting with the system to improve rating predictions and provide more relevant personalized recommendations. Adomavicius et al. [1,2] distinguish three main types of CARS: (1) those based on *pre-filtering*, which select the data that will be modeled according to the target recommendation context, (2) those based on *post-filtering*, which adjust the recommendation, e.g., by filtering or weighting the recommendations according to the target context, and (3) those based on *contextual modeling*, which incorporate contextual information into the model itself. Both pre- and post-filtering have the advantage of using any non-contextual algorithm while providing context-aware recommendations.

Pre-filtering techniques such as *item splitting* [5] and *micro-profiling* [3,21] generate contextual models based only on the relevant items or users data, which may be sparse. These techniques also require knowing the explicit context, while in some cases the context can only be derived implicitly.

Many works devise new contextual modeling algorithms. Some are for general purpose, such as *tensor factorization* [17,22], factorization machines [20], and context-aware factorization [4,14,16]. Others specifically aim at improving music recommendations [13] by extending the LDA model, improving Web search ranking via incorporating context into factorized models [23,25] or learning a variable length Hidden Markov Model [10]. These algorithms are often complex, introduce a large number of model parameters to be learned, and are mostly not available as off-the-shelf solutions.

Our work adopts a two-phase post-filtering approach. We compute an initial score while considering the current item context, and then weigh this score using the time of day as an additional context. The advantage of our approach is that it allows us to leverage available contextual information while using a standard non-contextual model.

Time-aware recommender system utilize the time as the main context of the recommendation. The main disadvantage of this approach is that it delineates a *static* mapping between users or groups of users, which we refer to as *entities*, and time slots. This is often not a valid assumption, as people may switch preferred times for watching TV. Our proposed framework supports a *dynamic* approach, which – in addition to the time of day – also relies on the currently watched show as context for recommending the subsequent show to watch. Our main assumption is that the currently watched show can implicitly serve as a dynamic entity-proxy. For example, consider a morning in a household where either a child or an adult is watching the TV. Whether the current show is *"Good Morning America"* or *"Dora the Explorer"* should affect the recommendation of what to watch next.

Xu et al. [26] uses LDA topic modeling in order to extract the browsing context of users. Their recommendation algorithm differs from ours as it only considers the similarity between the active browsing session (item context) and browsing access patterns (items topics) without considering the model of the active user.

Video and TV Recommendations. Video and TV content recommender systems tackle the multi-user problem in different ways. Some systems, like [6] ignore this problem, others [15,27] recognize that multi-user devices present a challenge for personalized recommendations, but do not try to address this issue. Netflix [11] and Android [18] suggest actively switching between users of a single device, other TV content recommnders [7,30] also ask users to log-in and collect their explicit feedback. Although it is technically possible for users to identify themselves by signing into the device, very few users currently log-in while watching TV.

Some recommendation systems for interactive TV platforms [29] utilize time as context by performing tensor factorization on user-item-context tensor. In contrast, our algorithm only trains simpler, non-contextual models, and applies both temporal and sequential context when serving the recommendation.

YouTube also recomends users which video to watch next [8,12]. The problem of video recommendation, however, is different as the inventory is not limited, and more importantly the input is less noisy, since users must be active in choosing the video, and the system can leverage explicit feedback by the users.

Recommending to a Multi-User Device. Television sets are a prime example of multi-user devices, however this setting is tackled also in other domains.

The work of Zhang et al. [28] focuses on the identification of users sharing an account. They developed a model of composite accounts as a union of linear subspaces, and then applied linear subspace clustering algorithms to identify the users. They also defined a statistical measure of the *compositeness* of an account, which is similar to our concept of device entropy (see Section 5.1). While their model is built on top of explicit ratings provided by users, our work is based on implicit signals.

A recent work by White et al. [24] explores personalized web-search on shared devices. They present methods for identifying shared devices, estimating the number of searchers on each device, and attributing new queries to individual searchers. The estimated number is used to guide a *k-means clustering* in segregating the device search log into logs of k searchers. A new query is assigned to a cluster by applying a similarity score that is based on features such as topic, time, and query length.

Table 1. Recommenders' Descriptions

Recommender	Description
GeneralPop	General popularity of item i
TemporalPop	Popularity of i at time-of-day t
SequentialPop	Popularity of i watched after c
DevicePop	Popularity of i within device d
DevicePop + X	*DevicePop* combined with a recommender X
LFM	Latent Factor Model with stochastic gradient descent
LDA	Latent Dirichlet Allocation applied as an LFM recommender
SequentialLFM/LDA	*LFM/LDA* with sequential context
TemporalLFM/LDA	*LFM/LDA* with temporal context
TempSeqLFM/LDA	*LFM/LDA* with both sequential & temporal contexts

3 Algorithms

This section describes the recommendation methods that were employed and compared in this work. We start by presenting in Section 3.1 several simple memory-based approaches that are popularity-oriented. Each of these approaches focuses on a different aspect of the data, such as the time-of-day, the current item being watched, and the history of a given device. The objective of these memory-based recommenders is to examine the potential power of various signals by putting them to use with basic and simple approaches. These methods serve as baselines in our evaluation section. We continue by describing in Section 3.2 how two collaborative-filtering methods are employed in order to produce personalized recommendations: one is a *Latent Factor Model* (LFM) and the other harnesses the *Latent Dirichlet Allocation* (LDA) algorithm. Finally, we describe in Section 3.3 how the contexts of the currently watched item and the time of day context are incorporated into these personalized methods. The way we utilize the context of the currently watched item is a main contribution of this work. It explains how we extend the standard factorization inner product into a *3-way* score calculation that exploits the available knowledge of what is being watched on top of a conventionally trained non-contextual model. Table 1 summarizes the recommenders we experimented with including their high level description.

3.1 Memory-Based Popularity Baselines

General Popularity. The General Popularity Recommender (*GeneralPop*) is a simple memory based method. This recommender takes the portion of devices that watched an item i to be i's score for all devices, which makes it a non-personalized recommender. The general popularity score for an item i is:

$$GeneralPop(i) = (\#devices\ that\ watched\ i)/\#devices$$

Temporal Popularity. A natural refinement to *GeneralPop* is considering the time when the recommendations are to be consumed. As different items are popular at different times of day, refining the popularity measurement to consider the time of recommendation is expected to improve the relevance of the recommended items. The Temporal Popularity Recommender (*TemporalPop*) measures the item's popularity in a specific time-of-day t, relative to its general popularity. It is a non-personalized recommender that for all devices calculates the score of item i at time t as:

$$TemporalPop(i,t) = Popularity(i|t)/GeneralPop(i),$$

where $Popularity(i|t) = \frac{\#devices\ that\ watched\ i\ @time(t)}{\#active\ devices\ @time(t)}$, and the $time(t)$ is an aggregation of a one-hour-granularity time slot (e.g. 8:00-9:00) over all days in the training data.

Sequential Popularity. The main contextual aspect of this work is what to watch next after the current show. The Sequential Popularity Recommender (*SequentialPop*) scores items according to their conditional popularity of being watched sequentially after a specific item. Given the currently watched item c, the score of an item i to be watched next, for any device is:

$$SequentialPop(i|c) = \frac{\#times\ i\ was\ watched\ after\ c}{\#times\ item\ c\ was\ watched}$$

This method is non-personalized as the score reflects the popularity of watching these two items in an ordered sequence independently of a specific device. We also experimented with a personalized version of *SequentialPop*, that only counts the number of times a given device d watched the two items consecutively, but found the training data to be too sparse.

An additional natural baseline is a recommender that simply recommends the next show on the same channel. However, since our data lacks the channel information, we cannot determine whether the user actively changed the channel or stayed on the same channel. The *SequentialPop* recommender is the closest approximation of this baseline, as a high probability of two shows being watched consecutively is a good indication of them airing on the same channel.

Device Popularity. The Device Popularity Recommender (*DevicePop*) relies on the assumption that users re-watch items they already watched in the past. As such, it fits well to the *habitual* setting that include recommendations of items previously watched by a given device. *DevicePop* is a personalized memory based method in which the score for device d and item i is:

$$DevicePop(d,i) = \#times\ device\ d\ watched\ item\ i$$

The weak spot of *DevicePop* is on items that were seldom or never watched before on a given device d, as *DevicePop* will induce score ties between large sets of such items. In order to overcome *DevicePop*'s inability to differentiate between such items, we often combined it with an additional recommender by adding their scores together. We denote such combinations as *DevicePop + X*, where X is the combined recommender. Note that *DevicePop* is clearly not suitable for the *exploratory* setting and was not tested in it.

3.2 Collaborative Filtering Methods

As basic personalized recommenders, we use two methods that rely on latent factors or topics: the first is a Latent Factor Model and the second is an application of Latent Dirichlet Allocation. Both methods output two matrices as their resulting models: a $|\mathcal{D}| \times n$ matrix $\mathcal{M}^{\mathcal{D}}$ and an $n \times |\mathcal{I}|$ matrix $\mathcal{M}^{\mathcal{I}}$, where $\mathcal{D}$ and $\mathcal{I}$ are the sets of all devices and items respectively. Each row of $\mathcal{M}^{\mathcal{D}}$ and each column of $\mathcal{M}^{\mathcal{I}}$ are vectors corresponding to a device and an item respectively. We denote the matrix vectors corresponding to device d and item i as $\mathcal{M}_d^{\mathcal{D}}$ (row) and $\mathcal{M}_i^{\mathcal{I}}$ (column) respectively. n is the selected latent dimension that represents n latent factors or topics. In our experiments we found the value $n=80$ produces the best results. The standard LFM inner product of $\mathcal{M}_d^{\mathcal{D}}$ and $\mathcal{M}_i^{\mathcal{I}}$ reflects the affinity between d and i. Formally, if $\bar{d} = \mathcal{M}_d^{\mathcal{D}}$ and $\bar{i} = \mathcal{M}_i^{\mathcal{I}}$, the recommendation score is calculated as:

$$R(d,i) = \langle \bar{d}, \bar{i} \rangle = \sum_{k=1}^{n} \bar{d}_k \cdot \bar{i}_k \qquad (1)$$

LFM with Stochastic Gradient Descent. As a state-of-the-art recommendation method we used LFM optimized using *Stochastic Gradient Descent*. The cost function we used for optimization is a log-sigmoid function that penalizes watched items with a low score and non-watched items with a high score:

$$cost(d,i) = \begin{cases} -log(Sig(R(d,i))), & \text{if } d \text{ watched } i, \\ -log(1 - Sig(R(d,i))), & \text{otherwise,} \end{cases}$$

where $Sig(x) = \frac{1}{1+e^{-x}}$ and $R(d,i)$ is the score depicted in Equation 1. To avoid overfitting, we used early-stop validation to determine the number of training iterations. This approach is denoted as *LFM* in the rest of the paper.

LDA as a Collaborative Filtering Recommender. To apply LDA on TV data, every device in the data is considered to be an input document for LDA and every item watched by that device is considered to be a word in that document. Multiple watches of the same item by a device correspond to a word appearing multiple times in the document. LDA's assumption that documents belong to multiple topics fits our case well – often multiple users, possibly with varying tastes, share the same TV set. Different combinations of the viewing users may have different "topics" of interest, or "tastes", that describe a given device. LDA models each device as a mixture of topics that relate to the combinations of entities that share the device.

We used an existing LDA implementation[2] to produce the $\mathcal{M}^{\mathcal{I}}$ matrix from which we inferred the $\mathcal{M}^{\mathcal{D}}$ matrix. Each row in $\mathcal{M}^{\mathcal{D}}$ corresponds to the probabilities of a given device to watch each of the n latent topics and each column in $\mathcal{M}^{\mathcal{I}}$ corresponds to the probabilities of a given topic to generate a view of each of the items in the data. In other words, $\mathcal{M}^{\mathcal{D}}_{d,k} = P(k|d)$ and $\mathcal{M}^{\mathcal{I}}_{i,k} = P(i|k)$. Thus, applying Equation 1 on $\bar{d} = \mathcal{M}^{\mathcal{D}}_d$ and $\bar{i} = \mathcal{M}^{\mathcal{I}}_i$ provides an estimation of the probability that item i will be watched on device d:

$$R(d,i) = \sum_{k=1}^n \bar{d}_k \cdot \bar{i}_k = \sum_{k=1}^n P(i|k) \cdot P(k|d) = P(i|d)$$

3.3   Contextual Personalization

Personalization Using Sequential Context. Equation 1 considers only the device d and the item i. Given that we also know the context of a currently watched item c, we can combine it into the recommendation score. Our assumption is that $\mathcal{M}^{\mathcal{I}}_c$, the latent representation of c, can provide information regarding the "taste" of the entity currently watching it. Intuitively, promoting agreement with c may refine the reliance on the latent representation of the device, which encapsulates together the "tastes" of multiple viewers. We thus extend the standard inner product calculation into a *3-way* calculation that takes c into account. This contextual recommendation score is the sum of the triple element-wise product of the vectors $\bar{d} = \mathcal{M}^{\mathcal{D}}_d$, $\bar{c} = \mathcal{M}^{\mathcal{I}}_c$, and $\bar{i} = \mathcal{M}^{\mathcal{I}}_i$:

$$R(d,i,c) = \sum_{k=1}^n \bar{d}_k \cdot \bar{c}_k \cdot \bar{i}_k \qquad (2)$$

As *LFM* model values are often negative, we normalized *LFM*'s model $\mathcal{M}^{\mathcal{I}}$ by adding the absolute value of $\mathcal{M}^{\mathcal{I}}$'s minimal entry to all entries, resulting with all non-negative item factors. The $\mathcal{M}^{\mathcal{D}}$ model is unaffected and may contain negative values. This normalization makes sure that for a *negative* user factor $\bar{d}_k$, the final score will decrease as the item factors $\bar{c}_k$ and $\bar{i}_k$ increase and vise versa. Without this normalization, the final score would have decreased when all three factors *agree* with negative values (multiplication of 3 negative values). Note that for using Equation 2 there is no need to change the training procedure or the existing models $\mathcal{M}^{\mathcal{D}}$ and $\mathcal{M}^{\mathcal{I}}$. The modification is only in how the recommendation score is computed. We denote applying the sequential context of the currently watched item on *LFM* and *LDA* as *SequentialLFM* and *SequentialLDA*.

Personalization Using Temporal Context. To apply temporal context into our personalized recommenders we take a post-filtering contextual approach. Basically, we combine a given recommender with the temporal context by multiplying the recommendation score $R(d,i)$ with the *TemporalPop* score of item

---

[2] LDA package by Daichi Mochihashi, NTT Communication Science Laboratories, *http://chasen.org/%7Edaiti-m/dist/lda/*

i at time t. Thus, we promote items in times of the day when they are more popular while maintaining the personalized aspect. In case $R(d, i) < 0$ we divide $R(d, i)$ by *TemporalPop* so that a high *TemporalPop* score will improve the recommendation score by making it "less negative". For example, for *LFM*:

$$TemporalLFM(d, i, t) = Temporal(R(d, i), i, t) = \ldots \tag{3}$$

$$= \begin{cases} R(d, i) \times TemporalPop(i, t), & \text{if } R(d, i) \geq 0, \\ R(d, i) \div TemporalPop(i, t), & \text{otherwise.} \end{cases}$$

where $R(d, i)$ is the non-contextual *LFM* score. We denote the *LFM* and *LDA* recommenders, combined with a temporal context as *TemporalLFM* and *TemporalLDA*.

Temporal and Sequential Context Composition. It is simple to combine the temporal context on top of a sequential context recommender. For example, applying both on the *LFM* recommender is a composition of the temporal context function over the *SequentialLFM* score:

$$R(d, i, c, t) = Temporal(SequentialLFM(d, i, c), i, t)$$

in accordance with Equations 2 and 3. The composition of temporal context over *SequentialLFM* and *SequentialLDA* is denoted as *TempSeqLFM* and *TempSeqLDA*, respectively.

4   Experimental Settings

4.1   The Data

Our analysis is based on broadcast viewership data that is collected from smart TV devices in households within the United States. The raw data is comprised of a set of device id, item id, timestamp triples. The item id uniquely identifies a series, not a specific episode. Neither the identity of the individual watching the show nor the channel on which the show is being broadcast are available to us.

Data was collected for 340,000 devices, which watched 19,500 unique items over a period of 4 months in early 2013 (exact numbers are reported in Table 2). We consider the first 3 months as training data – composed of more than 19 millions device-item pairs, while the last month is used as a test set. In the *habitual* setting, the test set includes all instances of consecutively watched items – a total of 3.8 million item pairs. The *exploratory* setting considers only consecutive items whose latter item was not watched by the device in the training data. There were 1.7 million consecutive item pairs in the test set of this setting.

Table 2. Data set numbers

	Total	Train	Test
Months	4	3	1
Devices	339,647	339,647	311,964
Items	19,546	17,232	11,640

4.2   Emulating Inventories

A TV recommender system ranks items from an inventory of shows available in a given context. However, the actual inventory available at each household depends on its location (for local channels), provider (cable, satellite), premium channel selections, subscriptions, and more. Our data lacks this information. We therefore need to emulate the unknown inventories.

In the evaluation stage, we go over pairs of items watched on a device in succession in the test set. Given a device d, a pair of consecutive items $\langle c, i \rangle$ watched by d in the test set, and the time t, (when the device finished watching c and started watching i), we emulate $\mathcal{I}_{c,t}$, the inventory of items available for watching after c at time t, as the set of all shows j that were watched by some device while d watched i. In addition, we require that for every show $j \in \mathcal{I}_{c,t}$, the pair $\langle c, j \rangle$ has been watched consecutively by some device in the *entire* data set (train and test portions). This procedure approximates the real TV lineup at time t on devices where show c is available. In the *habitual* setting, this procedure results in inventories comprising of 390 items on average. In addition, in the *exploratory* setting, items watched by the device in the training data are removed from the inventory for this specific device. This reduced the average size of inventories to 345 items.

4.3   Evaluation Metric

The metric used in our empirical evaluation is the *Average Rank Percentile* (ARP) metric. ARP measures how high (on average) the show actually watched next by the device was ranked by the recommender. Formally, given consecutive items $\langle c, i \rangle$ watched at time t by device d, we generate the inventory $\mathcal{I}_{c,t}$ and then rank all items $j \in \mathcal{I}_{c,t}$ using the model scoring function $R(d, j, c, t)$. The *rank percentile* is computed as $(|\mathcal{I}_{c,t}| - r(i) + 1)/|\mathcal{I}_{c,t}|$, where $r(i)$ is i's rank in the output of the model. ARP is the average rank percentile over all pairs of consecutive items.

Since each recommendation instance ranks an inventory containing exactly one "correct" answer (the show actually watched next), the per instance rank percentile is identical to the well-known *Area Under the ROC Curve* (AUC) metric on the ranked inventory. ARP is also somewhat similar to the well-known *Mean Reciprocal Rank* (MRR) metric; however, MRR values across multiple instances are comparable only when the lengths of the ranked lists are fixed. Since inventory sizes vary across different sequential and temporal contexts, the

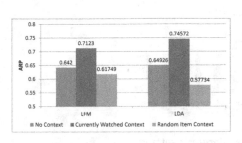

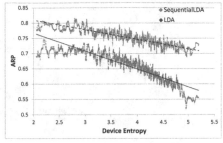

(a) Sequential context effect   (b) Profile diversity effect

Fig. 1. Attributing context to users (*exploratory* setting)

percentile in which the correct item is ranked is a better reflection of recommendation accuracy than its reciprocal rank.

From the ranked inventory list only k items are presented to the users. While in general these are the top-k items, the recommender system might apply a diversification policy or some filtering (like in the case the user asks only for a specific genre) that results in presenting items that are not at the top of the list. Therefore, we prefer the ARP metric for measuring the quality of the entire ranking over other popular metrics such as recall and precision @k that focus on the quality of its top.

5 Experimental Results

5.1 Attributing Context to Users

We begin by establishing the usefulness of the currently watched show as a context for recommending what to watch next. To this end, we compared several variations of *LFM* and *LDA* in sequential context (according to Equation 2): (a) using the currently watched item, (b) using a random item previously watched on the device, and (c) using no context. Figure 1a depicts the performance of these recommenders. The improvement in recommendation quality is clear when using the sequential context item, raising the ARP by almost 10% over *LDA*, and by almost 6% over *LFM*. Interestingly, using as a context an item randomly chosen from shows previously watched on the device performs worse than using no context at all. This demonstrates the importance of using the context item from the current viewing session and not an arbitrary item.

To further establish how accurately we attribute the context to a specific user we examine the performance of our recommenders as a function of the number of topics in the device, namely the *device topical entropy*. Intuitively, entropy gives an approximate measure of how diverse are the tastes of the users watching the device. Lacking the information on the number of users per device, we assume

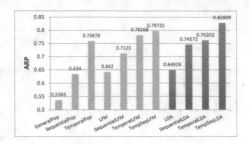

Fig. 2. ARP in the *exploratory* setting

a correlation between the taste diversity (number of topics) and the number of users sharing the device.

We use *LDA*'s model $\mathcal{M}^{\mathcal{D}}$ to compute the following entropy measure for a device d:

$$H_d = \sum_{0 \leq k \leq n} -P(k|d) \times \log P(k|d),$$

where $P(k|d) = \frac{\mathcal{M}^{\mathcal{D}}_{d,k}}{\sum_{0 \leq k' \leq n} \mathcal{M}^{\mathcal{D}}_{d,k'}}$. In other words, $P(k|d)$ is the probability that device d belongs to topic k. A zero entropy value means that the device's users span only a single topic with greater than 0 probability. On the other hand, a uniform distribution of viewed topics on a device maximizes the entropy. For example, when entropy is around 4.5, significant topic probabilities are distributed across more than $2^{4.5} = 22$ different topics. Figure 1b depicts the ARP of devices as a function of their topic distribution entropy. To reduce noise, the results are smoothed – each data point (x, y) represents a "sliding window", corresponding to the average ARP of 200 devices whose median entropy was x.

ARP results for non-contextual *LDA* demonstrate how hard the recommendation problem is across topical entropy ranges. Devices with low entropy values are more predictable and easier for recommendation generation. As entropy values rise, the taste variance (entropy) increases and the recommendation task becomes more difficult. As can be seen, *SequentialLDA* dominates *LDA* across all entropy ranges. In addition, as device entropy increases, the drop in performance is steeper for the non-contextual recommender. This implies that especially when device predictability is low (high entropy), relying on the currently watched item as context for recommendation increases precision. *SequentialLDA* maintains solid results (ARP > 0.7) even for the highest entropy values, where non-contextual *LDA* degenerates almost to a random recommender (ARP of ∼ 0.5). The results are shown only for *LDA*-based recommenders in the exploratory settings, but are consistent with their *LFM* and habitual settings counterparts.

5.2   The Exploratory Setting

In the *exploratory* setting, a recommender should rank items that were never watched on a given device. Figure 2 shows the performance of *LFM* and *LDA*

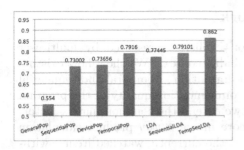

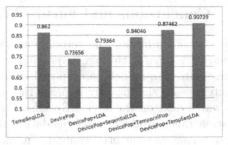

(a) Memory-based and LDA baselines with different contexts

(b) *DevicePop* combined with various recommenders

Fig. 3. ARP in the *habitual* setting

using different combinations of context, in comparison to the baseline recommenders (on the left-hand side).

One can notice that the performance of the *SequentialPop* baseline recommender are not so high. We conclude that a recommender that simply recommends to stay on the same channel will come up with similar results. Our non-contextual *LFM* and *LDA* recommenders are also comparable with *SequentialPop*. Sequential context is shown to improve performance for both *LFM* and *LDA*. Adding temporal context to non-contextual recommenders yields a greater improvement, surpassing the *TemporalPop* baseline. The clear winner in both cases is the recommender that combines both sequential and temporal context: *TempSeqLDA* displays a 27.5% ARP increase over the non-contextual *LDA*.

5.3   The Habitual Setting

Figure 3 displays ARP results for the *habitual* setting. Looking at Figure 3a, we find that adding sequential context to the *LDA* recommender gives a minor increase in accuracy, while adding both contexts (*TempSeqLDA*) gives a 11.3% increase over *LDA* and a 9% increase over *SequentialLDA*. In general, the *habitual* setting is considered to be easier as users exhibit repeated consumption patterns of TV shows. This can be observed in the relatively strong performance of non-contextual methods. Still, when considering these methods as baselines, the advantage of combining them with context is evident.

DevicePop is expected to be a solid recommender in the *habitual* setting. However, its main weakness is dealing with previously non-watched items. To counter this weakness, we experimented with combining *DevicePop* with other recommenders. Figure 3b shows results for four such combinations. Combining *DevicePop* with *LDA* and *SequentialLDA* increase ARP by 7.7% and 14% respectively. Combining *DevicePop* with *TemporalPop* yields a 18.7% increase in ARP, surpassing *TempSeqLDA*, which is given for reference on the left-most column. Finally, combining *DevicePop* with *TempSeqLDA* achieves an ARP > 0.9, which demonstrates the power of employing all contexts together.

6   Conclusions and Future Work

This work addresses the recommendation challenge presented by multi-user Smart TV devices through leveraging available context. We show how existing personalization models can tap real time information – temporal and sequential contexts – to significantly improve the recommendation quality of the program to watch next on the device. Specifically, we extend the common matrix factorization inner-product recommendation score into a *3-way* product calculation that considers sequential context – namely, the currently watched item – without changing the standard learning procedure or model.

Our experiments demonstrate that using context significantly improves recommendation accuracy, in both memory-based and collaborative filtering approaches; and that per context used, collaborative filtering schemes outperform the corresponding memory-based counterparts. The positive contribution of context is due to it "narrowing" the topical variety of the program to be watched next on the device. While tapping temporal context alone performs quite well in both the *exploratory* and *habitual* settings, enhancing it with sequential context results in significant additional accuracy gains. Finally, in the *habitual* setting, which is typical for TV, prediction accuracy improves when explicitly accounting for repeated item consumption patterns.

References

1. Adomavicius, G., Sankaranarayanan, R., Sen, S., Tuzhilin, A.: Incorporating contextual information in recommender systems using a multidimensional approach. ACM Transactions on Information Systems (TOIS) **23** (2005)
2. Adomavicius, G., Tuzhilin, A.: Context-aware recommender systems. In: Recommender Systems Handbook, pp. 217–253 (2011)
3. Baltrunas, L., Amatriain, X.: Towards time-dependant recommendation based on implicit feedback. In: CARS (2009)
4. Baltrunas, L., Ludwig, B., Ricci, F.: Matrix factorization techniques for context aware recommendation. In: RecSys, pp. 301–304 (2011)
5. Baltrunas, L., Ricci, F.: Context-based splitting of item ratings in collaborative filtering. In: RecSys, pp. 245–248 (2009)
6. Bambini, R., Cremonesi, P., Turrin, R.: A recommender system for an IPTV service provider: a real large-scale production environment. In: Recommender Systems Handbook, pp. 299–331 (2011)
7. Bellekens, P., Houben, G.-J., Aroyo, L., Schaap, K., Kaptein, A.: User model elicitation and enrichment for context-sensitive personalization in a multiplatform tv environment. In: EuroITV, pp. 119–128 (2009)
8. Bendersky, M., Pueyo, L.G., Harmsen, J.J., Josifovski, V., Lepikhin, D.: Up next: retrieval methods for large scale related video suggestion. In: KDD, pp. 1769–1778 (2014)
9. Blei, D.M., Ng, A.Y., Jordan, M.I.: Latent Dirichlet allocation. Journal of Machine Learning Research **3**, 993–1022 (2003)
10. Cao, H., Jiang, D., Pei, J., Chen, E., Li, H.: Towards context-aware search by learning a very large variable length hidden markov model from search logs. In: WWW, pp. 191–200 (2009)

11. Chaey, C.: (2013). http://www.fastcompany.com/3015138/fast-feed/now-you-can-have-multiple-user-profiles-on-one-netflix-account
12. Davidson, J., Liebald, B., Liu, J., Nandy, P., Vleet, T.V.: The youtube video recommendation system. In: RecSys, pp. 293–296 (2010)
13. Hariri, N., Mobasher, B., Burke, R.: Query-driven context aware recommendation. In: RecSys (2013)
14. Hidasi, B., Tikk, D.: Fast ALS-based tensor factorization for context-aware recommendation from implicit feedback. In: Flach, P.A., De Bie, T., Cristianini, N. (eds.) ECML PKDD 2012, Part II. LNCS, vol. 7524, pp. 67–82. Springer, Heidelberg (2012)
15. Hu, Y., Koren, Y., Volinsky, C.: Collaborative filtering for implicit feedback datasets. In: ICDM, pp. 263–272 (2008)
16. Karatzoglou, A.: Collaborative temporal order modeling. In: RecSys, pp. 313–316 (2011)
17. Karatzoglou, A., Amatriain, X., Baltrunas, L., Oliver, N.: Multiverse recommendation: n-dimensional tensor factorization for context-aware collaborative filtering. In: RecSys, pp. 79–86 (2010)
18. Kleinman, J.: (2013). http://www.technobuffalo.com/2013/09/16/android-4-2-multi-user-support-coming-to-some-samsung-tablets/
19. Koren, Y., Bell, R., Volinsky, C.: Matrix factorization techniques for recommender systems. Computer 42(8), 30–37 (2009)
20. Rendle, S., Gantner, Z., Freudenthaler, C., Schmidt-Thieme, L.: Fast context-aware recommendations with factorization machines. In: SIGIR, pp. 635–644 (2011)
21. Said, A., Luca, E.W.D., Albayrak, S.: Inferring contextual user profiles: improving recommender performance. In: RecSys Workshop on Context-Aware Recommender Systems (2011)
22. Shi, Y., Karatzoglou, A., Baltrunas, L., Larson, M., Hanjalic, A., Oliver, N.: TFMAP: optimizing map for top-n context-aware recommendation. In: SIGIR, pp. 155–164 (2012)
23. Weston, J., Wang, C., Weiss, R.J., Berenzweig, A.: Latent collaborative retrieval. In: ICML (2012)
24. White, R.W., Hassan, A., Singla, A., Horvitz, E.: From devices to people: attribution of search activity in multi-user settings. In: WWW, pp. 431–442 (2014)
25. Xiang, B., Jiang, D., Pei, J., Sun, X., Chen, E., Li, H.: Context-aware ranking in web search. In: SIGIR, pp. 451–458 (2010)
26. Xu, G., Zhang, Y., Yi, X.: Modelling user behaviour for web recommendation using LDA model. WI-IAT 3 (2008)
27. Xu, M., Berkovsky, S., Ardon, S., Triukose, S., Mahanti, A., Koprinska, I.: Catch-up tv recommendations: show old favourites and find new ones. In: RecSys, pp. 285–294 (2013)
28. Zhang, A., Fawaz, N., Ioannidis, S., Montanari, A.: Guess who rated this movie: identifying users through subspace clustering. In: UAI, pp. 944–953 (2012)
29. Zibriczky, D., Hidasi, B., Petres, Z., Tikk, D.: Personalized recommendation of linear content on interactive TV platforms: beating the cold start and noisy implicit user feedback. In: Workshop in Conference on User Modeling, Adaptation, and Personalization (2012)
30. Zimmerman, J., Kurapati, K., Buczak, A.L., Schaffer, D., Gutta, S., Martino, J.: TV personalization system: design of a TV show recommender engine and interface. In: Personalized Digital Television: Targetting Programs to Individual Viewers, pp. 27–51 (2004)

Nectar Track

Bayesian Hypothesis Testing
in Machine Learning

Giorgio Corani[✉], Alessio Benavoli,
Francesca Mangili, and Marco Zaffalon

Istituto Dalle Molle di Studi sull'Intelligenza Artificiale (IDSIA),
USI - SUPSI, Manno, Switzerland
{giorgio,alessio,francesca,zaffalon}@idsia.ch

Abstract. Most hypothesis testing in machine learning is done using the frequentist null-hypothesis significance test, which has severe drawbacks. We review recent Bayesian tests which overcome the drawbacks of the frequentist ones.

Keywords: Bayesian hypothesis testing · Null hypothesis significance testing

1 Introduction

Hypothesis testing in machine learning (for instance to establish whether the performance of two algorithms is significantly different) is usually performed using frequentist tests.

The highly-cited tutorial by [1] makes some important points. It recommends non-parametric rather than parametric tests for comparing multiple classifiers on multiple data sets. The advantages of the non-parametric approach are that they do *not* average measures taken on different data sets; they do *not* assume normality; they are *robust* to outliers. In particular, [1] recommends the signed-rank test for the pairwise comparison of *two* classifier over multiple data sets and the Friedman test for the comparison of *multiple* classifiers over multiple data sets. Modern procedures for the multiple comparisons are discussed in [1,2]. They control the family-wise error rate (FWER) while providing more power than the traditional Bonferroni correction. Both [1,2] assume the post-hoc analysis of the Friedman test to be based on the mean-ranks test. When comparing two algorithms A and B, the statistic of the mean-ranks test is proportional to the difference between the average rank of A and B, $\overline{R}_A - \overline{R}_B$. In a recent note [3], we recommend instead to *avoid* the mean-ranks test, as both $\overline{R}_A$ and $\overline{R}_B$ depend on the performance of the other algorithms included in the original experiment. This can make the results non-repeatable. For instance the difference between A and B could turn out to be significant if the pool comprises algorithms C, D, E and not significant if the pool comprises algorithms F, G, H. We instead recommend using the sign-test or the Wilcoxon signed-rank test, whose outcome only depends on the performance of A and B.

© Springer International Publishing Switzerland 2015
A. Bifet et al. (Eds.): ECML PKDD 2015, Part III, LNAI 9286, pp. 199–202, 2015.
DOI: 10.1007/978-3-319-23461-8_13

However such tests are based on the frequentist framework of the null-hypothesis significance tests (NHST). The NHST controls the Type I error, namely the probability of rejecting the null hypothesis when it is true. When multiple comparisons are performed, the NHST approach prescribes to control the family-wise error rate, namely the probability of finding at least one Type I error among the null hypotheses which are rejected. Yet null hypothesis significance testing has severe drawbacks.

Consider analyzing a data set of n observations with a NHST test. The sampling distribution used to determine the critical value of the test assumes that your intention was to collect exactly n observations. If your intention was different (for instance in machine learning you typically compare two algorithms on all the data sets that are available) the sampling distribution should be changed to reflect your actual sampling intentions [4]. This is never done, given the difficulty of formalizing one's intention and of devising an appropriate sampling distribution. This problem is thus important but generally ignored.

NHST can reject the null hypothesis or fail to reject it, but it cannot verify the null hypothesis. In other words, it does not provide any measure of evidence *for* the null hypothesis. Within the NHST framework accepting the null hypothesis is a weak decision: it does not mean that the null hypothesis is true.

NHST decisions are taken on the basis of the chosen significance α, namely the probability of rejecting the null hypothesis when it is true. Usually one sets $\alpha=0.01$ or 0.05, without having the possibility of a sound trade-off between Type I and Type II errors.

Bayesian hypothesis tests overcome these issues. The computation does not depend on the intention of the person who collected the data. The Bayesian test returns the posterior probability of the null and the alternative hypotheses. This allows to take decision which minimize the posterior expected value of the loss (posterior risk). For instance [5] reviews how to obtain Bayes-optimal decisions for a variety of different loss functions.

In [6] we proposed a Bayesian counterpart of the signed-rank test, which is the recommended test for comparing the score of two classifiers on multiple data sets. To devise this non-parametric test we adopted the Dirichlet process, which is often used in Bayesian non-parametrics. By means of simulations on artificial and real world data, we use our test to decide if a certain classifier is significantly better than another. The Bayesian and the frequentist signed-rank ($\alpha=0.05$) take the same decisions only when we assume the Type I error to be 19 times more costly than the Type II error. In this case, the optimal decision is to declare that classifier Y is better than classifier X when the posterior probability of this hypothesis is greater than $1-\alpha = 0.95$. For any other different cost setting the frequentist test is tied to control the Type I error, fixing $\alpha = 0.05$. Instead the Bayesian decision rule allows to minimize the posterior risk. The rule for optimal decisions (accepting or rejecting the null hypothesis) is equivalent to that of cost-sensitive classification [7]. For any other setting of the costs, the Bayesian test incurs *lower* costs than the frequentist test.

Assume now that the two classifiers have been assessed via cross-validation on a collection of data sets $D = \{D_1, D_2, \ldots, D_q\}$. One has to decide if the difference of accuracy between the two classifiers on the multiple data sets of D is significant. The signed-rank test both in its frequentist and Bayesian variant considers only the mean difference of accuracy measured on each data set, ignoring the associated uncertainty of the cross-validation estimates obtained on each data set.

In [8] we propose a test which performs inference on multiple data sets accounting for the correlation and the uncertainty of the estimates yielded by cross-validation on each data set. Our solution is based on two steps. First we develop a Bayesian counterpart of the correlated frequentist t-test [9], which is the standard test for analyzing cross-validation results. Under a specific matching prior the inferences of the Bayesian correlated t-test and of the frequentist correlated t-test are numerically equivalent. The meaning of the inferences is however different. The inference of the frequentist test is a p-value; the inference of the Bayesian test is a posterior probability. The posterior probabilities computed on the individual data sets can be combined to make further Bayesian inference on multiple data sets.

After having computed the posterior probabilities on each individual data set through the correlated Bayesian t-test, we merge them to make inference on D. We model each data set as a Bernoulli trial (borrowing the intuition of [10]), whose possible outcomes are the win of the first or the second classifier. The probability of success of the Bernoulli trial corresponds to the posterior probability computed by the Bayesian correlated t-test on that data set. The number of data sets on which the first classifier is more accurate than the second is a random variable which follows a Poisson-binomial distribution. We use this distribution to make inference about the difference of accuracy of the two classifiers on D. We are unaware of other approaches able to compare cross-validated classifiers on multiple data sets, accounting for the correlation and the uncertainty of the cross-validation estimates.

When comparing multiple classifiers, the recommended frequentist approach is the Friedman test. If it rejects the null hypothesis, one runs a procedure for multiple comparisons. A problem also of the modern procedures for multiple comparisons [1,2] is that they simplistically treat the multiple comparisons as independent from each other. But when comparing algorithms $\{a, b, c\}$, the outcome of the comparisons (a,b), (a,c), (b,c) are *not* independent.

In [11] we devised a Bayesian non-parametric procedure for comparing multiple classifiers. Adopting again the Dirichlet process (DP) [12] as a model for the prior, we first devised a Bayesian Friedman test. Then we designed a *joint* procedure for the analysis of the multiple comparisons which accounts for their dependencies. We analyze the posterior probability computed through the Dirichlet process, identifying statements of *joint* comparisons which have high posterior probability. The proposed procedure is a compromise between controlling the FWER and performing no correction of the significance level for the multiple comparisons. Our Bayesian procedure produces more Type I errors but fewer

Type II errors than procedures which control the family-wise error. In fact, it does not aim at controlling the family-wise error. We show the effectiveness of this approach in a simulation of sequential model selection among a large number of candidates (*racing*). Our procedure yields superior results compared to the traditional frequentist procedure thanks to both ability to manage dependencies among the multiple comparisons and to recognize equivalent models, narrowing down the pool of competing models. To recognize that the models have equivalent performance corresponds to verify the null hypothesis, which is impossible within NHST.

2   Software

The software for all our methods is available from http://ipg.idsia.ch/software/.

References

1. Demšar, J.: Statistical comparisons of classifiers over multiple data sets. Journal of Machine Learning Research **7**, 1–30 (2006)
2. García, S., Herrera, F.: An extension on statistical comparisons of classifiers over multiple data sets for all pairwise comparisons. Journal of Machine Learning Research **9**, 2677–2694 (2008)
3. Benavoli, A., Corani, G., Mangili, F.: Should we really use post-hoc tests based on mean-ranks? Journal of Machine Learning Research (2015) (in press)
4. Kruschke, J.: Doing Bayesian data analysis: A tutorial introduction with R. Academic Press (2010)
5. Müller, P., Parmigiani, G., Robert, C., Rousseau, J.: Optimal sample size for multiple testing: the case of gene expression microarrays. Journal of the American Statistical Association **99**(468), 990–1001 (2004)
6. Benavoli, A., Mangili, F., Corani, G., Zaffalon, M., Ruggeri, F.: A Bayesian Wilcoxon signed-rank test based on the Dirichlet process. In: Proceedings of the 31st International Conference on Machine Learning (ICML 2014), pp. 1026–1034 (2014)
7. Elkan, C.: The foundations of cost-sensitive learning. In: International Joint Conference on Artificial Intelligence, vol. 17, pp. 973–978 (2001)
8. Corani, G., Benavoli, A.: A Bayesian approach for comparing cross-validated algorithms on multiple data sets. Machine Learning (2015) (in press)
9. Nadeau, C., Bengio, Y.: Inference for the generalization error. Machine Learning **52**(3), 239–281 (2003)
10. Lacoste, A., Laviolette, F., Marchand, M.: Bayesian comparison of machine learning algorithms on single and multiple datasets. In: Proc. of the Fifteenth Int. Conf. on Artificial Intelligence and Statistics (AISTATS 2012), pp. 665–675 (2012)
11. Benavoli, A., Mangili, F., Corani, G., Zaffalon, M.: A Bayesian nonparametric procedure for comparing algorithms. In: Proceedings of the 32nd International Conference on Machine Learning (ICML 2015) (2015) (in press)
12. Ferguson, T.S.: A Bayesian analysis of some nonparametric problems. The Annals of Statistics **1**(2), 209–230 (1973)

Data-Driven Exploration
of Real-Time Geospatial Text Streams

Harald Bosch[✉], Robert Krüger, and Dennis Thom

Institute for Visualization and Interactive Systems,
University of Stuttgart, Universitätsstr. 38, 70569 Stuttgart, Germany
{bosch,krueger,thom}@vis.uni-stuttgart.de

Abstract. Geolocated social media data streams are challenging data
sources due to volume, velocity, variety, and unorthodox vocabulary.
However, they also are an unrivaled source of eye-witness accounts to
establish remote situational awareness. In this paper we summarize some
of our approaches to separate relevant information from irrelevant chatter using unsupervised and supervised methods alike. This allows the
structuring of requested information as well as the incorporation of unexpected events into a common overview of the situation. A special focus
is put on the interplay of algorithms, visualization, and interaction.

Keywords: Stream processing · Machine learning · Social media

1 Introduction and Problem Definition

Social media data comprises highly relevant information about events such as natural and technological disasters, crimes, and infectious diseases (e.g., see [1,2]).
Separating event-related data from noise such as chatter and speculations, however, is a challenging task. In this work, we summarize the contributions of our
former publications [3–6] regarding the interplay of interactive visualization and
(un)supervised machine learning approaches for gaining insights into massive,
real-time, geolocated message streams. From these contributions, a decision support system was built for (1) keeping an overview over the current situation,
(2) organize massive data streams while incorporating unexpected events, and
(3) adapt the selectivity and orchestration of filters to the current situation.
We further discuss on the synergistic effects between machine learning, information visualization, and human-computer-interaction. The following primarily
addresses Twitter as a data source and public safety as the application domain
but the approaches are applicable to others streams and applications as well.

2 Stream-Enabled X-means Clustering for Textual Data

As a means to incorporate unexpected incidents into situation awareness, an
unsupervised approach is used to find potential real-world events by identifying

© Springer International Publishing Switzerland 2015
A. Bifet et al. (Eds.): ECML PKDD 2015, Part III, LNAI 9286, pp. 203–207, 2015.
DOI: 10.1007/978-3-319-23461-8_14

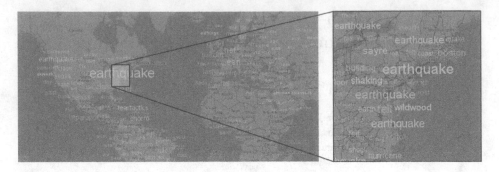

Fig. 1. Term usage clusters during a large earthquake in August 2011. The label size indicates the cluster's significance. Color indicates term relevance. By zooming in more space becomes available, clusters disaggregate and smaller clusters are shown.

accumulations of similar term usage in social media. The assumption is, that when people suddenly use similar words at the same location and the same time, they might have observed something unusual. We developed a stream-enabled variant of X-means to find such accumulations by clustering spatiotemporal coordinates for each observed term separately [3]. For each term a set of clusters is maintained. When a new message arrives, the cluster of each contained term is updated by adding the message coordinate to the nearest cluster. If the variance of the cluster exceeds a predefined threshold, it is split into two clusters. Because the temporal dimension increases monotonously, old clusters can become 'stale' when they will never be selected as the nearest cluster due to the large temporal difference. At this point, the stale cluster is evaluated for significance to dismiss cluster centroids that only cover sparse regions related to noise in the data. The criteria for significance are a low variance and the participation of multiple message authors as a means to exclude spamming users. This approach guarantees that spatiotemporally dense regions of similar term usage are covered by at least one, usually more cluster centroids and each centroid is related to a location instead of global chatter. The current state of the clustering is a list of locations (cluster centroid) for each term and a significance (the cluster's member count and variance) that can be used directly to place tags on a geographical map, using the significance for the font size (see Figure 1).

Overfitting as a Gift not a Curse. As indicated, our cluster splitting method tends to overfitting by creating too many centroids, especially if a cluster is spatially dense, but temporally elongated. This is an inherent problem of using a fixed variance threshold as splitting criterion because it cannot discriminate between, e.g., a sporting event in a stadium or a natural disaster covering a whole county. Luckily, the explorative data visualization can make use of this in order to offer a structural zooming interaction. In a global view all neighboring tags of a term can be aggregated into a larger tag by summing their significance, thus creating only one but larger tag. When users explore an area by zooming into it, the aggregation is less pronounced and reveals the geographical extent of

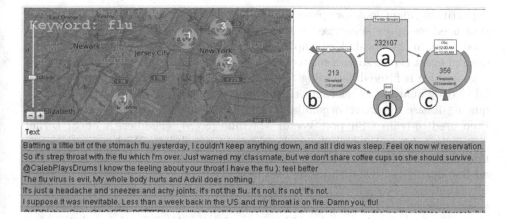

Fig. 2. New York influenza epidemic, January 2013. To filter the stream (a) for messages indicating illness we apply a pre-trained symptom classifier (b). Narrowing the results further with a term-based 'flu' filter (c) reveals all tweets indicating flu symptoms (d).

the event. The same benefit exists in the temporal domain. When the observed time frame is reduced, the amount of centroids used for creating tags is reduced alike.

Statistical Means for Relevance Indication. A problem for the clustering is, that there are 'stable' accumulations which are always dense enough to create significant centroids, most notably the names of touristic sights and cities. Social media authors will always use the terms 'London' and 'Eye' near the aptly named Ferris wheel. While being certainly significant accumulations, they are most often irrelevant due to missing novelty. Here, a temporal analysis can judge the novelty of a cluster. One approach takes the local history of the term usage and extracts the Seasonal and the Trend component using Loess (STL). If the remainder component, not explained by season or trend, is significantly higher than expected (z-score ≥ 2) the anomaly is considered novel [5]. Another approach is a pre-calculation of a smooth measure for location-dependent document densities for each term from a large collection of geolocated messages [4]. With this measure, the current term frequency can be contrasted with the expected value for this location. If deemed novel, the tag is highlighted in the tag map (Figure 1).

3   Orchestrating Filters Based on Classification

Unsupervised learning methods like the described clustering are powerful means to discover the unexpected. However, domain experts already have certain hypotheses or specific topics to search for. For example, a disaster prevention agency would be interested in catastrophe related information, while a health department might look for epidemics. To structure this a priori known information need we apply supervised learning algorithms [6]. Here, historical data is used to learn the vocabulary employed by social media authors to describe the desired event types. After a related event occurred, the constantly recorded message archive is filtered according to the location and time of the event and are labeled by analysts as relevant or irrelevant regarding the topic, e.g., an earthquake, disease outbreak, or similar. The labeled subset of past observations is then used to train a Support Vector Machine (SVM) classifier. A linear kernel is sufficient due to the high dimensionality of textual data. With a graph-based, interactive user interface the pre-trained classifiers can be loaded and combined ad-hoc during the live-monitoring of a new event to cover a multitude of possible situations (see Figure 2). Here, the combinations of classifiers can be used to assign messages to sets and track them throughout different views of the system to structure the analysis and avoid information overload. To this end, the classifiers' selectivity can be further adapted by the analysts through shifting the SVM decision boundary. This allows an ad-hoc trade-off between precision and recall.

4   Conclusion

Machine learning facilitates the exploration of massive spatiotemporal text streams. While unsupervised techniques are means for aggregated overviews and discovering the unexpected, supervised approaches can narrow down streams to specific topics. Interactive interfaces with suitable visualization can adjust classifications without opening the black-box and make these techniques accessible to domain experts without machine learning background. Moreover, interactive visual means can turn common drawbacks like overfitting and model inflexibility to strength—in our case by interactive structural zooming and model orchestration.

Acknowledgments. This work was supported by the BMBF project *VASA* project and the Horizon 2020 project *CIMPLEX*.

References

1. Sakaki, T., Okazaki, M., Matsuo, Y.: Earthquake shakes twitter users: Real-time event detection by social sensors. In: Int'l Conf. WWW, pp. 851–860 (2010)
2. Chew, C., Eysenbach, G.: Pandemics in the age of Twitter: content analysis of Tweets during the 2009 H1N1 outbreak. PLoS One 5(11) (2010)
3. Thom, D., Bosch, H., Koch, S., Wörner, M., Ertl, T.: Spatiotemporal Anomaly Detection through Visual Analysis of Geolocated Twitter Messages. In: IEEE Pacific Visualization Symposium, pp. 41–48 (2012)
4. Thom, D., Bosch, H., Ertl, T.: Inverse Document Density: A Smooth Measure for Location-Dependent Term Irregularities. In: COLING Conf., pp. 2603–2618 (2012)

5. Chae, J., Thom, D., Bosch, H., Jang, Y., Maciejewski, R., Ebert, D.S., Ertl, T.: Spatiotemporal Social Media Analytics for Abnormal Event Detection and Examination using Seasonal-Trend Decomposition. In: IEEE Conference on Visual Analytics Science and Technology, pp. 143–152 (2012)
6. Bosch, H., Thom, D., Heimerl, F., Püttmann, E., Koch, S., Krüger, R., Wörner, M., Ertl, T.: ScatterBlogs2: Real-Time Monitoring of Microblog Messages Through User-Guided Filtering. IEEE Trans. Vis. Comput. Graphics **19**(12), 2022–2031 (2013)

Discovering Neutrinos Through Data Analytics

Mathis Börner[✉], Wolfgang Rhode, Tim Ruhe,
for the IceCube Collaboration, and Katharina Morik

TU Dortmund University, Experimental Physics, Computer Science,
Otto-Hahn-Str. 12, 44227 Dortmund, Germany
{mathis.boerner,wolfgang.rhode,tim.ruhe,katharina.morik}@tu-dortmund.de
http://sfb876.tu-dortmund.de/SPP/sfb876-c3.html,
http://icecube.wisc.edu

Abstract. Astrophysical experiments produce Big Data which need efficient and effective data analytics. In this paper we present a general data analysis process which has been successfully applied to data from IceCube, a cubic kilometer neutrino detector located at the geographic South Pole.

The goal of the analysis is to separate neutrinos from atmospheric muons within the data to determine the muon neutrino energy spectrum. The presented process covers straight cuts, variable selection, classification, and unfolding. A major challenge in the separation is the unbalanced dataset. The expected signal to background ratio in the initial data (trigger level) is roughly $1:10^6$. The overall process was embedded in a multi-fold cross-validation to control its performance. A subsequent regularized unfolding yields the sought after neutrino energy spectrum.

Keywords: Neutrinos · IceCube · Machine learning · Random forest · Feature selection · Cross-validation · Signal and background separation

1 Introduction

IceCube is a neutrino detector located at the geographic South Pole with an instrumented volume of a cubic kilometer [1]. The detector consists of 86 strings at depths between 1450 m and 2450 m. Each string holds 60 digital optical modules (DOMs). The DOMs are designed to measure light and send a digitized signal to the surface. The purpose of the instrumentation is to measure Cherenkov light emitted by charged particles propagating through natural ice. The appearance of such a particle in the detector is referred to as an event. There are two types of events: events induced by neutrinos interacting in (or close to) the detector and events from muons which are produced in cosmic ray air showers in the atmosphere. Since only the neutrino spectrum is sought after the separation between atmospheric muon events and neutrino events is essential. Here, we summarize the a separation process implemented in RAPIDMINER [5] (based on [2]) and the subsequent unfolding.

This paper is based on work with the IceCube collaboration [3] and work in project C3 of the Collaborative Research Center SFB 876 which is funded by the DFG.

© Springer International Publishing Switzerland 2015
A. Bifet et al. (Eds.): ECML PKDD 2015, Part III, LNAI 9286, pp. 208–212, 2015.
DOI: 10.1007/978-3-319-23461-8_15

2   Selection of Neutrinos Events

Data for this analysis were taken, when the detector was under construction and consisted of 59 strings (IC59). The analysis faces two major challenges: In the initial data for each neutrino event 10^6 atmospheric muon events are detected and the rate of neutrinos decreases with the energy, proportional to $\sim E^{-3.7}$. The initial data rate is lowered, thereby complex reconstructions become feasible. This preselection also improves the signal to background ratio to roughly 1:1000.

The signature of atmospheric muons entering the detector shows no topological difference from an event induced by a muon neutrino. The approach is based on the fact that muons can only penetrate a few kilometers of ice while neutrinos can travel even through the earth's core. Hence, the presented approach only looks for events going upwards in the detector, towards the surface. Based on misreconstruction atmospheric muons dominate the data instead of muon neutrinos even for events with a reconstructed up-going track. To select only muon neutrino events a separation between well- and misreconstructed events needs to be conducted.

2.1   Data Preprocessing

The preprocessing consisted of two cuts. The first cut was applied on the reconstructed zenith angle to select up-going events. A second cut was applied on the *line fit* velocity [1] to reject spherical events, a topology that does not occur in high quality muon neutrino events. Both cuts were optimized simultaneously with respect to background rejection and signal efficiency. These two cuts rejected 91.4% of the background and retained 57.1% of the signal.

2.2   Variable Selection

Because not all variables are equally well suited for the event selection a representation in fewer dimensions needs to be found. Therefore, the Minimum Redundancy Maximum Relevance (MRMR) algorithm in the fast implementation of [6] was used for the selection of variables. Twenty-five variables were selected as this number shows a reasonable reduction of dimensions without losing too much information while showing stable behavior in the selection (detailed list of variables in [3]).

2.3   Performance of the Random Forest

From the machine learning point of view the event selection can be formalized in terms of a classification task with the classes *signal* (atmospheric neutrinos) and *background* (atmospheric muons). A Random Forest [8] was chosen as the machine learning algorithm. Training and testing were carried out in a standard five-fold cross-validation.

---

[1] Speed of the reconstructed event in the detector.

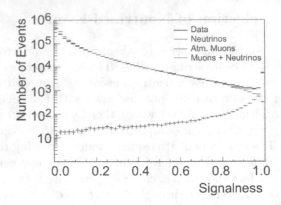

Fig. 1. Signalness (ratio of trees classifying an event as signal) distributions for data events (black) in comparison to the distributions of simulated events. Simulated signal events (Neutrinos) are depicted in blue, background events (atmospheric Muons) in magenta. The sum of the simulated events is shown in red.

The results of the Random Forest for simulated events and experimental data are shown in Fig. 1. For the analysis a strict cut of $S = 1$ was chosen. The good match between experimental data and simulated events indicates a stable performance of the forest. The errors for the distributions of the simulated events were obtained via cross-validation. The size of the errors is reasonably low and indicates a stable classification without any problems due to statistical fluctuations in the training events.

The purity of the final neutrino event sample was estimated to be $99.59^{+0.36}_{-0.37}\%$, while 18.2% of the signal is retained and 99.9999% of the background rejection is rejected.

3   Unfolding and the Resulting Energy Spectrum

The measurement of the neutrino energy is a so-called inverse problem. For this analysis the neutrino energy spectrum has to be reconstructed from measurements of the muons they induced. It can be expressed in an integral equation

$$g(y) = \int A(E,y)f(E)\,dE, \tag{1}$$

where $f(E)$ is the sought-after energy spectrum, $g(y)$ the distribution of measured variables and $A(E,y)$ a function describing the whole process from the production of the neutrino until the measurement in the detector.

To solve the integral equation, a regularized unfolding method was chosen (TRUEE [7]). The approach allows us to use up to three different variables. In this analysis the three variables were: the total amount of charge in the DOMs, number of unscattered photons and the length of the track from unscattered photons.

The resulting spectrum, related measurements, and theoretical predictions are shown in Fig. 2. It shows agreement both with related measurements and theoretical predictions.

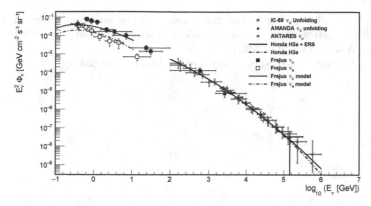

Fig. 2. The results of the analysis presented here are shown as red circles. Other measurements are depicted in black squares, hollow squares, black triangles, green triangles and blue. The curves shown originate from theoretical predictions. [3]

4   Summary and Results

This paper presents a data mining process that was successfully applied to and validated on data of the IceCube detector in its 59-string configuration. It was able to obtain 27771 atmospheric neutrino candidates in 346 days of IC59. The event selection method increased the neutrino rate from 49.3 neutrino events per day [4] to 80.3 neutrino events per day. The purity of the final neutrino sample was estimated to be $99.59^{+0.36}_{-0.37}\%$. The subsequent unfolding shows good agreement with prior measurements and extends the spectrum to energies never measured before.

References

1. Achterberg, A., et al.: First Year Performance of the IceCube Neutrino Telescope. Astroparticle Physics **26**, 155 (2006)
2. Ruhe, T., Morik, K., Rhode, W.: Application of RapidMiner in Neutrino Astronomy. In: Hofmann, M., Klinkenber, R. (eds.) RapidMiner: Data Mining Use Cases and Business Analytics Applications. CRC Press Book (2013)
3. Aartsen, M.G., et al.: Development of a general analysis and unfolding scheme and its application to measure the energy spectrum of atmospheric neutrinos with IceCube. The European Physical Journal C **3**, 75–116 (2015)
4. Abbasi, R., et al.: Measurement of the atmospheric neutrino energy spectrum from 100 GeV to 400 TeV with IceCube. Phys. Rev. D **83**(1), 012001(19) (2011)
5. Mierswa, I., Klinkenberg, R., Fischer, S., Ritthoff, O.: A Flexible Platform for Knowledge Discovery Experiments: YALE - Yet Another Learning Environment (2000)
6. Schowe, B., Morik, K.: Fast-ensembles of minimum redundancy feature selection. In: Okun, O., Valentini, G., Re, M. (eds.) Ensembles in Machine Learning Applications. SCI, vol. 373, pp. 75–95. Springer, Heidelberg (2011)

7. Mileke, N., et al.: Solving inverse problems with the unfolding program TRUEE: Examples in astroparticle physics. Nuclear Instruments and Methods in Physics Research A **697**, 133–147 (2013)
8. Breiman, L.: Random Forests. Mach. Learn. **45**(1), 5–32 (2001)

Listener-Aware Music Recommendation from Sensor and Social Media Data

Markus Schedl[(✉)]

Department of Computational Perception,
Johannes Kepler University, Linz, Austria
markus.schedl@jku.at
http://www.cp.jku.at

Abstract. Music recommender systems are lately seeing a sharp increase in popularity due to many novel commercial music streaming services. Most systems, however, do not decently take their listeners into account when recommending music items. In this note, we summarize our recent work and report our latest findings on the topics of tailoring music recommendations to individual listeners and to groups of listeners sharing certain characteristics. We focus on two tasks: *context-aware automatic playlist generation* (also known as serial recommendation) using sensor data and *music artist recommendation* using social media data.

1   Introduction

The importance of incorporating user characteristics and contextual aspects into recommender systems has been acknowledged many times [1,16,17]. Research that looks into this matter in the domain of music recommendation is scarce, though. Addressing this issue, we summarize our latest work on the tasks of (i) automatic music playlist generation incorporating contextual aspects of the listener and (ii) music artist recommendation tailored according to various user and listening characteristics. Both tasks are highly related to machine learning and data mining. In fact, we approach the former task by gathering a wide variety of listener-centric sensor data from a smart phone app and exploiting machine learning techniques to learn relationships between these features and music metadata (e.g. artist or track name). The latter task is related to data mining as we acquire and analyze huge amounts of listening events produced by users of social media and build recommendation algorithms that consider personal characteristics of the listeners and their listening behavior, also using novel features mined from user-generated data.

2   Automatic Playlist Adaptation Based on Sensor Data

Addressing the task of **automatic playlist generation**, we developed an *Android* app for smart devices, dubbed "Mobile Music Genius" (MMG) [6,8], which collects a variety of user-specific features during playback, ranging from time, location, and weather to ambient noise, light level, and motion. In addition, MMG gathers music metadata (artist, track, mood, and genre) and records

© Springer International Publishing Switzerland 2015
A. Bifet et al. (Eds.): ECML PKDD 2015, Part III, LNAI 9286, pp. 213–217, 2015.
DOI: 10.1007/978-3-s319-23461-8_16

player interaction (e.g. play, skip, pause events). Integrating a decision tree algorithm that is trained and retrained on the contextual features using track names as classes, MMG continuously monitors the context feature values of the listener and uses the classifier to suggest tracks suited to a given context, whenever the changes in context features exceed a threshold. These tracks are subsequently inserted into the playlist after the currently played one.

During a pilot study, we collected 7,628 data points (context features, music metadata, and interaction data) created by 48 students at JKU Linz. Based on this dataset, we investigated a variety of classifiers in cross-validation experiments for the task of **predicting music preferences from context features** [3]. Prediction was effected on four levels: artist, track, mood, and genre. Decision trees and random forests performed best for most prediction tasks, in particular for artist prediction (55% accuracy) and genre prediction (61% accuracy). We further analyzed which features are most important to predict contextual music preferences, and identified running tasks or apps, weather, time, and location, as well as general device properties as the most important ones for all four prediction levels [3]. For the prediction at track level, we found that player-related features, such as activation of repeat mode or the track belonging to a certain playlist, considerably contribute to a good performance, too.

3   Music Recommendation Based on Social Media Mining

In our research on music recommenders that are tailored to particular user characteristics, we consider two social media sources to acquire user and listening data: *Last.fm* and *Twitter*. While the former provides convenient API endpoints, we had to elaborate novel information extraction methods to identify listening events from streams of microblogs. In particular, we propose in [4] a hashtag-based filtering approach and a multi-stage rule-based method for matching microblog content and music metadata. Employing these methods yielded several datasets of *Twitter* users' listening events together with time stamps and spatial coordinates [4,7]. While it can be argued that exploiting social media data biases the results of recommendation experiments towards the user community of the respective platform, we show in [13] that the distribution of *Last.fm* listening events among major music genres is not far away from that reported in official music sales figures provided by the music industry. Exceptions are Classical music, which is underrepresented on *Last.fm* and even more on *Twitter*, and Metal music, which is overrepresented on *Last.fm* [5,13].

Based on the created social media datasets, in particular a set of 1 million geotagged listening events mined from *Twitter* (MMTD) [4] and a set of 200 million listening events mined from *Last.fm* (LFM-200m) [10], we model listeners by a variety of aspects and subsequently investigate the relationship between these user characteristics and the performance of stand-alone as well as hybrid recommendation algorithms [2,9,10]. We further propose new recommendation approaches for geotagged listening data [12,14].

Our overarching aim is to **identify the optimal music recommendation approach for a given listener** [9,10]. To this end, we categorize listeners according to age, gender, country, and genre preference [10], or novel listening-related features [2,9]. For each user group (e.g. listeners in a certain country), we then assess the performance of music artist recommenders, looking into several recommendation algorithms, their variants, and parameter settings (e.g. rating or score aggregation functions for nearest neighbor methods). Algorithms include user-based collaborative filtering (CF), popularity-based recommendation, and a CF algorithm extended by location-based filtering that we propose in [14]. In addition to these stand-alone recommenders, we create and evaluate hybrid systems, using different score normalization and fusion techniques. We investigate the performance of each algorithmic combination on each user category in cross-validation experiments using dataset LFM-200m. In addition to findings related to algorithmic tuning details, we made interesting observations specific to user categories: (i) recommendations are better when categorizing users according to age and country than according to gender or genre, (ii) younger people seem to be easier to satisfy by recommending overall popular music, whereas from age 41 upwards listeners tend to prefer CF-based recommendations, (iii) popularity-based recommendation works much better for women than for men, (iv) listeners in the investigated countries are served best by CF, except for Russia where mainstream recommendations seem to be more appreciated, and (v) Folk and Blues aficionados prefer our location-enhanced CF recommender.

In addition to the frequently exploited demographic features, we define a set of **novel user features that describe listening behavior**: play count, diversity, and mainstreaminess [2] as well as novelty [9]. While *play count* refers to features describing the overall amount of listening events of a user, *diversity* refers to the variety of genres listened to by a user, *mainstreaminess* is computed as the share of overall most popular music items among the user's listening events, and *novelty* as the share of music listened to for the first time by the user in a given time window. In an investigation similar to the one conducted in [10], we find on the MMTD dataset that all recommendation algorithms perform better in terms of F-measure for users with high play count, high diversity, and high mainstreaminess [2].

Comparing the results of our investigations of different user groups and recommendation algorithms [9,10] in terms of precision, recall, and F-measure, it turns out that performance varies strongly between user groups. Focusing on the recommendation algorithms, highest overall precision (6.35%) and F-measure (5.14%) is achieved by the combination of CF and popularity-based recommendation, while highest overall recall (13.85%) is achieved by further integrating content information based on similarity between collaborative music tags.

Our recent work further resulted in several **novel music recommendation approaches**. Exploiting GPS position information of listening events, we propose in [14] a recommendation algorithm that models each user's geospatial listening distribution via Gaussian mixture models and computes Earth mover's distances between these models, which are eventually integrated into a CF recommender. Another location-enhanced CF approach exploits geodesic distances between the users' spatial centroids of listening events, which are either linearly or exponentially weighted [12]. Eventually, we propose in [12] a hybrid music recommender that integrates our previous work on combining audio and web features for music similarity [11] and on rectifying similarity spaces to reduce hubs in recommendations [15].

4   Outlook

We are currently conducting experiments in which we consider a larger variety of user-specific factors, including features related to overall stylistic preferences, temporal aspects of music consumption, and openness to unknown music. We further plan to investigate the influence of high-level personal characteristics, such as affective states, personality traits, education, intelligence, and music sophistication, on music taste, and to exploit this knowledge to improve music recommendation algorithms.

Acknowledgments. This research is supported by the Austrian Science Fund (FWF): P25655 and by the EU-FP7 project no. 601166 ("PHENICX"). The author would further like to thank his colleagues and students who contributed to the work at hand.

References

1. Adomavicius, G., Tuzhilin, A.: Context-aware recommender systems. In: Recommender Systems Handbook, pp. 217–253. Springer (2011)
2. Farrahi, K., Schedl, M., Vall, A., Hauger, D., Tkalčič, M.: Impact of listening behavior on music recommendation. In: Proc. ISMIR (October 2014)
3. Gillhofer, M., Schedl, M.: Iron maiden while jogging, debussy for dinner? In: He, X., Luo, S., Tao, D., Xu, C., Yang, J., Hasan, M.A. (eds.) MMM 2015, Part II. LNCS, vol. 8936, pp. 380–391. Springer, Heidelberg (2015)
4. Hauger, D., Schedl, M., Košir, A., Tkalčič, M.: The million musical tweets dataset: what can we learn from microblogs. In: Proc. ISMIR (November 2013)
5. Lamere, P.: Social Tagging and Music Information Retrieval. New Music Research: Special Issue: From Genres to Tags - Music Information Retrieval in the Age of Social Tagging **37**(2), 101–114 (2008)
6. Schedl, M.: Ameliorating music recommendation: integrating music content, music context, and user context for improved music retrieval and recommendation. In: Proc. MoMM (December 2013)
7. Schedl, M.: Leveraging microblogs for spatiotemporal music information retrieval. In: Serdyukov, P., Braslavski, P., Kuznetsov, S.O., Kamps, J., Rüger, S., Agichtein, E., Segalovich, I., Yilmaz, E. (eds.) ECIR 2013. LNCS, vol. 7814, pp. 796–799. Springer, Heidelberg (2013)
8. Schedl, M., Breitschopf, G., Ionescu, B.: Mobile music genius: reggae at the beach, metal on a friday night? In: Proc. ACM ICMR (April 2014)
9. Schedl, M., Hauger, D.: Tailoring music recommendations to users by considering diversity, mainstreaminess, and novelty. In: Proc. ACM SIGIR (August 2015)
10. Schedl, M., Hauger, D., Farrahi, K., Tkalčič, M.: On the influence of user characteristics on music recommendation. In: Proc. ECIR (March-April 2015)
11. Schedl, M., Schnitzer, D.: Hybrid retrieval approaches to geospatial music recommendation. In: Proc. ACM SIGIR (July-August) (2013)

12. Schedl, M., Schnitzer, D.: Location-aware music artist recommendation. In: Gurrin, C., Hopfgartner, F., Hurst, W., Johansen, H., Lee, H., O'Connor, N. (eds.) MMM 2014, Part II. LNCS, vol. 8326, pp. 205–213. Springer, Heidelberg (2014)
13. Schedl, M., Tkalčič, M.: Genre-based analysis of social media data on music listening behavior. In: Proc. ACM Multimedia Workshop ISMM (November 2014)
14. Schedl, M., Vall, A., Farrahi, K.: User geospatial context for music recommendation in microblogs. In: Proc. ACM SIGIR (July 2014)
15. Schnitzer, D., Flexer, A., Schedl, M., Widmer, G.: Local and Global Scaling Reduce Hubs in Space. Journal of Machine Learning Research 13, 2871–2902 (2012)
16. Shi, Y., Larson, M., Hanjalic, A.: Collaborative Filtering Beyond the User-Item Matrix: A Survey of the State of the Art and Future Challenges. ACM Comput. Surv. 47(1), 3:1–3:45 (2014)
17. Zhang, Y.C., Seaghdha, D.O., Quercia, D., Jambor, T.: Auralist: Introducing Serendipity into Music Recommendation. In: Proc. WSDM, February 2012

Logic-Based Incremental Process Mining

Stefano Ferilli[1,2]($\boxtimes$), Domenico Redavid[3], and Floriana Esposito[1,2]

[1] Dipartimento di Informatica, Università di Bari, Bari, Italy
{stefano.ferilli,floriana.esposito}@uniba.it
[2] Centro Interdipartimentale per la Logica e Applicazioni,
Università di Bari, Bari, Italy
[3] Artificial Brain S.r.l., Bari, Italy
redavid@abrain.it

Abstract. Manually building process models is complex, costly and error-prone. Hence, the interest in process mining. Incremental adaptation of the models, and the ability to express/learn complex conditions on the involved tasks, are also desirable. First-order logic provides a single comprehensive and powerful framework for supporting all of the above. This paper presents a First-Order Logic incremental method for inferring process models. Its efficiency and effectiveness were proved with both controlled experiments and a real-world dataset.

1 Introduction

A *process* is a sequence of *events* associated to actions performed by agents. A *workflow* is a formal specification of how a set of tasks can be composed to result in valid processes, including sequential, parallel, conditional, or iterative execution. Each task may have preconditions and postconditions. An *activity* is the actual execution of a task. A *case* is a particular execution of actions compliant to a given workflow. Traces of cases may take the form of lists of events described by 6-tuples (T, E, W, P, A, O) where T is a timestamp, E is the type of the event (begin of process, end of process, begin of activity, end of activity), W is the name of the workflow the process refers to, P is a unique identifier for each process execution, A is the name of the activity, and O is the progressive number of occurrence of that activity in that process. A simple trace for an evening daily routine process case might be:

(201310151930,begin_of_process,evening,1,start,1).
(201310151930,begin_of_activity,evening,1,prepare_dinner,1).
(201310152005,end_of_activity,evening,1,prepare_dinner,1).
(201310152010,begin_of_activity,evening,1,watch_tv,1).
(201310152022,begin_of_activity,evening,1,have_dinner,1).
(201310152113,end_of_activity,evening,1,have_dinner,1).
(201310152240,end_of_activity,evening,1,watch_tv,1).
(201310152245,begin_of_activity,evening,1,use_bathroom,1).
(201310152358,end_of_activity,evening,1,use_bathroom,1).
(201310152358,end_of_process,evening,1,stop,1).

© Springer International Publishing Switzerland 2015
A. Bifet et al. (Eds.): ECML PKDD 2015, Part III, LNAI 9286, pp. 218–221, 2015.
DOI: 10.1007/978-3-s319-23461-8_17

Process mining aims at inferring workflow models from examples of cases. As reported in [4], previous works have encountered problems in dealing with concurrency or in considering different occurrences of the same activity. Using statistics about task frequency and mutual ordering yields less and less accurate models as long as the number of parallel tasks and/or nested loops increases. Genetic algorithms require very long times. Previous approaches in *Declarative Process Mining*, concerned with logic-based representations, including an incremental one, need both positive and negative examples, which is not standard in the process mining setting. Some works also tried to handle noise and probabilities and investigated the possibility of mining/inducing simple boolean conditions for the activities in a propositional setting [1,5]. Here we describe WoMan, an incremental process mining system based on First-Order Logic (FOL). FOL provides a great expressiveness potential to describe in a unified framework cases, models and contextual information.

2   A FOL-Based Proposal

WoMan [4] works in the declarative multi-relational learning setting. We translate case traces into FOL conjunctions based on two predicates:

$activity(S,T)$: at step S task T is executed;
$next(S',S'')$: step S'' follows step S'.

where the S's represent timestamps denoted by unique identifiers. The previous trace would translate to:

activity(s0,start), next(s0,s1), activity(s1,prepare_dinner), next(s1,s2),
activity(s2,watch_tv), next(s1,s3), activity(s3,have_dinner), next(s2,s4), next(s3,s4),
activity(s4,use_bathroom), next(s4,s5), activity(s5,stop)

The process model is described as a FOL conjunction based on two predicates:

$task(t,C)$: task t occurs in the multiset of cases C;
$transition(I,O,p,C)$: transition p, that occurs in the multiset of cases C,
   consists in ending all tasks in I and starting all tasks in O.

C can be exploited for computing statistics on the use of tasks and transitions, and thus to handle noise. A fragment of model accounting for the previous case would be:

task(prepare_dinner,[1,...]). | transition([start]-[prepare_dinner],1,[1,...]).
task(watch_tv,[1,...]). | transition([prepare_dinner]-[watch_tv,have_dinner],2,[1,...]).
task(have_dinner,[1,...]). | transition([watch_tv,have_dinner]-[use_bathroom],3,[1,...]).
task(use_bathroom,[1,...]). | transition([use_bathroom]-[stop],4,[1,...]).

This formalism is more expressive than Petri nets, in particular as regards the possibility of specifying invisible or duplicate tasks. It also permits to easily handle complex or tricky cases that require dummy or artificially duplicated task nodes, that cannot be handled by current approaches in the literature. This representation can be naturally extended by adding relevant contextual information expressed using user-defined, domain-specific predicates.

While learning the workflow structure, the FOL case descriptions, possibly extended with contextual information, can be also exploited as examples for learning task pre-conditions, using a FOL incremental learning system (e.g., InTheLEx [3]). In the previous case, a learned rule might be:

$$prepare\_dinner(X) :- day(X,D), saturday(D), bad\_weather(X).$$

(meaning that if at timestamp X it is bad weather, and it is saturday, then task prepare_dinner is enabled).

Differently from all previous approaches, WoMan is *fully incremental*: it can start with an empty model and learn from one case (while others need a large set of cases to draw significant statistics), and can refine an existing model according to new cases. This is a significant advance to the state-of-the-art, because continuous adaptation of the learned model to the actual practice can be carried out efficiently, effectively and transparently to the users. The learned model can be submitted to experts for analysis purposes, for improving their understanding of the process or for manually tailoring it. It can also be used to generate possible cases, or to supervise future behavior of the users and check whether it is compliant with the learned model, raising suitable warnings otherwise. The user's response to such warnings might be exploited to fix or refine the model.

3   Evaluation

A first evaluation of the proposed methodology used 11 artificial workflow models purposely devised to stress the learning methods, each involving different combinations of complexities and potentially tricky features. The experimental setting was as in [6]: 1000 training cases were randomly generated for each model and used to learn the model. This was repeated several times to ensure that the random generation did not affect the outcome. WoMan was able to learn the correct model in all cases within a few seconds and using less than 50 training cases. Even using 1000 examples, the technique in [6] was unable to learn 7 out of 11 test models. [2] learned the correct model only in 2 cases; nearly half of the times the wrong models did not even fulfill the syntactic requirements for Petri Nets; once it could not converge within a 6-hour limit.

WoMan was also tested on a real-world task concerning daily routines of people. Specifically, we used the Aruba dataset from the CASAS repository (http://ailab.wsu.edu/casas/datasets.html). It includes 220 cases involving 11 tasks, for a total of 13788 trace events. Learning showed a substantial convergence within the first 10 examples. A YAP Prolog 6.2.2 implementation of WoMan processed the whole dataset in 0.1 sec (including translation of traces into FOL). The learned model's average accuracy, evaluated by 10-fold cross-validation, was 92% (consider that each missed case costs nearly 5% accuracy). [6] and [2] returned formally wrong models according to Petri Net specifications.

Using as contextual information the status of the sensors installed in the Aruba home, task preconditions were learned as well. The Aruba dataset yielded 5976 training examples (27.16 per case on average) for learning task preconditions, but InTheLEx converged to the target theory using very few refinements (12 new clauses + 78 generalizations = 90 overall), taking only 10.97 sec per case on average. It was able to avoid overgeneralization, returning meaningful preconditions for 8 out of 11 tasks.

4 Conclusions

This paper presented WoMan, a process mining system based on First-Order Logic representations. It exploits a more expressive representation than previous proposals. Its incremental approach allows to learn from scratch and converge towards correct models using very few examples. It can also handle the context in which the activities take place, thus allowing to learn complex (and human-readable) preconditions for the tasks, using an First-Order Logic incremental learner. It can also handle noise in a straightforward way. Both controlled hard experiments and domain-specific ones, concerning people's daily routines, revealed that the method ensures quick convergence towards the correct model, using much less training examples than would be required by statistical techniques.

Acknowledgments. This work was partially funded by the Italian PON 2007-2013 project PON02_00563_3489339 'Puglia@Service'.

References

1. Agrawal, R., Gunopulos, D., Leymann, F.: Mining process models from workflow logs. In: Schek, H.-J., Saltor, F., Ramos, I., Alonso, G. (eds.) EDBT 1998. LNCS, vol. 1377, pp. 469–483. Springer, Heidelberg (1998)
2. de Medeiros, A.K.A., Weijters, A.J.M.M., van der Aalst, W.M.P.: Genetic process mining: an experimental evaluation. Data Min. Knowl. Discov. **14**(2), 245–304 (2007)
3. Esposito, F., Semeraro, G., Fanizzi, N., Ferilli, S.: Multistrategy theory revision: Induction and abduction in inthelex. Machine Learning Journal **38**(1/2), 133–156 (2000)
4. Ferilli, S.: Woman: Logic-based workflow learning and management. IEEE Transactions on Systems, Man and Cybernetics: Systems **44**(6), 744–756 (2013)
5. Herbst, J., Karagiannis, D.: Integrating machine learning and workflow management to support acquisition and adaptation of workflow models. In: Proceedings of the 9th International Workshop on Database and Expert Systems Applications, pp. 745–752. IEEE (1998)
6. Weijters, A.J.M.M., van der Aalst, W.M.P.: Rediscovering workflow models from event-based data. In: Hoste, V., De Pauw, G., (eds.) Proceedings of the 11th Dutch-Belgian Conference of Machine Learning (Benelearn 2001), pp. 93–100 (2001)

Mobility Mining for Journey Planning in Rome

Michele Berlingerio[1(✉)], Veli Bicer[1], Adi Botea[1], Stefano Braghin[1],
Nuno Lopes[1], Riccardo Guidotti[2], and Francesca Pratesi[2]

[1] IBM Research Ireland, Ballsbridge, Ireland
{mberling,velibice,adibotea,stefanob,nuno.lopes}@ie.ibm.com
[2] KDDLab Department of Computer Science, University of Pisa, Pisa, Italy
{guidotti,pratesi}@di.unipi.it

Abstract. We present recent results on integrating private car GPS routines obtained by a Data Mining module. into the PETRA (PErsonal TRansport Advisor) platform. The routines are used as additional "bus lines", available to provide a ride to travelers. We present the effects of querying the planner with and without the routines, which show how Data Mining may help Smarter Cities applications.

1 Introduction

Smart Cities applications are fostering research in many fields including Computer Science and Engineering. Data Mining is used to support applications such as optimization of a public urban transit network [3], event detection [2], and many more. Along these lines, the aim of the PErsonal TRansport Advisor (PETRA) EU FP7 project[1] is to develop an integrated platform to supply urban travelers with smart journey and activity advises, on a multi-modal network, while taking into account uncertainty:delays in time of arrivals, impossibility to board a (full) bus, walking speed, and so on. In this paper, we briefly describe the architecture of the PETRA platform, and present the results obtained by the embedded journey planner on thousands of planning requests, performed with and without the results coming from the Mobility Mining module. We show how, by integrating private transport routines into a public transit network, it is possible to devise better advises, measured both in terms of number of requests satisfied, and in terms of expected time of arrivals.

2 PETRA System Components

Figure 1 shows the diagram of a simplified system architecture for PETRA. We list and describe here the main modules used in this paper.

---

[1] http://www.petraproject.eu

© Springer International Publishing Switzerland 2015
A. Bifet et al. (Eds.): ECML PKDD 2015, Part III, LNAI 9286, pp. 222–226, 2015.
DOI: 10.1007/978-3-319-23461-8_18

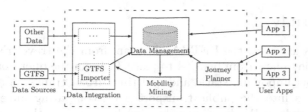

Fig. 1. Simplified PETRA architecture

2.1 Data Management

Handling large volumes of rich and heterogeneous urban data requires a tailored and scalable data management platform, from which we highlight the following modules: i) *data acquisition*, responsible for ingesting heterogeneous urban data; ii) *distributed data storage and indexing*, providing indexes designed for the different formats of data that can be handled by the system (relational, tabular, and graph data), and also their different types (geospatial, textual, etc); iii) *partitioning*, distributing the acquired data across the different nodes of the data storage; iv) *query and searching*, providing a combination of structural query processing and search techniques in order to answer different kinds of queries. The Data Manager (DM) exposes its data to the other PETRA components via a set of APIs, used, for example, by the Journey Planner (JP) to retrieve General Transit Feed Specification (GTFS) data from the DM's internal stored version.

2.2 Mobility Mining

This module fetches GPS data about individual private vehicle trajectories from the DM. We use a data mining process called *mobility profiling* to extract patterns from these traces. This process takes as input the users' trajectories and returns a set of individual *routines* describing their systematic movements [6]. Mobility patterns are expressed as sequences of GPS points with a temporal time stamp that can be exploited as "alternative bus routes" by the JP. These newly introduced routes represent an embedded carpooling service, transparently available in the PETRA application.

2.3 The Multi-modal Journey Planner

We deployed a multi-modal planner taking into account uncertainties related to the expected arrival time of the different modes of transport available in a city. The platform comprising the journey planner provides also functions such as plan execution monitoring, and replanning. The components used in our scenario are the multi-modal JP [4], which is used for the initial planning of journey, and a simulator for plan execution, which is used to monitor the validity of active journey plans. To better perform these tasks, the platform requires updated data. To achieve that, we created a connection between it and the DM, thus deploying the platform in Rome's use case.

3   Case Study

In the Rome's use case, the PETRA platform, from the traveler's perspective, provides journey plans from place A to place B. From the operator perspective, this is done by: importing static and real time urban transport data; merging private routines into the public transport data; computing unvertainty-aware multi-modal advises. We here describe the data used in this paper, how the import step works, and the results obtained with and without private routines.

3.1   Rome Data

The city of Rome, through the public agency Agenzia Mobilità, provides updated open data about its public transport systems. In particular, two main sources of information are offered via its website: i) Rome public transport GTFS, which is a snapshot of the entire public transport network updated every few weeks and ii) Rome public transport real time API. Also, Agenzia Mobilità is gathering a large collection of GPS traces from volunteers? private cars, used by the above described mobility mining module.

3.2   Importing Rome's Data

Importing Rome's data relies on an ad-hoc *data acquisition* module (named RDI, Rome Data Importer), that acts as a bridge between the different kinds of mobility data previously described and the internal DM. RDI performs two sub-tasks: the *daily update* and the *real time update*. The *daily update* consists of discovering *bus stops routines* and enforcing privacy over them. First the RDI transforms the private car routines into sequences of bus stops and combines them as bus lines: each GPS location is mapped to the closest bus stop within a given radius.In order to guarantee car drivers' privacy, the RDI checks if an external attacker could exploit the *bus stops routines* to discover their identity by analysing their vulnerability against the *linking attack* model [5]. Following the methodology in [1], the routines with an identification probability higher than a given *acceptable risk* are transformed into a safer version by removing some bus stops, otherwise they are deleted. Finally all the valid bus stop routines are added to the Rome GTFS data and sent to the DM. Each routine may be used by the JP like any other bus line, even for a portion of the trip. How to make sure the driver of the car can give a ride to the traveler is one of the challenges within the PETRA project. In the *real time update*, the RDI queries the Rome public transport real time API every t minutes, checking for updates (e.g. buses which have been delayed or cancelled) by comparing expected arrival times on the existing GTFS data with real time arrivals. Then it converts possible updates into GTFS format, and sends them to the DM.

3.3   Impact of Routines in Journey Planning

We ran the planning system in two different settings: NoRo, in which the planner uses all the public transport data available, but no routines; Ro, containing

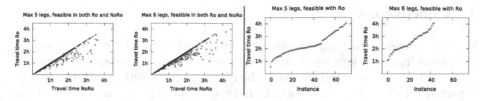

Fig. 2. Impact of routines on travel time.

both routines and public transport data. In each setting, we solved 2,000 queries (instances) with the origins and destinations chosen at random from the logs of the official journey planner of Agenzia Mobilità. In a query, users can set parameters such as the maximum walking time per journey m_w, and the maximum number of legs (i.e., segments) per journey m_l. We set m_w to 20 minutes, the default planner value. Half of the queries have m_l set to 5, and the other half is for $m_l = 6$. The public transport data we used has 8,896 stops and 391 routes. Each route is served by a number of trips, to a total of 39,422 trips per day. The Rome roadmap has 522,529 nodes and 566,400 links. In the GTFS data, we represent routines with a structure similar to public transport data. Each routine introduces a new route and a new trip. We started from 1,205,258 GPS trajectories from 262,657 users. After routine extraction, bus mapping, and anonymization, we ended up with 729 safe mapped routines from 641 users. This increases the number of bus routes to 1,120, for a total number of trips of 40,151.

Figure 2 illustrates the impact of adding routines as an additional mode. At the left, we compare the travel time in the Ro and NoRo settings. As expected, in a subset of cases, the travel time is the same. On the other hand, all points located below the main diagonal show instances where routines improve the time. In fact, routines can improve both the travel time and the number of legs per journey. The latter has two advantages. First, it makes a trip more convenient to the traveller, as it reduces the number of interchanges. Secondly, it helps increase the set of feasible instances (i.e., instances for which a solution exists). This is important because user-imposed constraints on m_l and m_w can restrict the set of feasible instances. For example, without using routines, in 29.3% of our queries (instances), it is impossible to complete the journey with at most 20 minutes of walking and at most 5 legs in the journey. Charts at the right in Figure 2 show instances that become feasible after adding routines. When m_l is set to 5, routines are part of the returned plan in 17.5% of the instances. Routines increase the percentage of feasible instances by 7.1%, to a total of 77.8%. In 9.6% of the instances, routines improve the travel time, the average savings per trip being equal to 25.5 minutes. When $m_l = 6$, routines become part of the plans in 22.3% of the instances. They increase the percentage of feasible instances from 84.5 to 88.9%. In 14.3% of the instances, routines improve the travel time, the average improvement amounting to 22.05 minutes per trip.

4   Conclusions

We have presented our results obtained by running the PETRA platform on the city of Rome for journey planning. Our results show an increased number of planning instances satisfied thanks to the routines, along with a reduced average expected travel time. Future works include: i) exploiting the uncertainty of the routines for more robust advises; ii) devising the platform for tourism activity planning; iii) extending the mobility mining to crowd patterns.

Acknowledgement. This work has been partially supported by the EC under the FET-Open Project n. FP7-ICT-609042, PETRA.

References

1. Basu, A., et al.: A privacy risk model for trajectory data. In: Zhou, J., Gal-Oz, N., Zhang, J., Gudes, E. (eds.) Trust Management VIII. IFIP AICT, vol. 430, pp. 125–140. Springer, Heidelberg (2014)
2. Berlingerio, M., Calabrese, F., Lorenzo, G.D., Dong, X., Gkoufas, Y., Mavroeidis, D.: Safercity: A system for detecting and analyzing incidents from social media. In: (Demo paper) IEEE 2013 Workshops, pp. 1077–1080 (2013)
3. Berlingerio, M., Calabrese, F., Di Lorenzo, G., Nair, R., Pinelli, F., Sbodio, M.L.: AllAboard: a system for exploring urban mobility and optimizing public transport using cellphone data. In: Blockeel, H., Kersting, K., Nijssen, S., Železný, F. (eds.) ECML PKDD 2013, Part III. LNCS, vol. 8190, pp. 663–666. Springer, Heidelberg (2013)
4. Botea, A., Nikolova, E., Berlingerio, M.: Multi-modal journey planning in the presence of uncertainty. In: ICAPS (2013)
5. Monreale, A., Andrienko, G.L., Andrienko, N.V., Giannotti, F., Pedreschi, D., Rinzivillo, S., Wrobel, S.: Movement data anonymity through generalization. Transactions on Data Privacy 3(2) (2010)
6. Trasarti, R., Pinelli, F., Nanni, M., Giannotti, F.: Mining mobility user profiles for car pooling. In: KDD 2011, pp. 1190–1198 (2011)

Pattern Structures and Concept Lattices for Data Mining and Knowledge Processing

Mehdi Kaytoue[1]([⊠]), Victor Codocedo[1], Aleksey Buzmakov[2],
Jaume Baixeries[3], Sergei O. Kuznetsov[4], and Amedeo Napoli[2]

[1] Université de Lyon, CNRS, INSA-Lyon, LIRIS, UMR5205, 69621 Lyon, France
mehdi.kaytoue@insa-lyon.fr
[2] LORIA (CNRS - Inria Nancy Grand Est - Université de Lorraine),
B.P. 239, 54506 Vandœuvre-lès-Nancy, France
[3] Universitat Politècnica de Catalunya, 08032 Barcelona, Catalonia, Spain
[4] National Research University Higher School of Economics (HSE),
Kochnovski pr.3, Moscow 125319, Russia

Abstract. This article aims at presenting recent advances in Formal
Concept Analysis (2010-2015), especially when the question is dealing
with complex data (numbers, graphs, sequences, etc.) in domains such
as databases (functional dependencies), data-mining (local pattern discovery), information retrieval and information fusion. As these advances
are mainly published in artificial intelligence and FCA dedicated venues,
a dissemination towards data mining and machine learning is worthwhile.

1 Pattern Structures in Formal Concept Analysis

Formal Concept Analysis (FCA) is a branch of applied lattice theory that
appeared in the 1980's [11]. Starting from a binary relation between a set of
objects and a set of attributes, formal concepts are built as maximal sets of
objects in relation with maximal sets of attributes, by means of derivation operators forming a Galois connection. Concepts form a partially ordered set that
represents the initial data as a hierarchy, called the concept lattice. This conceptual structure has proved to be useful in many fields, e.g. artificial intelligence, knowledge management, data-mining and machine learning, morphologi-
cal mathematics, etc. In particular, several results and algorithms from itemset
and association rule mining and rule-based classifiers were already characterized in terms of FCA [17,20]. For example, the set of frequent closed itemsets
is an order ideal of a concept lattice; association rules and functional dependencies can be characterized with the derivation operators; jumping patterns were
defined as hypotheses, etc, not to mention efficient polynomial-delay algorithms
for building all closed itemsets such as *CloseByOne* [18].

The goal of this communication is to present our recent advances in FCA over
the period 2010–2015, especially when the question is dealing with complex data,
thanks to the rich formalism of *pattern structures* [10]. This general approach
translates FCA to any partially ordered data descriptions to deal elegantly with

© Springer International Publishing Switzerland 2015
A. Bifet et al. (Eds.): ECML PKDD 2015, Part III, LNAI 9286, pp. 227–231, 2015.
DOI: 10.1007/978-3-319-23461-8_19

non binary, say *complex, heterogeneous* and *structured* data. Pattern structures also allow new ways of solving problems in several applications (see next section).

The key idea relies on defining so-called *similarity operators* which induce a semi-lattice on data descriptions. Several alternative attempts were made for defining such semi-lattices on sets of graphs and logical formulas (see, e.g., the works of Chaudron&Maille, Ferré&Ridoux, Polaillon&Brito cited in [16]). Formally, a pattern structure is a triple $(G, (D, \sqcap), \delta)$ where G is a set of objects, $(D, \sqcap)$ is a meet-semi-lattice of potential object descriptions and $\delta : G \longrightarrow D$ is a mapping associating each object with its description. Elements of D are called patterns and are ordered with a subsumption relation $\sqsubseteq$: $\forall c, d \in D$, $c \sqsubseteq d \iff c \sqcap d = c$. For any $A \subseteq G$ and $d \in (D, \sqcap)$, two derivation operators are defined: as $A^\square = \bigsqcap_{g \in A} \delta(g)$ and $d^\square = \{g \in G | d \sqsubseteq \delta(g)\}$. These operators form a Galois connection between $(\wp(G), \subseteq)$ and $(D, \sqcap)$. Pattern concepts of $(G, (D, \sqcap), \delta)$ are pairs of the form (A, d), $A \subseteq G$, $d \in (D, \sqcap)$, such that $A^\square = d$ and $A = d^\square$. For a pattern concept (A, d), d is a pattern intent and is the common description of all objects in A, the pattern extent. When partially ordered by $(A_1, d_1) \leq (A_2, d_2) \Leftrightarrow A_1 \subseteq A_2$ ($\Leftrightarrow d_2 \sqsubseteq d_1$), the set of all concepts forms a complete lattice called pattern concept lattice.

Pattern structures offer a concise way to define closed patterns. They also allow efficient polynomial-delay algorithms (modulo complexity of computing $\sqsubseteq$ and $\sqcap$) [18]. In presence of large datasets, they offer natural approximation tools (*projections*, detailed below) and achieve lazy classification [19].

Data Heterogeneity. When D is the power set of a set of items I, $\sqcap$ and $\sqsubseteq$ are the set intersection and inclusion resp.: pattern intents are closed itemsets and we fall back in standard FCA settings. Originally, pattern structures were introduced to handle objects described by labeled graphs [18]. We developed the general approach in various ways for handling objects described by: numbers and intervals [16], partitions [4], sequences [5] and trees [22].

Data Approximation. Pattern structure projections simplify computation and reduce the number of concepts [10]. For example, a set of labeled graphs can be projected as a set of k-chains [21], while intervals can be enlarged [12]. A projection ψ associates any pattern to a *more general* pattern covering more objects. A projection is $\sqcap$-preserving: $\forall c, d \in D, \psi(c \sqcap d) = \psi(c) \sqcap \psi(d)$. We studied how numerical data can be projected when a similarity relation between numbers (symmetric, reflexive but not transitive relation) is considered and showed that a projection can be performed as a pre-processing task. We also introduced a wider class of projections [6]: while projections can only modify object descriptions, *o-projections* modifies the semi-lattice of descriptions.

Data Representation. For any pattern structure, a *representation* context can be built, which is a binary relation encoding the pattern structure. Concepts in both data representations are in 1-1-correspondence. We studied this aspect for several types of patterns, designing the transformation procedures and evaluating in which conditions one data representation prevails [4,15]. We also showed that the bijection does not hold in general for minimal generators (qualified as *free* or *key* in pattern mining) [15]. The impact of projections on representation contexts are investigated with the new class of *o*-projections [6].

2 Applications

Database and Functional Dependencies. Characterizing and computing functional dependencies (FDs) are an important topic in database theory (see e.g. references in [4]). In FCA, Ganter & Wille proposed a first characterization of FDs as implications in a formal context (binary relation) obtained after a transformation of the initial data [11]. However, n^2 objects are created from the n initial tuples. To overcome this problem, we present a characterization of functional dependencies in terms of (partition) pattern structures that offers additional benefits for the computation of dependencies [4]. This method can be naturally generalized to other types of FDs (multi-valued and similarity dependencies [3]).

Pattern Mining and Biclustering. Biclustering aims at finding local patterns in numerical data tables. The motivation is to overcome the limitation of standard clustering techniques where distance functions using all the attributes may be ineffective and hard to interpret. Applications are numerous in biology, recommendation, etc. (see references in [7,13]). In FCA, formal concepts are maximal rectangles of *True* values in a binary data-table (modulo columns/rows permutations). Accordingly, concepts are binary biclusters with interesting properties: maximality (via a closure operator), overlapping and partial ordering of the local patterns. Such properties are key elements of a mathematical definition of numerical biclusters and the design of their enumeration algorithms. We highlight these links for several types of biclusters with interval [14] and partition pattern structures [7] and their representation contexts. Next investigations concern dimensionality: a bijection between n-clusters and $n + 1$-concepts is proven [13].

Information Retrieval. FCA has been used in a myriad of ways to support a wide variety of information retrieval (IR) techniques and applications [9]: the concept lattice represents concisely the document and the query space which can be used as an index for automatic retrieval. In the last years, the Boolean IR model (and consequently, FCA) has been considered as too limited for modern IR requirements, such as large datasets and complex document representations. Pattern structures have shown a great potential to reuse the body of work of FCA-based IR approaches by providing support to complex document representations, such as numerical and heterogeneous indexes [8]. In the context of semantic web, a noticeable application of this model is RDF data completion [1].

Information Fusion for Decision Making. Merging information given by several sources (databases, experts...) into an interpretable and useful format is a tricky task. Fusion results may not be in suitable form for being used in decision analysis. This is due to the fact that information sources are heterogeneous, noisy and inconsistent. We investigated how FCA and pattern structures can be used in decision making when fusion is required: pattern concept lattices (based on intervals) provide an information fusion space where maximal subsets of information can be detected and support decision making [2].

References

1. Alam, M., Buzmakov, A., Codocedo, V., Napoli, A.: An approach for improving RDF data with formal concept analysis. In: Int. Joint Conf. on Artif. Intell. (2015)
2. Assaghir, Z., Napoli, A., Kaytoue, M., Dubois, D., Prade, H.: Numerical information fusion: lattice of answers with supporting arguments. In: Int. Conf. on Tools with Artificial Intelligence (ICTAI), pp. 621–628. IEEE (2011)
3. Baixeries, J., Kaytoue, M., Napoli., A.: Computing similarity dependencies with pattern structures. In: Int. Conf. on Concept Lattices and Their Applications (CLA), CEUR 1062, pp. 33–44 (2013)
4. Baixeries, J., Kaytoue, M., Napoli, A.: Characterizing functional dependencies in formal concept analysis with pattern structures. Ann. Math. Artif. Intell. **72**(1–2), 129–149 (2014)
5. Buzmakov, A., Egho, E., Jay, N., Kuznetsov, S., Napoli, A., Raïssi, C.: On Mining Complex Sequential Data by Means of FCA and Pattern Structures. International Journal of General Systems (2015)
6. Buzmakov, A., Kuznetsov, S.O., Napoli, A.: Revisiting pattern structure projections. In: Baixeries, J., Sacarea, C., Ojeda-Aciego, M. (eds.) ICFCA 2015. LNCS (LNAI), vol. 9113, pp. 200–215. Springer, Heidelberg (2015)
7. Codocedo, V., Napoli, A.: Lattice-based biclustering using partition pattern structures. In: European Conf. on Artificial Intelligence (ECAI) (2014)
8. Codocedo, V., Napoli, A.: A proposition for combining pattern structures and relational concept analysis. In: Glodeanu, C.V., Kaytoue, M., Sacarea, C. (eds.) ICFCA 2014. LNCS (LNAI), vol. 8478, pp. 96–111. Springer, Heidelberg (2014)
9. Codocedo, V., Napoli, A.: Formal concept analysis and information retrieval – a survey. In: Baixeries, J., Sacarea, C., Ojeda-Aciego, M. (eds.) ICFCA 2015. LNCS (LNAI), vol. 9113, pp. 61–77. Springer, Heidelberg (2015)
10. Ganter, B., Kuznetsov, S.O.: Pattern structures and their projections. In: Delugach, H.S., Stumme, G. (eds.) ICCS 2001. LNCS (LNAI), vol. 2120, pp. 129–142. Springer, Heidelberg (2001)
11. Ganter, B., Wille, R.: Formal Concept Analysis. Springer, Berlin (1999)
12. Kaytoue, M., Assaghir, Z., Napoli, A., Kuznetsov, S.O.: Embedding tolerance relations in fca: an application in information fusion. In: CIKM. ACM (2010)
13. Kaytoue, M., Kuznetsov, S.O., Macko, J., Napoli, A.: Biclustering meets triadic concept analysis. Ann. Math. Artif. Intell. **70**(1–2), 55–79 (2014)
14. Kaytoue, M., Kuznetsov, S.O., Napoli, A.: Biclustering numerical data in formal concept analysis. In: Jäschke, R. (ed.) ICFCA 2011. LNCS (LNAI), vol. 6628, pp. 135–150. Springer, Heidelberg (2011)
15. Kaytoue, M., Kuznetsov, S.O., Napoli, A.: Revisiting numerical pattern mining with formal concept analysis. In: Int. Joint Conf. on Art. Intell. (IJCAI) (2011)
16. Kaytoue, M., Kuznetsov, S.O., Napoli, A., Duplessis, S.: Mining gene expression data with pattern structures in formal concept analysis. Inf. Sci. **181**(10) (2011)
17. Kuznetsov, S.O.: Galois connections in data analysis: contributions from the soviet era and modern russian research. In: Ganter, B., Stumme, G., Wille, R. (eds.) Formal Concept Analysis. LNCS (LNAI), vol. 3626, pp. 196–225. Springer, Heidelberg (2005)

18. Kuznetsov, S.O.: Learning of simple conceptual graphs from positive and negative examples. In: Żytkow, J.M., Rauch, J. (eds.) PKDD 1999. LNCS (LNAI), vol. 1704, pp. 384–391. Springer, Heidelberg (1999)
19. Kuznetsov, S.O.: Fitting pattern structures to knowledge discovery in big data. In: Cellier, P., Distel, F., Ganter, B. (eds.) ICFCA 2013. LNCS (LNAI), vol. 7880, pp. 254–266. Springer, Heidelberg (2013)
20. Kuznetsov, S.O., Poelmans, J.: Knowledge representation and processing with formal concept analysis. Wiley Interdisc. Rew.: Data Mining and Knowledge Discovery 3(3), 200–215 (2013)
21. Kuznetsov, S.O., Samokhin, M.V.: Learning closed sets of labeled graphs for chemical applications. In: Kramer, S., Pfahringer, B. (eds.) ILP 2005. LNCS (LNAI), vol. 3625, pp. 190–208. Springer, Heidelberg (2005)
22. Leeuwenberg, A., Buzmakov, A., Toussaint, Y., Napoli, A.: Exploring pattern structures of syntactic trees for relation extraction. In: Baixeries, J., Sacarea, C., Ojeda-Aciego, M. (eds.) ICFCA 2015. LNCS (LNAI), vol. 9113, pp. 153–168. Springer, Heidelberg (2015)

Privacy Preserving Blocking and Meta-Blocking

Alexandros Karakasidis[1]([✉]), Georgia Koloniari[2], and Vassilios S. Verykios[1]

[1] School of Science & Technology, Hellenic Open University, Patras, Greece
{a.karakasidis,verykios}@eap.gr
[2] Department of Applied Informatics, University of Macedonia, Thessaloniki, Greece
gkoloniari@uom.gr

Abstract. Record linkage refers to integrating data from heterogeneous sources to identify information regarding the same entity and provides the basis for sophisticated data mining. When privacy restrictions apply, the data sources may only have access to the merged records of the linkage process, comprising the problem of privacy preserving record linkage. As data are often dirty, and there are no common unique identifiers, the linkage process requires approximate matching and it renders to a very resource demanding task especially for large volumes of data. To speed up the linkage process, privacy preserving blocking and meta-blocking techniques are deployed. Such techniques derive groups of records that are more likely to match with each other. In this nectar paper, we summarize our contributions to privacy preserving blocking and meta-blocking.

Keywords: Privacy · Record linkage · Blocking · Meta-blocking

1 Introduction

Considering the data explosion we experience the last decade, we seek ways to boost the results of data mining. As related data are highly scattered, i.e., in different organizations databases, on the web, etc., integrating large volumes of data comprises an indispensable first step towards mining more useful information that could not be discovered if we consider each separate database in isolation.

This process of identifying and linking information across multiple databases, that refers to the same real world entity, is known as the problem of *record linkage*. When privacy concerns arise, the record linkage problem is augmented to its privacy preserving version, where the participants should not gain any additional information regarding each other's data, apart from the linkage results.

For instance, let us consider a medical researcher who wishes to perform a study on the interactions between certain prescribed medicine over the last decade, using data from hospitals and clinics from all over Europe. This comprises a data mining problem, where the additional requirement of privacy is posed due to the sensitive nature of the data. These data are not all stored in a single database which may be mined, but originate from multiple health care

© Springer International Publishing Switzerland 2015
A. Bifet et al. (Eds.): ECML PKDD 2015, Part III, LNAI 9286, pp. 232–236, 2015.
DOI: 10.1007/978-3-319-23461-8_20

Fig. 1. Privacy preserving record linkage workflow.

units from different countries each of them using its own database. Consequently, these data should be merged, using a private record linkage protocol.

The lack of global unique identifiers deems necessary the use of common attributes, in most cases textual, for identifying the matching records. As attribute values are usually the result of user input, they are most often dirty, requiring methods for approximate matching. Taking into account the large volumes of available data, we confront a very resource demanding task. To deal with this, *privacy preserving matching* (PPM) is often preceded by a *privacy preserving blocking* (PPB) phase. PPB speeds up matching by organizing candidate records that are more likely to match into blocks, based on the values of their attributes. The attributes selected for PPB and PPM are respectively called blocking and matching attributes. Lately, privacy preserving meta-blocking was introduced which, applied after blocking, aims at reorganizing the way records within a block should be matched, so as to further improve performance. The phases of the overall linkage process are depicted in Fig.1.

Blocking imposes an additional filtering step to the matching process, thus increasing its precision. On the other hand, blocking may eliminate matching record pairs, thus decreasing recall. Therefore, some blocking techniques compromise result quality [9], while others rely on efficiency-privacy tradeoffs failing to significantly improve performance for large scale data without sacrificing their privacy [8]. Finally, there are approaches that, though efficient, are limited to specific data types, either numerical or nominal [2]. In this nectar paper, we present our contribution on privacy preserving blocking and meta-blocking methods. Our aim is to boost performance while maintaining high levels of matching quality without compromising privacy.

2   Privacy Preserving Blocking and Meta-Blocking

We first present three privacy preserving blocking techniques and then, the only work up to now on privacy preserving meta-blocking. The first blocking technique is designed for textual data, while all others may be adopted for both textual and numerical data using appropriate distance and similarity measures.

Phonetic Code Based Private Blocking. A phonetic code is a hash produced by a phonetic algorithm for matching words based on their pronunciation. The main feature of phonetic algorithms is their fault tolerance against typographical errors. In [3], we present a two-party phonetic based privacy preserving blocking protocol. The two parties (data sources) agree on the use of a set of phonetic algorithms and blocking attributes. Each party then encodes the blocking fields with each of the algorithms. To increase the entropy of each dataset and consequently reduce the ability of predicting its values, fake phonetic codes are injected. Next, all phonetic codes are encrypted using a secure hash function and records are grouped into blocks according to their hashes. Each record is assigned to multiple blocks according to each of the blocking attributes encoded by each of the phonetic algorithms.

High matching quality is assured by using multiple phonetic codes per blocking attribute, thus overcoming through redundancy certain weak points of phonetic algorithms, such as in Soundex [7], where an error in the first letter produces a different code. Privacy is assured as phonetic algorithms are one-way functions which apply information suppression, and improved with the use of fake codes and encryption. With respect to efficiency, phonetic codes are very fast to compute, and moreover, matching on identical phonetic hashes enables us to deploy indexes to further speed up matching, achieving up to 61.4% speedup with respect to plain matching and recall at 0.67 [3].

Reference Table Based Private Blocking. Reference tables are publicly available datasets used to provide privacy by avoiding direct comparison between the two databases, using instead reference values as a comparison basis. Our contribution comprises of two methods, that employ a third party. The two sources individually cluster the same reference values. Each record is classified to a cluster (class) based on some distance or similarity measure, associating one of the reference values to its blocking attribute. Records classified at the same class comprise a block. The two sources send the classified record ids to the third party who merges blocks belonging to the same class. Final blocks are returned to the sources only when they contain record ids from both sources. Matching may then be performed either at the sources or at the third party.

Reference table based k-anonymous private blocking [5] is the first work using this concept for privacy preserving blocking. Nearest Neighbor clustering is used to form clusters of at least k-elements, thus ensuring k-anonymity as each record is assigned to a class based on its similarity with one of at least k-elements. However, while ensuring privacy and result quality, the method incurs high complexity as each blocking attribute should be looked up against all reference values. Experiments show that linkage with our method requires half the execution time of plain matching with recall up to 0.78 [5].

Multidimensional private blocking [4] improves the performance of [5], by using k-Medoids for clustering. Each blocking attribute is checked only against cluster medoids, thus reducing the method's complexity. Moreover, the use of an edit distance negates the need for a reference table to contain values similar to the ones contained in the datasets. For numerical fields, bins are created based on numerical reference values. This work introduces the concept of multidimensional blocking. In blocking, when more than one blocking fields are used, the same procedure has to be repeated and a record may fall within numerous blocks.

With multidimensional blocking, the two sources locate the class of each record for each of the blocking attributes used. Then, they calculate the intersection of classes each record belongs to. As such, a record is associated with less blocks resulting in reduced matching operations. Execution time drops further to 11% of plain matching, while recall is 0.73 [4].

Privacy Preserving Meta-Blocking. *Sorted neighborhood on encrypted fields* (SNEF) [6], based on [1], is to the best of our knowledge, the only privacy preserving meta-blocking method. Multidimensional private blocking is extended by associating each record within each block with a score derived by an objective function that uses the edit distance between each blocking attribute and the cluster medoid of its class. After the third party merges the blocks, the party who performs the privacy preserving matching sorts the records within each block based on their scores. A sliding window of size w slides over the records, and each record is checked against the next w records in the block, rendering the matching complexity within each block from quadratic to linear. SNEF does not compromise privacy since each record is associated with a single number which cannot be factorized due to the properties of the objective function. There is a tradeoff between matching quality and time efficiency, depending on window size which, nevertheless, remains linear. As experiments show, SNEF further improves multidimensional blocking's time by 20% with a recall around 0.70 [6].

3   Conclusion and Future Work

We presented our work on privacy preserving blocking techniques, which are efficient while assuring privacy and result quality. Next, we plan to accelerate our blocking methods by using random samples instead of clustering.

References

1. Hernández, M.A., Stolfo, S.J.: Real-world data is dirty: Data cleansing and the merge/purge problem. Data Min. Knowl. Discov. **2**(1), 9–37 (1998)
2. Inan, A., Kantarcioglu, M., Ghinita, G., Bertino, E.: Private record matching using differential privacy. In: ACM EDBT (2010)
3. Karakasidis, A., Verykios, V.S.: Secure blocking + secure matching = secure record linkage. JCSE **5**(3), 223–235 (2011)
4. Karakasidis, A., Verykios, V.S.: A highly efficient and secure multidimensional blocking approach for private record linkage. In: IEEE ICTAI (2012)
5. Karakasidis, A., Verykios, V.S.: Reference table based k-anonymous private blocking. In: ACM SAC (2012)

6. Karakasidis, A., Verykios, V.S.: A sorted neighborhood approach to multidimensional privacy preserving blocking. In: IEEE ICDMW (2012)
7. Odell, M., Russell, R.: The Soundex coding system. US Patents 1261167 (1918)
8. Vatsalan, D., Christen, P., Verykios, V.S.: Efficient two-party private blocking based on sorted nearest neighborhood clustering. In: ACM CIKM (2013)
9. Yakout, M., Atallah, M.J., Elmagarmid, A.K.: Efficient private record linkage. In: IEEE ICDE (2009)

Social Data Mining and Seasonal Influenza Forecasts: The FluOutlook Platform

Qian Zhang[1], Corrado Gioannini[2],
Daniela Paolotti[2], Nicola Perra[1], Daniela Perrotta[2], Marco Quaggiotto[2],
Michele Tizzoni[2], and Alessandro Vespignani[1,2(✉)]

[1] MOBS, Northeastern University, Boston, MA, USA
{qi.zhang,n.perra,a.vespignani}@neu.edu
[2] ISI Foundation, Turin, Italy
{corrado.gioannini,daniela.paolotti,daniela.perrotta,
marco.quaggiotto,michele.tizzoni}@isi.it

Abstract. FluOutlook is an online platform where multiple data sources are integrated to initialize and train a portfolio of epidemic models for influenza forecast. During the 2014/15 season, the system has been used to provide real-time forecasts for 7 countries in North America and Europe.

Keywords: Real-time forecasting · Epidemic modeling · Data mining

1 Introduction

The real-time monitoring and modeling of infectious disease is being redefined by the novel availability of large scale social media and digital surveillance data. Several methods use social data, like search engine queries and tweets, as inputs for time series analysis; Google Flu Trends (GFT) [1] being probably the most known example. Unfortunately, most of the current approaches are unable to capture the disease transmission dynamics and its long-term trends, and suffer from several issues related to biases and statistical sampling [2]. Here we present FluOutlook (http://fluoutlook.org/), an online platform exposing real-time seasonal influenza forecasts. It integrates current and historical surveillance data, social data mining and several forecast models. Along with standard regression statistical models, FluOutlook includes stochastic generative models simulating the disease progression at the level of single individuals. The platform reports in real-time the influenza intensity with a lead time of up to four weeks, as well as main indicators of the epidemic season at its early stages. FluOutlook provides a description of the seasonal influenza that could be used by public health agency to guide their decision making process, as well as to compare and assess the performance of different forecast approaches.

A. Bifet et al. (Eds.): ECML PKDD 2015, Part III, LNAI 9286, pp. 237–240, 2015.
DOI: 10.1007/978-3-319-23461-8_21

2   Methodology

The FluOutlook platform consists of two parts: a computational framework that provides predictions and a user-friendly website that provides their visualization. The system architecture, shown in Fig. 1, is made by three main components. The first component mines and assimilates the social and surveillance data needed to initialize the modeling approaches. The second component is the computational system that generates the numerical output of the modeling approaches. The third component is the statistical pipeline that compares the models' output with the current ground truth, available to define the forecast ensemble that is eventually exposed on the platform. The website of the platform runs as a Python Flask application with a PostgreSQL database, served through the Apache web service. In the landing page, maps show the current influenza activity level in each country and indicate the observed trend. The forecasting page provides more detailed predictions for each country.

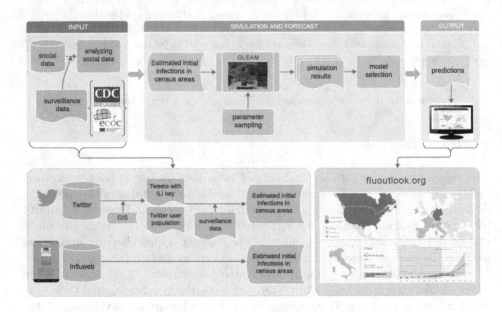

Fig. 1. The FluOutlook system architecture and the landing page of the Fluoutlook website.

2.1   Simulation Engine

The FluOutlook platform has at its core a computational modeling and simulation engine. We use different forecast methodologies based on statistical regression approaches and the GLEAM (GLobal Epidemic And Mobility model) [3,4]. GLEAM is a spatial, stochastic and individual based epidemic model based on three layers.The population layer is based on the high-resolution population database of the Gridded Population of the World project by the Socio-Economic Data and Applications Center (SEDAC) that estimates population with a granularity given by a lattice of cells covering the whole planet at a resolution of 15×15 minutes of arc. The Mobility layer integrates short-range and long-range transportation data. Long-range air travel mobility is based on travel flow data obtained from the Official Airline Guide (OAG) databases. The model identifies 3, 362 subpopulations in 220 different countries. The model simulates the short and long-range mobility of individuals between these subpopulations using a stochastic procedure. The infection dynamics takes place within each subpopulation and considers different infectious disease dynamic and the intervention measures being considered. GLEAM has been successfully used during the 2009 H1N1 pandemic to provide short and long term predictions of its course [5,6].

2.2 Tracking Seasonal Flu with Social Data Mining

One of the key and novel components in FluOutlook is the estimation of the initial cases necessary to run the GLEAM model (see Fig. 1). While many different sources of data can be integrated, currently FluOutlook adopts two sources of geolocalized data: Twitter and a participatory system for digital surveillance called Influweb [8].

Inferring Initial Infections from Twitter. By filtering geolocalized tweets with a set of ILI-related keywords, we can obtain spatial and temporal information about ILI. We use 40-50 keywords for each given language. Not all ILI keywords have the same relevance. Moreover, tweets containing an ILI keyword may not necessarily contain information related with influenza. Although it is still challenging to filter noisy information, we simplify this process by ranking the ILI keywords with consideration of the correlation between the time series of the surveillance data and the volume of tweets for a given key word. In each country, the volume of geolocalized tweets, Twitter user population, and actual population allow the estimation of the relative number of initial infections for a given week in each subpopulation area.

Inferring Initial Infections from Influweb. Open source indicators, such as Twitter, search engine queries, Wikipedia, may be mixed with irrelevant information and over-represent some particular demographic [2]. Novel data sources can overcome these issues. For instance, online self-reporting platforms are designed to provide more accurate in-time indicators of disease activity. Influenzanet [7] is one of such online platforms. It monitors ILI activities using self-reported information coming from volunteers across several countries in Europe. Influweb, a part of Influenzanet project, is a system to collect information on influenza-like-illness in Italy [8]. Voluntary participants across the country register on the website and submit to the system information about their locations, demographic and influenza-related health status. To better monitor the ILI activity, the volunteers are invited to weekly update their health conditions. The high quality and reliability of such data allow to infer initial infections in a given week in any census area directly. In FluOutlook, we use Influweb data to initialize GLEAM simulations in Italy.

2.3 Generative Model Selection and Forecast Output

The simulation module in the platform performs a Latin hypercube sampling of a parameter space used in the GLEAM model, and generates for each sampled point P a statistical ensemble of the epidemic profiles. From each statistical ensemble, the model selection module estimates the likelihood function $L(P|X)$, where $X = x_0, x_1, ..., x_{N-1}$ indicates the ILI surveillance data in a given fitting window of length N. By considering the likelihood region defined by relative likelihood function in defining the parameters' range, the module selects a set of models. The selected models provide both long-term predictions for epidemic peak time and intensity, and short-term predictions for the epidemic profiles in the future four weeks. In Fig. 1, we show predicted epidemic profiles for Italy in week 2, 2015 with initial infections inferred from Twitter and Influweb, as well as other time-series predicting methods. In the forecasting page, the predicted statistical confidence intervals for peak time and intensity are given on the bottom of the web page.

3 Conclusion

Since the fall 2014 FluOutlook platform has provided real-time forecast of seasonal flu for the United States, Canada, Italy, France, Netherlands, Spain and Ireland The platform can incorporate surveillance data from any other countries or regions and is able to provide forecast of seasonal influenza for countries with high quality social data.

References

1. Ginsberg, J., et al.: Detecting influenza epidemics using search engine query data. Nature **457**, 1012–1014 (2009)
2. Lazer, D., Kennedy, R., King, G., Vespignani, A.: Big data: The parable of Google Flu: traps in big data analysis. Science **343**, 1203–1205 (2014)
3. Balcan, D., et al.: Multiscale mobility networks and the spatial spreading of infectious diseases. Proc. Natl. Acad. Sci. **106**, 21484–21489 (2009)
4. Balcan, D., et al.: Modeling the spatial spread of infectious diseases: The GLobal Epidemic and Mobility computational model. J. Comput. Sci. **1**, 132–145 (2010)
5. Balcan, D., et al.: Seasonal transmission potential and activity peaks of the new influenza A(H1N1): a Monte Carlo likelihood analysis based on human mobility. BMC Medicine **7**, 45 (2009)
6. Tizzoni, M., et al.: Real-time numerical forecast of global epidemic spreading: case study of 2009 A/H1N1pdm. BMC Medicine **10**, 165 (2012)
7. https://www.influenzanet.eu/
8. https://www.influweb.it/

Star Classification Under Data Variability: An Emerging Challenge in Astroinformatics

Ricardo Vilalta[1]([✉]), Kinjal Dhar Gupta[1], and Ashish Mahabal[2]

[1] Department of Computer Science, University of Houston, Houston, TX 77204, USA
{vilalta,kinjal13}@cs.uh.edu
[2] Department of Astronomy,
California Institute of Technology, Pasadena, CA 91125, USA
aam@astro.caltech.edu

Abstract. Astroinformatics is an interdisciplinary field of science that applies modern computational tools to the solution of astronomical problems. One relevant subarea is the use of machine learning for analysis of large astronomical repositories and surveys. In this paper we describe a case study based on the classification of variable Cepheid stars using domain adaptation techniques; our study highlights some of the emerging challenges posed by astroinformatics.

Keywords: Astroinformatics · Domain adaptation · Variable star classification

1 Introduction

The recent emergence of a new field of study named *astroinformatics*, comes as a response to the rapid growth of data volumes corresponding to a variety of astronomical surveys. Data repositories have gone from gigabytes into terabytes, and we expect those repositories to reach the petabytes in the coming years. This massive amount of data stands in need of advanced computational solutions. The general aim in astroinformatics is the application of computational tools to the solution of astronomical problems; key issues involve not only an efficient management of data resources, but also the design of new computational tools that efficiently capture the nature of astronomical phenomena.

Recent work reports on successful applications of machine learning for analysis of large astronomical repositories and surveys [2]. Machine learning is already an indispensable resource to modern astronomy, serving as an instrumental aid during the process of data integration and fusion, pattern discovery, classification, regression and clustering of astronomical objects. And we expect machine learning to produce high-impact breakthroughs when large (petabyte) datasets become available. As an illustration, LSST (Large Synoptic Survey Telescope), will survey the sky to unprecedented depth and accuracy at an impressive temporal cadence [3]; it will generate an expected 30 terabytes of data obtained each night to provide a complete new view of the deep optical universe in the realm of time domain

© Springer International Publishing Switzerland 2015
A. Bifet et al. (Eds.): ECML PKDD 2015, Part III, LNAI 9286, pp. 241–244, 2015.
DOI: 10.1007/978-3-319-23461-8_22

astronomy. Other projects, such as CRTS (Catalina Realtime Transient Survey), have already begun to yield impressive scientific results in time-domain astronomy. And projects such as GAIA[1] and DES (Dark Energy Survey) promise to illuminate unprecedented amounts of the time-varying Universe.

2   Cepheid Variable Star Classification

One important challenge in the analysis of astronomical data is that as we move from one survey to another, the nature of the light-curves changes drastically. As an example, some surveys contain rich sources of data in terms of temporal coverage, but the depth is shallow. Other surveys capture objects at extreme depths but for a short time only. And even if we remain within the same survey, analyzing objects that belong to different regions of the sky can bring substantial differences in measurements. All these factors lead to different aspects of data variability.

We now describe a case study where we address the analysis of a rich variety of large surveys under data variability (previous reports can be found in [5,6]). The problem we address is characterized by an original source surveys where class labels for astronomical objects abound, and by a target survey with few class labels, and where feature descriptions may differ significantly (i.e., where marginal probabilities may differ). The problem is also known as *domain adaptation*, or *concept shift*, in machine learning [1,4]. A solution to this common problem in astronomy carries great value when dealing with large datasets, as it obviates the compilation of class labels for the new target set.

Our study is confined to the context of Cepheid variable star classification [5,6], where the goal is to classify Cepheids according to their pulsation modes (we focus on the two most abundant classes, which pulsate in the fundamental and first-overtone modes). Such classification can in fact be attained for nearby galaxies with high accuracy (e.g., Large Magellanic Cloud) under the assumption of class-label availability. The high cost of manually labeling variable stars, however, suggests a different mode of operation where a predictive model obtained on a data set from a source galaxy T_{tr}, is later used on a test set from a target galaxy T_{te}. Such scenario is not straightforwardly attained, as shown in Fig. 1 (left), where the distribution of Cepheids in the Large Magellanic Cloud LMC galaxy (source domain, top sample), deviates significantly from that of M33 galaxy (target domain, bottom sample). In this example, we employ two features only: apparent magnitude in the y-axis, and log period in the x-axis, but our solution is general and allows for a multi-variate representation. Both the offset along apparent magnitude[2], and the significant degree of sample bias, are mostly due to the fact that M33 is $\sim 16\times$ farther than the LMC. Our assumption is then

---

[1] Satellite mission launched in 2013 by the European Space Agency to determine the position and velocity of a billion stars, creating the largest and most precise 3D map of the Milky Way.

[2] Apparent magnitude m, is defined as $m = -2.5 \times \log_{10} \frac{L}{d^2}$, where d is the distance from Earth to the star measured in parsecs, and L is the star luminosity. Hence, smaller numbers correspond to brighter magnitudes (higher fluxes).

that the difference in the joint input-output distribution between the target and source surveys is mainly due to a systematic shift of sample points.

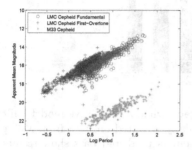 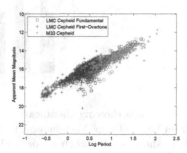

Fig. 1. Left. The distribution of Cepheids along the Large Magellanic Cloud LMC (top sample), deviates significantly from M33 (bottom sample). Right. M33 is aligned with LMC by shifting along mean magnitude.

Our proposed solution shows evidence of the usefulness of domain adaptation in star classification [6]. The main idea consists of shifting T_{te} using maximum likelihood. As an example, if we assume the marginal distribution from which the training is drawn, $P_{\mathrm{tr}}(\mathbf{x})$, is a mixture of Gaussians, we can then estimate parameters directly from our sample T_{tr}, since we know all class labels (i.e., we know which vector belongs to each component or Gaussian). This enables us to have a complete characterization of the marginal distribution: $P_{\mathrm{tr}}(\mathbf{x}) =$

$$\sum_{i=1}^{c} \phi_i \; g_i(\mathbf{x}|\mu_i, \Sigma_i), \; g_i(\mathbf{x}|\mu_i, \Sigma_i) = \frac{1}{(2\pi)^{n/2}|\Sigma_i|^{1/2}} \exp\{-\tfrac{1}{2}(\mathbf{x} - \mu_i)^T \Sigma_i^{-1}(\mathbf{x} - \mu_i)\},$$

where ϕ_i, μ_i, and Σ_i are the mixture coefficient (i.e., prior probability), mean and covariance matrix of the ith component respectively, n is the number of features, and c is the number of components. We can then define a new testing set $T_{\mathrm{te}}' = \{\mathbf{x}'\}$, where $\mathbf{x}' = (x_1 + \delta_1, x_2 + \delta_2 + ... + x_n + \delta_n)$, since we know a shift has occurred along our input features. Our approach is then to find the set of shifts $\Delta = \{\delta_i\}$ that maximizes the log likelihood of T_{te}' with respect to distribution $P_{\mathrm{tr}}(\mathbf{x})$: $\mathcal{L}(\Delta|T_{\mathrm{te}}') = \log \prod_{k=1}^{q} P_{\mathrm{tr}}(\mathbf{x}^k) = \sum_{k=1}^{q} \log P_{\mathrm{tr}}(\mathbf{x}^k)$.

To solve this optimization problem, we used an iterative gradient ascent approach; we search the space of values in Δ for which the log-likelihood function reaches a maximum value. Fig. 2 shows our results; we used Cepheid variables from Large Magellanic Cloud (LMC) as the source domain, and M33 as the target domain. There is a significant increase in accuracy with the data alignment step, which serves as evidence to support our approach.

3   Conclusions and Remarks

The variability of surveys in terms of depth and temporal coverage in astronomy calls for specialized techniques able to learn, adapt, and transfer predictive models from source light-curve surveys to target light-curve surveys. In this paper we

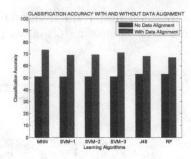

Fig. 2. Plot bars showing classification accuracy when a predictive model built using data from LMC (Large Magellanic Cloud) galaxy is tested on data from M33 galaxy. Blue bars show results when no data alignment is used; red bars show results using data alignment.

show a methodology along this direction that accounts for a data misalignment caused by a systematic data shift.

To end, we point to the importance of exploiting contextual information when modeling astronomical phenomena. This is because the surroundings of a variable source are essential to determine the nature of the object under study. For example, a supernova is easiest to distinguish from other variable and normal objects because it exhibits one brightening episode and then it fades away over weeks. However, if the context reveals the presence of a galaxy nearby, the supernova interpretation becomes much more plausible. A radio source in close proximity to a transient, in contrast, suggests a Blazar classification and is evidence against a supernova. Such contextual information is key to attain accurate predictions, and will become increasingly accessible with the advent of extremely large astronomical surveys.

References

1. Ben-David, S., Blitzer, J., Crammer, K., Kulesza, A., Pereira, F., Wortman, J.: A Theory of Learning from Different Domains. Machine Learning, Special Issue on Learning From Multiple Sources **79**, 151–175 (2010)
2. Ivezic, Z., Connolly, A.J., VanderPlas, J.T., Gray, A.: Statistics, Data Mining, and Machine Learning in Astronomy: A Practical Python Guide for the Analysis of Survey Data, Princeton Series in Modern Observational Astronomy. Princeton University Press (2014)
3. LSST Science Book, version 2.0, 245 authors. http://www.lsst.org/lsst/scibook
4. Quinonero-Candela, J., Sugiyama, M., Schwaighofer, A., Lawrence, N. D.: Dataset Shift in Machine Learning. MIT Press (2009)
5. Vilalta, R., Dhar Gupta, K., Macri, L.: A Machine Learning Approach to Cepheid Variable Star Classification using Data Alignment and Maximum Likelihood. Astronomy and Computing **2**, 46–53 (2013). Elsevier
6. Vilalta, R., Dhar Gupta, K., Macri, L.: Domain adaptation under data misalignment: an application to cepheid variable star classification. In: The 22nd International Conference on Pattern Recognition (ICPR 2014), Stockholm, Sweden (2014)

The Evolution of Social Relationships and Strategies Across the Lifespan

Yuxiao Dong[1], Nitesh V. Chawla[1(✉)], Jie Tang[2], Yang Yang[2], and Yang Yang[1]

[1] Interdisciplinary Center for Network Science and Applications,
Department of Computer Science and Engineering,
University of Notre Dame, Notre Dame, USA
{ydong1,nchawla,yyang1}@nd.edu
[2] Department of Computer Science and Technology,
Tsinghua University, Beijing, China
jietang@tsinghua.edu.cn, yyang.thu@gmail.com

Abstract. In this work, we unveil the evolution of social relationships across the lifespan. This evolution reflects the dynamic social strategies that people use to fulfill their social needs. For this work we utilize a large mobile network complete with user demographic information. We find that while younger individuals are active in broadening their social relationships, seniors tend to keep small but closed social circles. We further demonstrate that opposite-gender interactions between two young individuals are much more frequent than those between young same-gender people, while the situation is reversed after around 35 years old. We also discover that while same-gender triadic social relationships are persistently maintained over a lifetime, the opposite-gender triadic circles are unstable upon entering into middle-age. Finally we demonstrate a greater than 80% potential predictability for inferring users' gender and a 73% predictability for age from mobile communication behaviors.

Our study [1] is based on a real-world large mobile network of more than 7 million users and over 1 billion communication records, including phone calls and text messages (CALL and SMS). Previous work shows that human social strategies used by people to meet their social needs indicate complex, dynamic, and crucial social theories [2]. This work unveils the significant social strategies and social relationship evolution across one's lifespan in human communication. Specifically, we investigate the interplay of demographic characteristics and three types of social relationships, including social ego, social tie, and social triad.

The social strategies that people use to build their ego social networks are observed from Figure 1. The X-axis represents central users' age from 18 to 80 years old and the Y-axis represents the demographic distribution of users' friends, in which positive numbers denote female friends' age and negative numbers denote male friends'. The spectrum color, which extends from dark blue

---

This work was originally published in the 20th ACM SIGKDD International Conference on Knowledge Discovery and Data Mining ($KDD'14$) [1]. This extended abstract has been largely extracted from the publication.

© Springer International Publishing Switzerland 2015
A. Bifet et al. (Eds.): ECML PKDD 2015, Part III, LNAI 9286, pp. 245–249, 2015.
DOI: 10.1007/978-3-319-23461-8_23

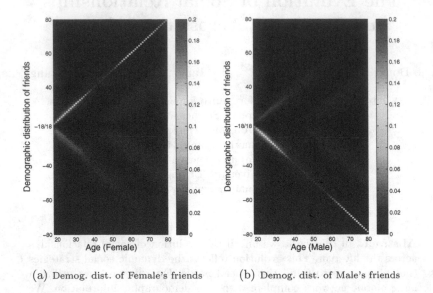

(a) Demog. dist. of Female's friends   (b) Demog. dist. of Male's friends

Fig. 1. Friends' demographic distribution in ego social networks. X-axis: (a) female age; (b) male age. Y-axis: age of friends (positive: female friends, negative: male friends). The spectrum color represents the friends' demographic distribution.

(low) to red (high), represents the probability of one's friends belonging to the corresponding age (Y-axis) and gender (positive or negative). First, the highlighted diagonal lines indicate that people tend to communicate with others of both similar age and gender, i.e., age homophily and gender homophily. Furthermore, we can see that young and middle-age people put increasing focus on the same generation and decreasing focus on the older generation, while the seniors devote more attention on the younger generation even along with the sacrifice of age homophily. Third, we observe that young people are active in broadening social circles (high degree and low clustering coefficient centralities), while seniors tend to keep small but stable connections (low degree and high clustering coefficient centralities).

We further study the social strategies by which people maintain their social tie relationships. The heat maps in Figure 2 visualize the communication frequencies—the number of calls per month between two people with different demographic profiles. Four sub-figures detail the average numbers of calls between two individuals, two males, two females, and one male and one female, respectively. We can see that the interactions between two young males are more frequent than those between two young females (Cf. Figures 2(b) and 2(c)), and moreover, opposite-gender interactions between one young female and male are much more frequent than those between same-gender individuals (Cf. Figure 2(d)). However, reversely, same-gender interactions between two middle-age individuals are more frequent than those between opposite-gender

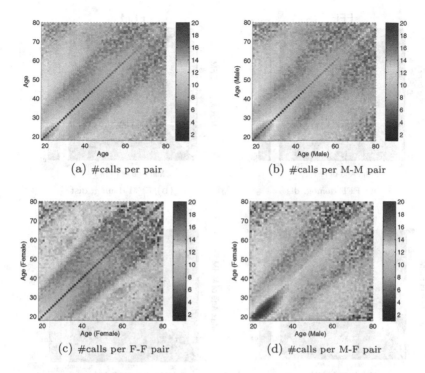

(a) #calls per pair

(b) #calls per M-M pair

(c) #calls per F-F pair

(d) #calls per M-F pair

Fig. 2. Strength of social tie. XY-axis: age of users with specific gender. The spectrum color represents the number of calls per month. (a), (b), and (c) are symmetric.

individuals. In addition to the diagonal lines in each sub-figure, the cross-generation areas that extend from green to yellow indicate that on average 13 calls per month have been made between people aged 20–30 years old and those aged 40–50 years old. The frequent cross-generation interactions are maintained to bridge the age gaps between different generations, such as parents and children, managers and subordinates, and advisors and advisees, etc.

More interestingly, we highlight the social strategies on triadic relationships unveiled from Figure 3, wherein the X-axis and Y-axis denote the minimal and maximal age of three users within a closed social triad. Sub-figures 3(a) and 3(d) show the distributions of same-gender triads: 'FFF' (Female-Female-Female) and 'MMM' (Male-Male-Male), and sub-figures 3(b) and 3(c) present distributions for users' age in opposite-gender triads: 'FFM' and 'FMM'. From heat-map visualization, we observe that people expand both the same-gender and opposite-gender triadic relationships during the dating active period. However, people's attention to opposite-gender circles quickly disappears after entering into middle-age (Cf. Figures 3(b) and 3(c)) and the same-gender triadic relationships are persistent over a lifetime (Cf. Figures 3(a) and 3(d)). To the best of our knowledge, we are the first to discover the instability of opposite-gender triadic relationships and the persistence of same-gender triadic relationships over a lifetime in a large

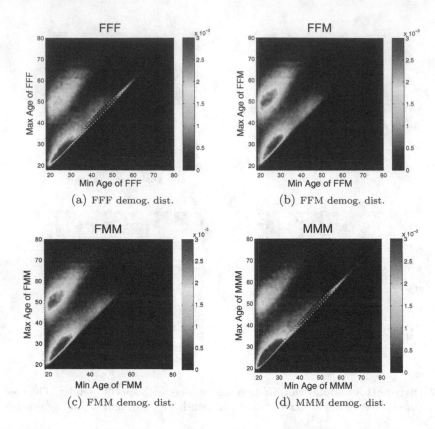

(a) FFF demog. dist.

(b) FFM demog. dist.

(c) FMM demog. dist.

(d) MMM demog. dist.

Fig. 3. Demographic distribution in social triadic relationships. X-axis: minimum age of three users in a triad. Y-axis: maximum age of three users. The spectrum color represents the distributions.

population, which demonstrates the evolution of social strategies that are used by people to meet their social needs in different life stages.

Based on these discovered social strategies, we further study to what extent users' demographic information can be inferred from mobile communication behaviors. The objective is to infer users' gender and age simultaneously by leveraging their interrelations. We present the WhoAmI framework—a Multiple Dependent-Variable Factor Graph model, whereby the social interrelations between users with different demographic profiles can be modeled. On both CALL and SMS networks, the *WhoAmI* method can achieve an accuracy of 80% for predicting users' gender and 73% for users' age according to their daily mobile communication patterns, significantly outperforming several alternative data mining methods.

Acknowledgments. Nitesh V. Chawla, Yuxiao Dong, and Yang Yang are supported by the Army Research Laboratory under Cooperative Agreement Number W911NF-09-2-0053, and the U.S. Air Force Office of Scientific Research (AFOSR) and the Defense Advanced Research Projects Agency (DARPA) grant #FA9550-12-1-0405. Jie Tang and Yang Yang are supported by the National High-Tech R&D Program of China (No. 2014AA015103), Natural Science Foundation of China (No. 61222212), and National Basic Research Program of China (2014CB340506).

References

1. Dong, Y., Yang, Y., Tang, J., Yang, Y., Chawla, N.V.: Inferring user demographics and social strategies in mobile social networks. In: KDD 2014, pp. 15–24. ACM (2014)
2. Palchykov, V., Kaski, K., Kertész, J., Barabási, A.-L., Dunbar, R.I.M.: Sex differences in intimate relationships. Scientific Reports **2**, 370 (2012)

Understanding Where Your Classifier Does (Not) Work

Wouter Duivesteijn[✉] and Julia Thaele

Technische Universität Dortmund, Dortmund, Germany
{wouter.duivesteijn,julia.thaele}@tu-dortmund.de

Abstract. FACT, the First G-APD Cherenkov Telescope, detects air showers induced by high-energetic cosmic particles. It is desirable to classify a shower as being induced by a gamma ray or a background particle. Generally, it is nontrivial to get any feedback on the real-life training task, but we can attempt to understand how our classifier works by investigating its performance on Monte Carlo simulated data. To this end, in this paper we present the SCaPE (Soft Classifier Performance Evaluation) model class for Exceptional Model Mining, which is a Local Pattern Mining framework devoted to highlighting unusual interplay between multiple targets. The SCaPE model class highlights subspaces of the search space where the classifier performs particularly well or poorly. These subspaces arrive in terms of conditions on attributes of the data, hence they come in a language a human understands, which should help us understand where our classifier does (not) work.

1 Introduction

The FACT telescope [1,2] is an Imaging Air Cherenkov Telescope, designed to detect light emitted by secondary particles, generated by high-energetic cosmic particles interacting with the atmosphere of the Earth. For astrophysical reasons, it is important to classify the light as resulting from the atmosphere being hit by a gamma ray or a proton; the latter occur much more frequently, but the former are the more interesting in gamma astronomy. Currently, one of the used classifiers is a random forest, whose performance needs our detailed attention.

The problem with training a classifier on real astrophysical data is that there is no clear feedback. Based on the observed light, we could deduce whether the inducing particle is a gamma ray or a proton. Then, we can look in the direction from which the particle originated, and strive to find an astrophysical source generating gamma rays. But even if we find such a source, there is no certain way of telling what kind of particle induced the original observation. Effectively, we are dealing with a feedbackless learning task, and it is typically hard to finetune a classifier without feedback.

This Nectar Track submission presents the paper [4]. A significantly longer version of that paper appeared as a technical report [5].

© Springer International Publishing Switzerland 2015
A. Bifet et al. (Eds.): ECML PKDD 2015, Part III, LNAI 9286, pp. 250–253, 2015.
DOI: 10.1007/978-3-319-23461-8_24

To study our learning performance, we turn to Monte Carlo data. We simulate particle interactions with the atmosphere, as well as reflections of the resulting Cherenkov light with telescope mirrors on the one hand and the FACT camera electronics on the other hand. This gives us a dataset of camera images that is equivalent in form to a dataset we would get from real astrophysical observations, except that we also know the true label of our classification task. By training our random forest on this dataset, we obtain the soft classifier probabilities for each record. Through studying the interaction between the binary ground truth that we already knew and the soft classifier probabilities we learned from the data, we can understand where our classifier performs exceptionally.

We study this interaction with Exceptional Model Mining (EMM) [3,7]: a Local Pattern Mining framework, seeking coherent subsets of the dataset where multiple targets interact in an unusual way. We present the SCaPE (Soft Classifier Performance Evaluation) model class for EMM, seeking subgroups for which a soft classifier represents a ground truth exceptionally well or poorly. This provides us with insight where our classifier does (not) work.

2   Related Work

Previous work exists on discovering subgroups displaying unusual interaction between multiple targets, for instance in the previously developed model classes for EMM: correlation, regression, Bayesian network, and classification (cf. [3,7]). The last of these model classes is particularly related to the SCaPE model class, with some major differences. Most notably, the classification model class *investigates* classifier *behavior* in the *absence* of a ground truth, whereas the SCaPE model class *evaluates* classifier *performance* in the *presence* of a ground truth. Hence, the two model classes are different means to achieve different ends.

Automated guidance to improve a classifier has been studied in the data mining subfield of meta-learning: how can knowledge about learning be put to use to improve the performance of a learning algorithm? In almost all of the existing meta-learning work, the focus is on letting the machine learn how the machine can perform better. By contrast, the SCaPE model class for EMM focuses on providing *understanding* to the domain expert where his/her classifier works well or fails. As such, SCaPE provides progress on the path sketched by Vanschoren and Blockeel [10, Section 5]: "We hope to advance toward a meta-learning approach that can explain not only *when*, but also *why* an algorithm works or fails [...]". Vilalta and Drissi [11, Section 4.3.1] do devote a subsubsection to "Finding regions in the feature space [...]", but this is in the context of algorithm selection.

A very recent first inroad towards peeking into the classifier black box is the method by Henelius et al. [6], who strive to find groups of attributes whose interactions affect the predictive performance of a given classifier. This is more akin to the classification model class for EMM. While Henelius et al. study hard classifiers, the SCaPE model class is designed for soft classifiers.

3   The SCaPE Model Class for EMM

Exceptional Model Mining (EMM) [3,7] is a framework within *Pattern Mining* [8]: the broad subfield of data mining where only a part of the data is described at a time, ignoring the coherence of the remainder. EMM is a supervised variant of Pattern Mining, typically invoked in a multi-target setting: there are several attributes $t_1, \ldots, t_m$ that are singled out as the *targets* of EMM. The goal of EMM is to find subgroups of the datasets where these targets display an unusual *interaction*. This interaction is captured by the definition of a *model class*, and subgroups are deemed interesting when their model is exceptional, which is captured by the definition of a *quality measure*.

In the SCaPE model class for EMM, we assume a dataset Ω, which is a bag of N records of the form $x = (a_1, \ldots, a_k, b, r)$. We call $\{a_1, \ldots, a_k\}$ the *descriptive attributes*, or *descriptors*, whose domain is unrestricted. The remaining two attributes, b and r, are the *targets*. The first, b, is the *binary target*; we will denote its values by 0 and 1. The second, r, is the *real-valued target*, taking values in $\mathbb{R}$. The goal of the SCaPE model class is to find subgroups for which the soft classifier outputs, as captured by r, represent the ground truth, as captured by b. In [4] and [5], we define a quality measure to assess this quality in a subgroup. Conceptually, the real-valued target r imposes a total order on the records of the dataset. The quality measure considers the ranking of the values of the binary target b under this order, and computes an average ranking loss [9]. This average ranking loss is computed for the entire dataset, and for each subgroup under consideration; the quality of a subgroup is compared to the overall quality in the dataset at hand. Subgroups with a higher-than-usual average ranking loss highlight areas of poor classifier performance, and subgroups with a lower-than-usual average ranking loss highlight areas of good classifier performance.

4   Experimental Results

In [4], we have presented subgroups found on the Monte-Carlo simulated FACT data, along with astrophysical interpretations. Additionally, in [5], we have presented results on nine UCI datasets. These results showcase what the SCaPE model class can unearth in your dataset, and describe problematic areas of the search space for these well-known datasets, which forms an interesting resource for any data miner striving to evaluate their methods on these datasets.

5   Conclusions

In gamma ray astronomy, the separation of gamma and proton showers marks an important step in the analysis of astrophysical sources. Better classifier performance leads to less dilution of the interesting physics results and improves the statement of results of the astrophysical source. The result set will more frequently contain the infrequently appearing gamma showers, which should increase the effective observation time. Due to the importance of the separation in this field, understanding why the classifier does not perform as desired is extremely valuable. The SCaPE model class for EMM helps to understand how to improve the classifier performance.

Beyond its importance within astrophysics, SCaPE is agnostic of the domain of the dataset it analyzes. In fact, it can be used to assess the performance of any soft classifier on any dataset when a ground truth is available. This makes SCaPE an invaluable tool for any data miner who wants to learn where his/her classifier works well and where its performance can be improved. What one could practically *do* with this knowledge depends on the task at hand. Our FACT experiments teach us at which settings the telescope delivers the best results, which allows us to improve the effectiveness of future observations. One could imagine the benefits of oversampling difficult regions, or learning a more expressive classifier on only the difficult regions of the input space. SCaPE provides you with the understanding where your classifier does (not) work: feel free to reap the benefits of that knowledge in any way you see fit.

Acknowledgments. This research is supported in part by the Deutsche Forschungsgemeinschaft (DFG) within the Collaborative Research Center SFB 876 "Providing Information by Resource-Constrained Analysis", project C3.

References

1. Anderhub, H., Backes, M., Biland, A., et al.: Design and Operation of FACT - the First G-APD Cherenkov Telescope, arXiv:1304.1710 (astro-ph.IM)
2. Bretz, T., Anderhub, H., et al.: FACT – The First G-APD Cherenkov Telescope: Status and Results, arXiv:1308.1512 (astro-ph.IM)
3. Duivesteijn, W.: Exceptional Model Mining, Ph.D. thesis, Leiden University (2013)
4. Duivesteijn, W., Thaele, J.: Understanding where your classifier does (Not) work – the SCaPE model class for EMM. In: Proc. ICDM, pp. 809–814 (2014)
5. Duivesteijn, W., Thaele, J.: Understanding Where Your Classifier Does (Not) Work – the SCaPE Model Class for Exceptional Model Mining, technical report 09/2014 of SFB876 at TU Dortmund (2014)
6. Henelius, A., Puolamäki, K., Boström, H., Asker, L., Papapetrou, P.: A peek into the black box: exploring classifiers by randomization. Data Mining and Knowledge Discovery **28**(5–6), 1503–1529 (2014)
7. Leman, D., Feelders, A., Knobbe, A.J.: Exceptional model mining. In: Daelemans, W., Goethals, B., Morik, K. (eds.) ECML PKDD 2008, Part II. LNCS (LNAI), vol. 5212, pp. 1–16. Springer, Heidelberg (2008)
8. Morik, K., Boulicaut, J.F., Siebes, A. (eds.): Local Pattern Detection. Springer, New York (2005)
9. Tsoumakas, G., Katakis, I., Vlahavas, I.P.: Mining multi-label data. In: Data Mining and Knowledge Discovery Handbook, pp. 667–685. Springer (2010)
10. Vanschoren, J., Blockeel, H.: Towards understanding learning behavior. In: Proc. BENELEARN, pp. 89–96 (2006)
11. Vilalta, R., Drissi, Y.: A Perspective View and Survey of Meta-Learning. Artificial Intelligence Review **18**(2), 77–95 (2002)

Visual Analytics Methodology for Scalable and Privacy-Respectful Discovery of Place Semantics from Episodic Mobility Data

Natalia Andrienko[1,2(✉)], Gennady Andrienko[1,2], Georg Fuchs[1], and Piotr Jankowski[3,4]

[1] Fraunhofer Institute IAIS, Sankt Augustin, Germany
{natalia.andrienko,gennady.andrienko,
georg.fuchs}@iais.fraunhofer.de
[2] City University London, London, UK
[3] San Diego State University, San Diego, USA
pjankows@mail.sdsu.edu
[4] Institute of Geoecology and Geoinformation,
Adam Mickiewicz University, Poznan, Poland

Abstract. People using mobile devices for making phone calls, accessing the internet, or posting georeferenced contents in social media create episodic digital traces of their presence in various places. Availability of personal traces over a long time period makes it possible to detect repeatedly visited places and identify them as home, work, place of social activities, etc. based on temporal patterns of the person's presence. Such analysis, however, can compromise personal privacy. We propose a visual analytics approach to semantic analysis of mobility data in which traces of a large number of people are processed simultaneously without accessing individual-level data. After extracting personal places and identifying their meanings in this privacy-respectful manner, the original georeferenced data are transformed to trajectories in an abstract semantic space. The semantically abstracted data can be further analyzed without the risk of re-identifying people based on the specific places they attend.

1 Introduction

The topic of this presentation, based on [1], is semantic modeling and semantic analysis of mobility data (trajectories of people). Currently, the main approach to attaching semantics to mobility data is comparing the locations of points from trajectories with locations of known places of interest (POI) [2]. This approach, however, cannot identify places having personal meanings, such as home and work. Identifying personal places is a challenging problem requiring scalable methods that can cope with numerous trajectories of numerous people while respecting their personal privacy [3, 4]. Our contribution consists of such an approach and a method for semantic abstraction of mobility data enabling further analyses without compromising personal privacy.

© Springer International Publishing Switzerland 2015
A. Bifet et al. (Eds.): ECML PKDD 2015, Part III, LNAI 9286, pp. 254–258, 2015.
DOI: 10.1007/978-3-319-23461-8_25

A special focus of the paper is episodic mobility data [5, 6], where large temporal and spatial gaps can exist between consecutive records, but the proposed approach also works for data with fine temporal resolution. It adheres to the visual analytics paradigm [7], which combines computational analysis methods, such as machine learning techniques, with interactive visual tools supporting human reasoning.

2   Problem Statement and Methodology Overview

The input **data** are episodic human mobility traces, such as records about the use of mobile phones or other mobile devices at various locations. Each record includes a person's (user's) identifier, time stamp, and location specification, which may be geographic coordinates or a reference to some spatial object with known coordinates, such as mobile network antennas. The **goal** is to obtain "semantic trajectories" [2], in which the geographic locations are substituted by semantic labels denoting the meanings of the visited places or types of activities performed there, e.g., 'home', 'work', 'eating', 'recreation', etc. The transformation needs to be done for a large set of individuals in such a manner that their geographic positions are hidden from the analyst. The resulting semantic trajectories are devoid of geographic positions and thus can be viewed and further analyzed without compromising individuals' location privacy.

For checking the plausibility of places to have this or that meaning, land use (LU) data are utilized. For example, when a set of places is going to be labelled as 'home', it is checked, based on LU data, whether most of them are in residential areas. A possible alternative is data about POI, such as public transport stops, schools, shops, and restaurants, which can be retrieved from geographic databases or obtained from map feature services, such as OpenStreetMap (www.openstreetmap.org). Having POI data, it is possible to derive counts of different POI types inside places or within a specified distance threshold. Before assigning some meaning to a set of places, the compatibility of this meaning with the POI types occurring in these places is checked.

The analytical workflow consists of the following steps:

1. Extract repeatedly visited personal and public places.
2. For each place, compute a time series of visits by hourly intervals within the weekly cycle, i.e., ignoring the specific dates.
3. Attach LU or POI attributes to the places.
4. For each target meaning ('home', 'work', 'eating', 'shopping', etc.):
    4.1. Derive relevant attributes (criteria) from the time series of place visits.
    4.2. Based on the attribute values, select candidate places for the target meaning.
    4.3. Validate the place selection with LU or POI data. Iteratively modify the selection for maximizing the proportion of relevant land uses or POI types.
    4.4. Assign the target meaning to the selected places. Exclude from further analysis the places that have already received meanings.
5. Replace the geographic positions in the input data with the semantic labels (meanings) of the places containing the positions.
6. Create a "semantic space", i.e., a spatial arrangement of the set of place meanings, and treat the transformed data as trajectories in this semantic space.
7. Apply movement analysis methods to the semantic space trajectories.

For place extraction, we have developed a special algorithm that groups position records by spatial proximity. To extract personal places, the positions of each individual are clustered separately; to extract public places, the positions of all people are clustered together. Places are defined by constructing boundaries (spatial convex hulls or buffers) around the clusters of positions. The tool works automatically. It takes input data from the database, processes the data, and puts the resulting place boundaries back in the database without showing them to the analyst.

The task of identifying place meanings requires utilization of a human analyst's background knowledge and cognitive capabilities. This task is supported by interactive visual techniques that only show data aggregated over either the whole set or groups of places and do not allow access to individual data. Multi-criteria ranking is used for identifying the most probable home and work places. The places with the best ranks are considered as candidates for receiving the target meaning. The fitness of the candidates is checked using the statistics of the associated LU or POI classes. The criteria weights can be interactively modified to maximize the proportion of relevant LU or POI classes and minimize the proportion of irrelevant classes. For target meanings other than 'home' and 'work', candidate places are selected through interactive filtering based on relevant temporal attributes and land use or POI information.

3   Feasibility Studies

The feasibility of the approach is demonstrated using two case studies: one with simulated tracks from the VAST Challenge 2014 [8], for which ground truth is available, and the other with real traces built from georeferenced tweets posted during one year within the metropolitan area encompassing San Diego (USA) and the surrounding communities. The datasets contain positions of 35 personal cars and 4,286 Twitter users, respectively. We extracted 202 personal and 41 public places from the VAST Challenge data and 38,225 personal and 9,301 public places from the San Diego data.

By applying our methodology to the VAST Challenge data, we were able to identify the meanings of 170 personal places (84%) and 40 public places out of 41 (97.5%). The results match the available ground truth information. For the San Diego test case, we managed to attach semantic labels to 65% of the personal places and 55.3% of the public places. We were able to identify the probable home places of 3,873 persons (90.4%) and the probable work or study places for 2,171 persons (50.7%). For 1,950 persons (45.5%), it was possible to find both home and work places. The largest class of personal places is 'shopping' (4,695 places). Other large classes include 'eating' (2,194), 'social life' (1,497), which includes places with many visits in the evening and night hours and on the weekend, and 'transport' (1,315). No ground truth is available for checking these results; however, further analysis of the semantically abstracted data (semantic trajectories) corroborates the plausibility of place meaning assignments. Fig. 1 shows an example of a possible avenue to further analysis.

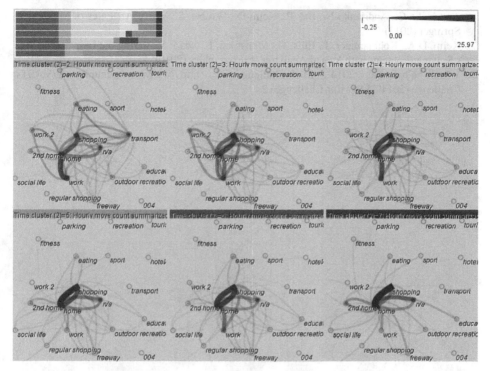

Fig. 1. The background of the maps is a "semantic space", i.e., a 2D spatial arrangement of the set of semantic labels of places. The San Diego data transformed to semantic trajectories have been aggregated into flows between the semantic space locations by hourly time intervals of the weekly cycle. The intervals have been clustered based on the similarity of the sets of the flows; the cluster membership is represented by colors in the calendar view on the top left (each cluster has a specific color). The maps show the flows summarized by the time clusters.

References

1. Andrienko, N., Andrienko, G., Fuchs, G., Jankowski, P.: Scalable and Privacy-respectful Interactive Discovery of Place Semantics from Human Mobility Traces. Information Visualization (2015). doi:10.1177/1473871615581216, Appendix: http://geoanalytics.net/and/papers/placeSemantics/
2. Parent, C., Spaccapietra, S., Renso, C., et al.: Semantic Trajectories Modeling and Analysis. ACM Computing Surveys **45**(4), article 42 (2013)
3. Giannotti, F., Pedreschi, D. (eds.): Mobility, Data Mining and Privacy - Geographic Knowledge Discovery. Springer, Berlin (2008)
4. Cuellar, J., Ochoa, M., Rios, R.: Indistinguishable regions in geographic privacy. In: Ossowski, S., Lecca, P. (eds.) Proc. 27th Annual ACM Symposium Applied Computing (SAC 2012), pp. 1463–1469. ACM, March 26–30, 2012
5. Andrienko, G., Andrienko, N., Stange, H., Liebig, T., Hecker, D.: Visual Analytics for Understanding Spatial Situations from Episodic Movement Data. Künstliche Intelligenz **26**(3), 241–251 (2012)

6. Andrienko G., Andrienko N., Bak P., Keim D., Wrobel, S.: Visual Analytics of Movement. Springer (2013)
7. Keim, D.A., Kohlhammer, J., Ellis, G., Mansman, F. (eds.): Mastering the Information Age - Solving Problems with Visual Analytics, Eurographics (2010)
8. VAST Challenge 2014: Mini-Challenge 2. http://www.vacommunity.org/VAST+ Challenge+2014%3A+Mini-Challenge+2

Will This Paper Increase Your h-index?

Yuxiao Dong, Reid A. Johnson, and Nitesh V. Chawla$^{(\boxtimes)}$

Interdisciplinary Center for Network Science and Applications,
Department of Computer Science and Engineering,
University of Notre Dame, Notre Dame, IN, USA
{ydong1,rjohns15,nchawla}@nd.edu

Abstract. A widely used measure of scientific impact is citations. However, due to their power-law distribution, citations are fundamentally difficult to predict. Instead, to characterize scientific impact, we address two analogous questions asked by many scientific researchers: "How will my h-index evolve over time, and which of my previously or newly published papers will contribute to it?" To answer these questions, we perform two related tasks. First, we develop a model to predict authors' future h-indices based on their current scientific impact. Second, we examine the factors that drive papers—either previously or newly published—to increase their authors' predicted future h-indices. By leveraging relevant factors, we can predict an author's h-index in five years with an R^2 value of 0.92 and whether a previously (newly) published paper will contribute to this future h-index with an F_1 score of 0.99 (0.77). We find that topical authority and publication venue are crucial to these effective predictions, while topic popularity is surprisingly inconsequential. Further, we develop an online tool that allows users to generate informed h-index predictions. Our work demonstrates the predictability of scientific impact, and can help researchers to effectively leverage their scholarly position of "standing on the shoulders of giants."

Scientific impact plays a pivotal role in the evaluation of the output of scholars, departments, and institutions. A widely used measure of scientific impact is citations, with a growing body of literature focused on predicting the number of citations obtained by any given publication. The effectiveness of citation prediction, however, is fundamentally limited by their power-law distribution, whereby publications with few citations are extremely common and publications with many citations are relatively rare. In light of this limitation, we instead investigate scientific impact by addressing two analogous questions [1], both related to the measure of h-index [2] and asked by many academic researchers: *"How will my h-index evolve over time, and which of my previously and newly published papers will contribute to my future h-index?"*

Y. Dong and R.A. Johnson—Provided equal contribution to this work.

This work was published at the 8th ACM International Conference on Web Search and Data Mining (*WSDM'15*) [1]. This extended abstract has been largely extracted from the publication.

© Springer International Publishing Switzerland 2015
A. Bifet et al. (Eds.): ECML PKDD 2015, Part III, LNAI 9286, pp. 259–263, 2015.
DOI: 10.1007/978-3-319-23461-8_26

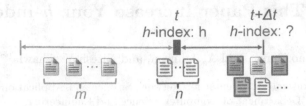

Fig. 1. **Illustrative example of scientific impact prediction.** Before time t, a scholar published m papers and had an h-index of h. Our prediction problems are targeted at answering two questions: 1) First, what is the scholar's future h-index, h', at time $t+\Delta t$? 2) Second, which of his/her papers, both (a) those m papers previously published before t and (b) those n new papers published at t, will contribute to h'?

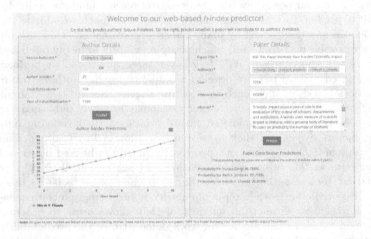

Fig. 2. **Prototype h-index prediction tool.** The left side may be used to predict the development of authors' h-indices and the right side may be used to predict whether a paper will contribute to its authors' h-indices.

To tackle these questions, we formulate two scientific impact prediction problems, as shown in Figure 1. Our primary problem is to determine whether a given previously or newly published paper will, after a predefined timeframe, influence a particular author's predicted *future* h-index. As a secondary problem, we predict authors' future h-indices based on their current scientific impact. These predicted future h-indices are then used as the future h-indices in our primary task, with the purpose of accounting for the change in the author's h-index over the prediction timeframe. Besides addressing these problems, we have also developed and deployed an online tool (see Figure 2) that allows users to generate h-index predictions informed by our findings.

Using a large-scale academic dataset with over 1.7 million authors and 2 million papers from the premier online academic service ArnetMiner [3], we demonstrate a high level of predictability for scientific impact as measured by our two problems in Figure 3. Accordingly, we find strong performance for our first task of predicting an author's future h-index. We can predict an author's h-index

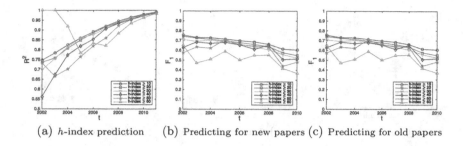

(a) h-index prediction   (b) Predicting for new papers (c) Predicting for old papers

Fig. 3. Scientific Impact Predictability. (a) Predicting authors' future h-indices; (b) Predicting whether newly published papers can increase their primary authors' future h-indices; (c) Predicting whether previously published papers can increase their primary authors' future h-indices.

in five years with an R^2 value of 0.9197. This performance generally increases as the prediction timeframe is shortened, with a prediction of ten years achieving an R^2 of 0.7461. We can predict whether a previously or newly published paper will contribute to an author's future h-index in five years with respective F_1 scores of 0.99 and 0.77, improvements of +130% and +160% over random guessing. Predictive performance for newly published papers generally decreases as the prediction timeframe is shortened, but is consistently high for previously published papers. Our results also indicate that authors with low h-indices are easier to predict for than those with high ones.

We also assess the factors that influence our predictive results. For our secondary problem, we find that the author's current h-index is most telling, followed by the number of publications and co-authors. For our primary problem, we investigate six groups of factors that drive a paper's citation count to become greater than its primary author's h-index, including the paper's author(s), content, published venue, and references, as well as social and temporal effects related to its author(s). Figure 4 shows the response curve of the most important factor (as evaluated by correlation coefficients) for each group of factors. We find that topical authority is the most telling factor for newly published papers, while the existing citation information is most telling for previously published ones, followed by the authors' influence and the publication venue. We also find that publication venue and the author collaborations are moderately significant factors for longer prediction periods, but inconsequential for shorter ones. Finally, we are surprised to find that topic popularity is insignificant for both previously and newly published papers.

Overall, our findings unveil the predictability of scientific impact, deepen the understanding of scientific impact measures, and provide scholars with concrete suggestions for expanding their scientific influence. Salient points include:

- A scientific researcher's authority on a topic is the most decisive factor in facilitating an increase in his or her h-index. This coincides with the fact that the society fellows or lifetime honors are typically awarded for contributions to a topic or domain.

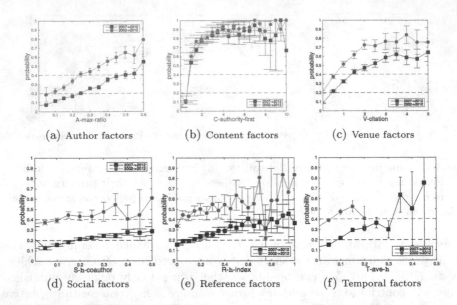

(a) Author factors   (b) Content factors   (c) Venue factors

(d) Social factors   (e) Reference factors   (f) Temporal factors

Fig. 4. Factor response curves with $\Delta t = 5$ or 10 $(t + \Delta t = 2012)$. x-axis: factor value; y-axis: probability that a given paper published in t increase its primary author's h-index in 2012. *A-max-ratio*: the ratio between max-h-index and #papers attributed to the primary author; *C-authority-first*: the consistence between the first author's authority and this paper; *V-citation*: the #average-citations of papers published in this venue; *S-h-coauthor*: the average h-index of co-authors of the paper's authors; *R-h-index*: the references' h-index; *T-ave-h*: the average Δh-indices of the authors between now and three years ago. All response probabilities are observed at a 95% confidence interval.

– The level of the venue in which a given paper is published is another crucial factor in determining the probability that it will contribute to its authors' h-indices. The suggestion here lies in the every scholar's aim: *Target and publish influential scientific results in top venues.*
– Publishing on an academically "hot" but unfamiliar topic is unlikely to further one's scientific impact, at least as measured by an increase in one's h-index. This reminds us that *one should not turn to follow the vogue topics that are beyond his or her expertise.*

We strongly believe that our findings can help lead to the improved use of scientific impact measures, though we caution that *in no way should our research be construed as advocating the use of the h-index or any other measure as a deciding factor in one's research pursuits.*

Acknowledgments. This work is supported by the Army Research Laboratory under Cooperative Agreement Number W911NF-09-2-0053, the U.S. Air Force Office of Scientific Research (AFOSR) and the Defense Advanced Research Projects Agency (DARPA) grant #FA9550-12-1-0405, and the National Science Foundation (NSF) Grant OCI-1029584.

References

1. Dong, Y., Johnson, R.A., Chawla, N.V.: Will this paper increase your h-index?: Scientific impact prediction. In: WSDM 2015, pp. 149–158. ACM, New York (2015)
2. Hirsch, J.E.: An index to quantify an individual's scientific research output. PNAS **102**(46), 16569–16572 (2005)
3. Tang, J., Zhang, J., Yao, L., Li, J., Zhang, L., Su, Z.: ArnetMiner: Extraction and mining of academic social networks. In: KDD 2008, pp. 990–998 (2008)

Demo Track

CardioWheel: ECG Biometrics on the Steering Wheel

André Lourenço[1,2]([⊠]), Ana Priscila Alves[1,2], Carlos Carreiras[1,2],
Rui Policarpo Duarte[1], and Ana Fred[1,2]

[1] CardioID Technologies Lda, Lisbon, Portugal
[2] Instituto de Telecomunicações, Lisbon, Portugal
arlourenco@lx.it.ptz

Abstract. Monitoring physiological signals while driving is a recent trend in the automotive industry. We present CardioWheel, a state-of-the-art machine learning solution for driver biometrics based on electrocardiographic signals (ECG). The presented system pervasively acquires heart signals from the users hands through sensors embedded in the steering wheel, to recognize the driver's identity. It combines unsupervised and supervised machine learning algorithms, and is being tested in real-world scenarios, illustrating one of the potential uses of this technology.

Keywords: Electrocardiographic signals (ECG) · Biometrics · Automotive industry · Personalization · Security

1 Introduction

Automatic personalization of car settings, based on the driver's identity, is becoming a standard in the automotive industry. Examples of these adjustable settings include seat and mirror positions, favorite radio stations, address lists, among others. Traditionally, this problem has been tackled by resorting to something the user has, like a personal physical key. Recently, with the arrival of connected car systems, such as Android Auto and Apple Carplay, driver authentication is routed through the smartphone, which typically has an associated user identity. Moreover, integration with cloud services increases the range of personalization systems, leveraging the available digital footprint of the user. Biometric recognition, which makes use of intrinsic measurable properties of the user, is an alternative authentication method with undisputed advantages. The inclusion of biometrics into cars is not new [2,10], though it has not been widely adopted, mainly due to usability issues.

Our system uses the heart signal to enable driver recognition, in a similar fashion as with a fingerprint. The electrocardiographic (ECG) signal is acquired from the driver's hands with sensors embedded in the steering wheel, continuously and unobtrusively while the user is driving. This allows for biometrics to be integrated in cars in an innovative way.

A. Bifet et al. (Eds.): ECML PKDD 2015, Part III, LNAI 9286, pp. 267–270, 2015.
DOI: 10.1007/978-3-319-23461-8_27

2 Target Users

According to a recent report [1], half of all surveyed consumers purchase cars based on the brand's technological reputation. In the survey, consumers demonstrated how information and technology are crucial throughout the car experience. Additionally, consumers are willing to disclose personal information for customization, security and savings, with 60% of the population willing to provide biometric information, such as fingerprints and DNA samples, in return for personalized security or car security.

Major brands are starting to introduce biometric technology in their products. For example, Ford [3], in a patent approved in January 2015, outlines a system that uses a smartphone to connect to a car's controller over either Bluetooth or Wi-Fi that allows to lock and unlock the doors, in combination with a biometric capture device (including a retinal scan, a fingerprint sensor, voice recognition, or face recognition).

One example of application is car-sharing, where the use of biometrics makes driver swapping easier, nixing the reliance on a physical key to open and start the vehicle. Additionally, it could also be used to restrict certain features based on the driver, e.g. restricting the speed for a young driver, or allowing only to drive at specific times.

3 System Overview

The CardioWheel solution monitors the ECG while driving, extracting relevant information related with the driver's identity and health state. System developments were made both in Hardware and Software. The Hardware needs to handle two main tasks: i) ECG acquisition and analog filtering; ii) Signal processing and classification.

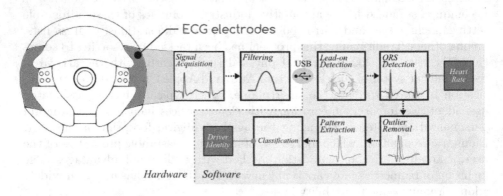

Fig. 1. System Architecture

The acquisition of physiological data in the steering wheel of the car represents a big challenge. We followed recent trends towards an off-the-person sensing

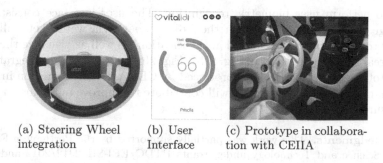

(a) Steering Wheel integration   (b) User Interface   (c) Prototype in collaboration with CEIIA

Fig. 2. System integration in the steering wheel

approach [9], designing our system for ECG acquisition using conductive fabric electrodes. The signals thus acquired have a lower Signal to Noise Ratio (SNR) than when compared to traditional approaches. However, the main advantage of this method is that it can be easily integrated in the steering wheel, without impacting on the driver's usual behavior. One additional challenge are the artifacts introduced by the highly dynamic setting, as, for example, when the driver may have one of the hands off of the steering wheel (e.g. when changing gears). Commonly used state-of-the-art ECG signal processing algorithms are not designed to handle such dynamic settings, having been developed for clinical-grade equipment in stationary environments.

We have designed and produced a custom-made PCB board focused on noisy ECG signals, which includes a one-lead ECG sensor at the hands [7], an ARM processor for signal acquisition and filtering, and an isolation stage to guarantee user protection. Signal processing and classification are performed on an Intel Edison computational unit.

Our work follows a partially fiducial framework, and in Fig. 1 we depict the block diagram of the proposed biometric system. After signal acquisition and filtering, the system detects the presence of the driver's hands on the sensor, with a windowing approach. Signal windows with hands present are then forwarded to a QRS detection block [4], which segments the signal into heartbeats and also estimates the heart rate. Anomalous heartbeats are discarded with an unsupervised method [5]. The pattern extraction block takes the preprocessed input heartbeats to compute a mean template from 5 consecutive segments [8]. The templates are then fed to a classifier, based on support vector machines (SVM), following the approach proposed in [6].

4   Demonstration Prototype

The CardioWheel integration on a real-world scenario was made by sewing a conductive fabric to a standard steering wheel cover, as illustrated in Fig. 2(a). Each electrode is connected to our custom PCB board, where the ECG acquisition takes place. The ECG board is connected via USB to the Intel Edison board to perform user biometric recognition. The user interface consists of a web page that communicates with the Edison board via websockets, allowing the user to perform enrollment and authentication, as well as visualize the heart rate in real time, as depicted in Fig. 2(b). Fig. 2(c) represents the integration of our prototype in an electrical car produced by CEIIA. This integration in a real life scenario produces data, which will contribute to improve our solution in the near future.

Acknowledgments. This work was partially supported by the Portuguese Foundation for Science and Technology under grants PTDC/EEI-SII/2312/2012 and PEst-OE/EEI/LA0008/2013. We also want to thanks Hugo Silva and José Guerreiro for their colaboration on the development of the first prototypes.

References

1. Cisco: Consumers Desire More Automated Automobiles, According to Cisco Study (2014). http://newsroom.cisco.com/press-release-content?type=web content&articleId=1184392
2. Gilpin, L.: Ford moves toward facial recognition and gesture UI in the car (2014). http://www.zdnet.com/article/ford-moves-toward-facial-recognition-and-gesture-ui-in-the-car/
3. Lavrinc, D.: Ford Wants To Let You Open Your Car With Your Eyeballs (2015). http://jalopnik.com/ford-wants-to-let-you-open-your-car-with-your-eyeballs-1683328257
4. Lourenço, A., Silva, H., Lourenço, R.L., Leite, P.L., Fred, A.L.N.: Real time electrocardiogram segmentation for finger based ECG biometrics. In: Proc International Conf. on Bio-inspired Systems and Signal Processing (BIOSIGNALS), pp. 49–54, February 2012
5. Lourenço, A., Silva, H., Carreiras, C., Fred, A.: Outlier detection in non-intrusive ECG biometric system. In: Kamel, M., Campilho, A. (eds.) ICIAR 2013. LNCS, vol. 7950, pp. 43–52. Springer, Heidelberg (2013)
6. Lourenço, A., Silva, H., Fred, A.L.N.: ECG-based biometrics: A real time classification approach. In: IEEE International Workshop on Machine Learning for Signal Processing (MLSP), pp. 1–6, September 2012
7. Silva, H., Lourenço, A., Lourenço, R., Leite, P., Coutinho, D., Fred, A.: Study and evaluation of a single differential sensor design based on electro-textile electrodes for ECG biometrics applications. In: Proc. IEEE Sensors Conference, pp. 1764–1767, October 2011
8. Silva, H., Gamboa, H., Fred, A.: Applicability of lead v_2 ECG measurements in biometrics. In: Proc. of the Int. eHealth, Telemedicine and Health ICT Forum (Med-e-Tel), pp. 177–180, April 2007
9. Silva, H., Lourenço, A., Canento, F., Fred, A., Raposo, N.: ECG biometrics: principles and applications. In: Proc. of the 6th Conf. on Bio-Inspired Systems and Signal Processing (BIOSIGNALS) (2013)
10. Volkswagen: Biometric Driver Identification (2015). http://www.volkswagenag.com/content/vwcorp/content/en/innovation/communication_and_networking/Biometric.html

Data Patterns Explained with Linked Data

Ilaria Tiddi [✉], Mathieu d'Aquin, and Enrico Motta

Knowledge Media Institute, The Open University, Milton Keynes, UK
{ilaria.tiddi,mathieu.daquin,enrico.motta}@open.ac.uk

Abstract. In this paper we present the system Dedalo, whose aim is to generate explanations for data patterns using background knowledge retrieved from Linked Data. In many real-world scenarios, patterns are generally manually interpreted by the experts that have to use their own background knowledge to explain and refine them, while their workload could be relieved by exploiting the open and machine-readable knowledge existing on the Web nowadays. In the light of this, we devised an automatic system that, given some patterns and some background knowledge extracted from Linked Data, reasons upon those and creates well-structured candidate explanations for their grouping. In our demo, we show how the system provides a step towards automatising the interpretation process in KDD, by presenting scenarios in different domains, data and patterns.

1 Introduction

In Knowledge Discovery in Databases (KDD), patterns are defined as "statements or expressions describing an interesting relationship among a subset of the analysed data, which is typically resulted from a data mining process (classification, cluster, sequence-pattern mining, association rules mining and so on)" [1]. Our work focuses on the KDD step following the data mining one, i.e. the process of pattern interpretation.

Let us imagine that we aim at explaining why a term such as *A Song of Ice and Fire* is searched over the Web only at specific times of the year: this is shown in Figure 1a, where one can observe regular popularity peaks. Such a pattern can only be explained by someone who, having background knowledge about the fantasy novels, can explain that the popularity increases in those periods in which a new Game of Thrones TV season or a new novel is released. In many other real-world contexts, the revealed patterns are generally provided to experts that analyse, refine and interpret them in order to reuse them for further purposes. For instance, patterns can be used in Business Intelligence for decision making, in E-commerce for item recommendation, in Learning Analytics for assisting people's learning. Producing pattern explanations becomes then an intensive and time-consuming activity, particularly when the background knowledge needs to be gathered from different domains and sources.

With that said, we state that the Web knowledge in the form of Linked Data can facilitate the problem of interpreting Knowledge Discovery patterns. Linked

© Springer International Publishing Switzerland 2015
A. Bifet et al. (Eds.): ECML PKDD 2015, Part III, LNAI 9286, pp. 271–275, 2015.
DOI: 10.1007/978-3-319-23461-8_28

Data refer to a set of best practices for publishing and connecting structured data on the Web [2], with the purpose of fostering reuse, linkage and consumption of data. Thanks to their well-established principles (use of HTTP URIs for naming, provision of useful information about data, and inclusion of links to connect to external resources), Linked Data consist nowadays in a globally available knowledge graph, where several datasets are represented in RDF standards, can be accessed and understood by machines and, most importantly, are connected across disciplines. In our example, it is possible to use information about events (e.g., times and topics) to detect that the peaks of popularity correspond to moments where there have been events somehow related to the book series *A Song of Ice and Fire*.

In this demo, we present Dedalo, a tool to generate Linked Data candidate explanations, as in Figure 1b, from data mining patterns such as the one of Figure 1a. We will show how Dedalo can be applied to patterns and scenarios of different nature, thanks to the variety of domains existing within Linked Data. Our work aims at proving that Linked Data can help turning the interpretation process into an automatic process relieving the manual effort of the experts.

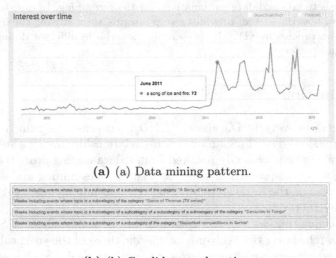

(a) (a) Data mining pattern.

(b) (b) Candidate explanations.

Fig. 1. Dedalo Explanation Visualizer. In (a) the pattern *A Song of Ice and Fire* chosen by the user. In (b) Dedalo gives the best candidate explanations based on the Linked Data knowledge.

2   Dedalo's Implementation

We implemented Dedalo as a system that integrates the following modules: an *Explanation Generator*, which produces explanations based on an Inductive Logic Programming (ILP) strategy; a *Background Knowledge Builder*, which builds the background knowledge using an A* strategy to traverse the Linked

Data graph and collect the salient information, and an *Explanation Visualiser* which finally presents the users both the pattern to explain and the generated candidate explanations.

Explanation Generator. This module is designed with the idea that it is possible to learn why some items, considered as the positive examples, belong to a pattern, while some others, the negative examples, do not. Following common Machine Learning approaches, the initial data are therefore organised in positive and negative observations to learn from. More specifically, given a pattern to be explained which is selected by a user, the items belonging to it will be considered as positive examples, while the ones not belonging to it will be the negative examples. In the example of the search term *A Song of Ice and Fire*, each search rate evaluated as a peak is considered as positive example, while the remaining are considered as negative examples. The aim of this module is to derive candidate explanations which cover a maximum number of positive examples and a minimum number of negative examples, e.g., high search rates correspond to events somehow related to the fantasy series. As in Inductive Logic Programming, explanations are derived by reasoning upon the background knowledge about both the positive and negative examples, which is built using statements extracted directly from Linked Data.

Background Knowledge Builder. This module automatically and iteratively builds the background knowledge from Linked Data. The assumption here is that it is not feasible to import the whole knowledge represented in Linked Data (also, most of the knowledge might indeed not be relevant). On the other hand, it is possible to iteratively extend the background knowledge about the data,

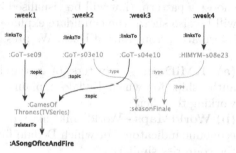

Fig. 2. Dedalo's built graph.

with the aim of deriving explanations which represent a bigger portion of positive examples (i.e., the pattern to explain). Starting from the URI representation of the items in the dataset, which in our case consists in weeks of a year in which the term is searched, a graph is iteratively built by following the URIs links and exploring ("dereferencing") the new discovered entities. In this way, no a priori knowledge is introduced in the process: The graph exploration is simply carried out by following existing links between Linked Data URIs. The Linked Data traversal relies on the assumption that data are connected and therefore data sources can be easily and naturally spanned to gather new, unknown knowledge to reason upon. For instance in Figure 2 many of the weeks are linked to aired TV episodes (through the relation *:linksTo*), and some of those are further linked to the Game of Thrones TV series (following the relation *:topic*), which in turn is linked to *A Song of Ice and Fire* through *:relatesTo*.

Explanation Visualiser. The final module consists in presenting to the user the candidate explanations that have been found for the pattern he had initially chosen. Candidate explanations consists in a path of RDF properties and one final entity that a subset of items have in common: In the example of Figure 2, one of the explanations we can derive is shown as e_1=⟨:linksTo.:topic.:relatesTo→:ASo- ngOfIceAndFire⟩. The evaluation of the candidate explanations is shaped as in a classification task, where the classifier prediction is represented by the number of positive examples that the explanation covers, and the external judgment consists in the whole set of positive examples. The closer those two sets are, the better the explanation represents the pattern to explain, and the better it is evaluated. The best explanations are then visualised and presented to the user in natural language, as in Figure 1b, where we show the candidate explanations generated for the search term *A Song of Ice and Fire*.

3   Demo Scenarios

During the demonstration we will present to the audience scenarios of different nature, following use-cases of our previous works [3]. Users will be allowed to choose a pattern, that will be visualised in the way it is provided to Dedalo, and will be also shown the candidate explanations that Dedalo has derived with the information from Linked Data. We present them below.

(a) KMiData - Clusters of researchers grouped according to their co-authorship, for which Dedalo explains the reasons for which the authors are working together;
(b) WorldMaps - Worldbank[1] maps of countries grouped according to different economic indicators, for which Dedalo finds socio-economical reasons explaining the countries similarity;
(c) Education - Clusters of words grouped according to their semantic similarity, for which Dedalo is able to find which topic relates them;
(d) Trends - Trends of topics searched over the last 10 years, for which Dedalo explains the peaks of popularity.

We will encourage the audience in understanding the efforts required to give an explanation for a pattern and will show the benefits of using Linked Data to assist the explanation process. By using different scenarios we intend to make a step forward in the automatisation of the pattern interpretation process.

References

1. Fayyad, U., Piatetsky-Shapiro, G., Smyth, P.: From data mining to knowledge discovery in databases. AI Magazine **17**(3), 37 (1996)

---

[1] http://www.worldbank.org/

2. Heath, T., Bizer, C.: Linked data: Evolving the web into a global data space. Synthesis Lectures on the Semantic Web: Theory and Technology 1(1), 1–136, (2011). Semantic Web: Research and Applications (pp. 560–574). Springer, Berlin Heidelberg
3. Tiddi, I., d'Aquin, M., Motta, E.: Dedalo: Looking for clusters explanations in a labyrinth of linked data. In: Presutti, V., d'Amato, C., Gandon, F., d'Aquin, M., Staab, S., Tordai, A. (eds.) ESWC 2014. LNCS, vol. 8465, pp. 333–348. Springer, Heidelberg (2014)

GAZOUILLE: Detecting and Illustrating Local Events from Geolocalized Social Media Streams

Pierre Houdyer[1], Albrecht Zimmerman[2], Mehdi Kaytoue[2(✉)], Marc Plantevit[3], Joseph Mitchell[1], and Céline Robardet[2]

[1] Tapastreet Ltd., 36 Dame Street, Dublin, Ireland
{pierre,joe}@tapastreet.com
[2] INSA-Lyon, CNRS, LIRIS UMR5205, 69621 Lyon, France
{albrecht.zimmerman,mehdi.kaytoue,celine.robardet}@liris.cnrs.fr
[3] Université Lyon 1, CNRS, LIRIS UMR5205, 69622 Lyon, France
marc.plantevit@liris.cnrs.fr

Abstract. We present GAZOUILLE, a system for discovering local events in geo-localized social media streams. The system is based on three core modules: (i) social networks data acquisition on several urban areas, (ii) event detection through time series analysis, and (iii) a Web user interface to present events discovered in real-time in a city, associated to a gallery of social media that characterize the event.

1 Introduction

Social networks (such as Twitter, Instagram, ...) are rich sources of information that can be used to build a huge number of applications and services for certain end-users (b2c), for companies, e.g. with analytics platforms (b2b), but also to help governments and charitable organizations. Through several public APIs, one can access streams of messages, often provided with text (including hashtags, user mentions and URIs), media (images or video) and geo-tags indicating the position of the user emitting the message (called *post* in the sequel).

One way of exploiting such data is to discover global trends and detecting events in the streams of posts. The motivations are manifold: disaster detection, epidemic surveillance, identification of newsworthy events that traditional media are slow to pick up, identification of trends, monitoring of brand perception, etc. The question of how to identify events in streams of text data has been a research topic for more than a decade now, starting from e-mail, via blog posts, to location-based social networks data [2]. The general idea underlying most of that work is identifying "bursty" topics (mentioned significantly more often during a time period than in the period preceding it).

Whereas most of the existing systems identify global trends, only a few take into account the geo-localization of the posts for detecting local events [3]. This is actually the goal of GAZOUILLE: harvesting data from urban areas, the system is able to detect spatially circumscribed events in real-time and to intelligibly

This work was supported by the project GRAISearch (FP7-PEOPLE-2013-IAPP).

A. Bifet et al. (Eds.): ECML PKDD 2015, Part III, LNAI 9286, pp. 276–280, 2015.
DOI: 10.1007/978-3-319-23461-8_29

characterize them (with their periodicity, users, key-words, and via a media gallery, e.g. in Figure 2). In what follows, we present an overview of GAZOUILLE (Section 2) and a use-case on the city of New-York (Section 3).

2   System Overview

The system architecture is illustrated in Figure 1: a crawler engine harvests social networks (e.g. Twitter and Instagram) in specific locations (cities) and populates a database with posts. An event detection module is running continuously with a sliding window, and enters detected events into the database. Finally, a Web user interface allows to choose a time window, and explore the most highly expressed events with an intuitive characterization (see elements in Figures 2 and 3). We explain the different modules now.

Data acquisition. For acquiring data, trackers are set on a selected city, on which a grid is defined: each cell gives rise to a geo-tagged query on each social network every 5 minutes (default refresh rate). The back-end is realized in *Ruby*, connects to data providers' official APIs, and stores data in a *PostgreSQL* DBMS.

Data preparation. From each geo-tagged post, we extract meaningful keywords (e.g. hashtags and user mentions from *tweets*). Natural language processing tools could also be used here for stemming, lematization, etc. In the end, a city gives rise to a single stream of pairs $(timestamp, \{word_1, ..., word_n\})$.

Event detection. The detection is performed in a window of a given size. Each time the trackers refresh, the window is right-slid and a new detection is performed. Based on our review of the state-of-the-art, we have selected a lightweight method for bursty term detection [1] and implemented it with several modifications to handle texts from social networks (the method was originally designed analyzing news corpus). The method described in [1] transforms each terms' *document frequency - inverse document frequency* (DFIDF) scores over time via a Discrete Fourier Transform and derives its *periodicity* and *strength of expression* from the resulting periodogram. Originally, stopwords are used to identify "irrelevant" terms but since those are not available in our settings (and difficult to derive for Twitter in general), we classify all terms with less than average expression thus (denoted as 'L' for *Low* in Figures 2 and 3, 'H' for *high* otherwise). Individual bursts are modeled as *Gaussians*.

Front-end. The user selects a window of time in which events are detected. Bursty terms are ranked w.r.t. their strength of expression. The user selects the

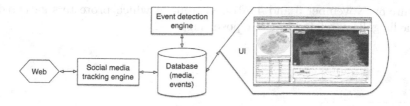

Fig. 1. System architecture

terms he is interested in and the rest of the interface updates (details hereafter). We used *Google Charts*[1] and *magic wall*[2] UI components.

3   Use Case: New York City

We collected tweets posted from October the 31st to November the 4th 2014 in New York City, covering the districts of Brooklyn, Manhattan, Queen's, and Staten Island (geo-tagged queries to the Twitter API every five minutes). We accordingly tracked $200,000$ geo-localized tweets in a period containing the 2014 NYC marathon (November 2). Our goal is to validate that this event will be discovered in time and space, but we should also be able to identify several other events that happened in this area and time frame. The event detector engine is run with a sliding window of 128 time stamps, each corresponding to ten minutes.

As an end user, the window in which discovered events should be given can be selected and moved. We set the window to the day of the marathon (note that dates are in Dublin GMT timezone), see Figure 2. The ordered list of bursty terms is then refreshed automatically. In this case the most strongly expressed terms are #nycmarathon and @nycmarathon. Expression is given in percentage of expression w.r.t. the maximum, i.e. the most bursty term. We then select those two first terms and all tweets gathered by our initial trackers that involve these terms are displayed on the map. The shape of these geo-localized tweets strongly resembles the known marathon course (left map bordered in black). The third bursty terms also concerns the marathon. The fourth term is the user mention of a singer from Harlem that released a song that day, freely available on the Web, thus shared and discussed on Twitter. Finally, the last given bursty term concerns @CRinQC who we identified as a Republican tweeter whose criticism of President Obama was re-tweeted by noted New York business and former Republican presidential hopeful Donald J. Trump, amplifying his expression. The user can then discover other terms in the list.

When shifting the time window one day later (see Figure 3), the interface updates. We now have a new list of bursty terms. Whereas tags related to the New York marathon are still present, they are not the burstiest terms anymore. We select the four most strongly expressed terms (*@dalailamatruth*, *#dalailam-abeacon*, *#religiousfreedom*, and *#dalailama*) since one may assume they concern the Dalai Lama. Plotting the concerned geo-localized tweets on the map result in a very concentrated area: the Beacon theater where the Daila Lama was giving a lecture on November 3 and 4 2014, in front of which protesters gathered, as the media gallery related to these posts suggests.

---

[1] https://developers.google.com/chart/interactive/docs/gallery/timeline
[2] http://teefouad.com/plugins/magicwall

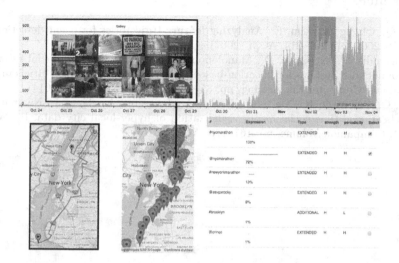

Fig. 2. Detecting events in New-York on November 2 with selected bursty terms *#nyc-marathon* and *@nycmarathon*, corresponding geo-localized tweets (right map), and the known NYC 2014 marathon course (left map). In the table, expression is given in percentage with respect to the top expressed term, while expression and periodicity are given as high ('H') and low ('L'), i.e. above/below their average value.

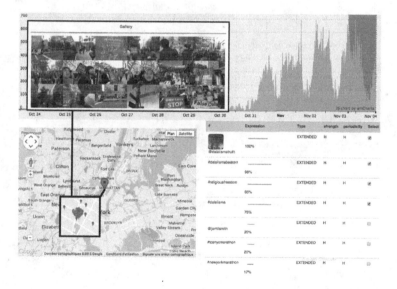

Fig. 3. Detecting events in New-York on November 3 with selected bursty terms *@dalailamatruth*, *#dalailamabeacon*, *#religiousfreedom*, and *#dalailama* and corresponding geo-localized tweets all concentrated around the Beacon Theater. A gallery of the corresponding media is also given.

References

1. He, Q., Chang, K., Lim, E.: Analyzing feature trajectories for event detection. In: SIGIR 2007, pp. 207–214. ACM (2007)
2. Symeonidis, P., Ntempos, D., Manolopoulos, Y.: Location-based social networks. Recommender Systems for Location-based Social Networks. Springer Briefs in Electrical and Computer Engineering, pp. 35–48. Springer, New York (2014)
3. Xia, C., Schwartz, R., Xie, K.E., Krebs, A., Langdon, A., Ting, J., Naaman, M.: Citybeat: real-time social media visualization of hyper-local city data. In: WWW 2014, Companion Volume, pp. 167–170. ACM (2014)

HiDER: Query-Driven Entity Resolution for Historical Data

Bijan Ranjbar-Sahraei[1]([⊠]), Julia Efremova[2], Hossein Rahmani[1],
Toon Calders[3], Karl Tuyls[4], and Gerhard Weiss[1]

[1] Maastricht University, Maastricht, The Netherlands
b.ranjbarsahraei@maastrichtuniversity.nl
[2] Eindhoven University of Technology, Eindhoven, The Netherlands
[3] Université Libre de Bruxelles, Brussels, Belgium
[4] University of Liverpool, Liverpool, UK

Abstract. Entity Resolution (ER) is the task of finding references that
refer to the same entity across different data sources. Cleaning a data
warehouse and applying ER on it is a computationally demanding task,
particularly for large data sets that change dynamically. Therefore, a
query-driven approach which analyses a small subset of the entire data
set and integrates the results in real-time is significantly beneficial. Here,
we present an interactive tool, called HiDER, which allows for query-
driven ER in large collections of uncertain dynamic historical data. The
input data includes civil registers such as birth, marriage and death
certificates in the form of structured data, and notarial acts such as
estate tax and property transfers in the form of free text. The outputs
are family networks and event timelines visualized in an integrated way.
The HiDER is being used and tested at BHIC center(Brabant Historical
Information Center, https://www.bhic.nl); despite the uncertainties of
the BHIC input data, the extracted entities have high certainty and are
enriched by extra information.

1 Introduction

In the domain of historical research vast amount of historical data exists. Digiti-
zation and correction of data is an everyday process in historical centers. Additionally, some projects such as Ancestory.com[1] are using crowdsourcing and vol-
unteering efforts to improve the quality of their database on census records and
civil registers. This results in many dynamically changing large data corpora,
requiring efficient ER.

   This work develops, based on the work of [1], a query-driven tool for Historical
Data Entity Resolution called HiDER. HiDER has the following advantages: (a)
HiDER allows for ER across different data sources; (b) the changes in input data
and ER algorithms can be incorporated in generating outcomes in real time; (c)
by using *Lucene*'s inverted indexing, both structured and unstructured data are
handled, and fuzzy search allows for compensating missing data and spelling

---

[1] Ancestry.com Inc., http://www.ancestry.com

© Springer International Publishing Switzerland 2015
A. Bifet et al. (Eds.): ECML PKDD 2015, Part III, LNAI 9286, pp. 281–284, 2015.
DOI: 10.1007/978-3-319-23461-8_30

variations, and (d) graph-based ER allows for detecting and visualizing "family networks".

2   The HiDER System

The HiDER system is developed on an Apache web server, equipped with Solr search platform. HiDER works as follows: a user gives a query which consists of at least a family name, but can also contain names of a couple, date and location and relatives' names. Subsequently, HiDER searches for relevant records existing in different sources and presents them in an integrated way. To do this, HiDER uses an inverted index data structure to retrieve a subset of records from multiple corpora, and applies an ER process, developed previously by the authors in [2,3], on this subset on the fly. As such, the system is flexible in the sense that it adapts with minimal effort to changes in the corpus. In Fig. 1, different modules of HiDER are shown. Next we introduce each of these modules, in detail.

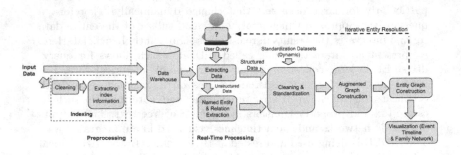

Fig. 1. The HiDER query-driven ER process.

Preprocessing: The **input data** consists of historical documents of the 18th and 19th centuries in the form of structured civil registers and unstructured notarial acts. We refer to each civil register or notarial act as a *record* and each person mentioned in a record as a *reference*. Upon arrival of a new record or when an existing record is updated, the important information of the record is **cleaned** and stored in an **inverted index**. For structured data, the names, locations, date and type of the record are the indexed information, and a *general text* field is used to generate an inverted index for every term which appears in the record. For the unstructured data the *general text* field is used to generate the inverted index for every term in the text of document. The indexing procedure is computationally light and still captures every information in the record.

Real-Time Processing: Real-Time processing is the main part of the HiDER system. Depending on the user query, HiDER uses the available indexes in the data warehouse for **Extracting Data**. The available *faceting feature* guides a user to drill into his/her target data (see left column in Fig. 2). Furthermore, the user can choose between strict and fuzzy search, where the latter one allows for compensation of spelling errors and missing data. The retrieved unstructured data is then further processed for **Named Entity** and **Relation Extraction**; for more information we refer to [2]. Additional **Cleaning** and **Standardization** is applied to the outputs of previous modules. For instance, extra symbols are removed from names, and names with spelling variations are standardized. The standardization databases[2] are continuously updated based on the user feedback and experts knowledge and updates are incorporated in answering future queries.

In the **Augmented Graph Construction** phase, the contextual information available in each record is translated to a graph component. The graph consisting of these components is then augmented by adding so called *block nodes* which capture the important features of each name such as its first and last few letters and its length (see [4] for examples). Once the augmented graph is constructed, a *random walk*-based entity detection approach is used to detect all references with highly similar contextual information (i.e., similar neighbor nodes in the augmented graph), indicating that they all refer to the same entity. Once the entities are detected, the **Entity Graph Construction** is accomplished by merging each set of references that correspond to the same entity (this technique is elaborated by the authors in [3]).

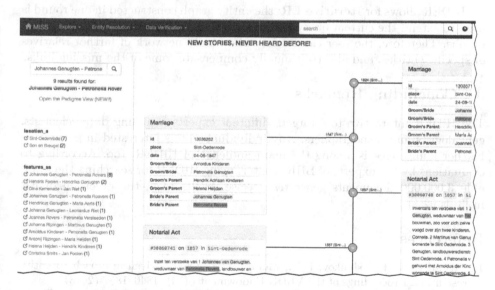

Fig. 2. Part of the HiDER interface upon arrival of a query: searching tool and faceting are shown on the left, and the event timeline is shown on the right.

Visualization serves as an indispensable tool to evaluate the entity graph manually, and is also a way to deliver the results to the user. HiDER is capable of visualizing the entity graph in the form of *event timelines*. In event timelines

---

[2] e.g., http://www.meertens.knaw.nl/cms/en/collections/databases

the information of each record is shown in the form of a floating card, while the important entities are highlighted, and the cards are sorted based on the date of the records (see Fig. 2). To visualize the *family networks*, due to complexity of the generated entity graphs, we propose a novel visualization scheme for the genealogical data by combining every two individuals with marriage relations into single couple nodes, and use graph traversing algorithms to categorize nodes into different generations (see Fig. 3).

Fig. 3. The HiDER visualization of a family network: each link connects parents, on the left, to a child and his/her spouse, on the right. Users can interactively focus on the nodes and expand them. Coloring of nodes adapts to the mouse position.

HiDER allows for **Iterative ER**; the entity graph constructed in one round is used to extend the current query and as such to iteratively construct new entity graphs. Therefore, the user can retrieve the family network of farther relatives of specific entities, and also to manually compensate some of the missing links.

3   Concluding Remarks

The HiDER interactive tool targets different experts including data scientists, genealogists and demographers. Any individual who is interested in generating his/her family tree is among the main audience of HiDER, too. According to evaluations by the experts of BHIC center, using HiDER for searching the available 3,000,000 documents generates precise results (e.g., the precision of ER in [3] is 92%).

References

1. Altwaijry, H., Kalashnikov, D.V., Mehrotra, S.: Query-driven approach to entity resolution. Proceedings of the VLDB Endowment **6**(14), 1846–1857 (2013)
2. Efremova, J., Ranjbar-Sahraei, B., Rahmani, H., Oliehoek, F.A., Calders, T., Tuyls, K.: Multi-source entity resolution for genealogical data. In: Population Reconstruction. Springer (2015) (in press)
3. Rahmani, H., Ranjbar-Sahraei, B., Weiss, G., Tuyls, K.: Entity resolution in disjoint graphs: an application on genealogical data. Intelligent Data Analysis **20**(2) (2016) (in press)
4. Rahmani, H., Ranjbar-Sahraei, B., Weiss, G., Tuyls, K.: Contextual entity resolution approach for genealogical data. In: Workshop on Knowledge Discovery, Data Mining and Machine Learning, Aachen, Germany (2014)

IICE: Web Tool for Automatic Identification of Chemical Entities and Interactions

Andre Lamurias[1,2](✉), Luka A. Clarke[1], and Francisco M. Couto[2]

[1] Faculdade de Ciências, BioISI: Biosystems & Integrative Sciences Institute,
Universidade de Lisboa, Lisboa, Portugal
[2] LaSIGE, Departamento de Informática, Faculdade de Ciências,
Universidade de Lisboa, 1749-016 Lisboa, Portugal
alamurias@lasige.di.fc.ul.pt, laclarke@fc.ul.pt, fcouto@di.fc.ul.pt

Abstract. Automatic methods are being developed and applied to transform textual biomedical information into machine-readable formats. Machine learning techniques have been a prominent approach to this problem. However, there is still a lack of systems that are easily accessible to users. For this reason, we developed a web tool to facilitate the access to our text mining framework, IICE (Identifying Interactions between Chemical Entities). This tool annotates the input text with chemical entities and identifies the interactions described between these entities. Various options are available, which can be manipulated to control the algorithms employed by the framework and to the output formats.

Keywords: Text mining · Machine learning · Ontologies · Named entity recognition · Relation extraction

1 Introduction

The amount of information about chemical compounds that is published in the form of scientific literature is growing at an unprecedented rate [1]. To update the chemical interactions described in databases, such as DrugBank [4] and IntAct [3], relies on manual reading and parsing the literature. This means that this update will always lag behind scientific publications, as experts extract the relevant information from the papers. For this reason, there is a growing need for automatic methods that transform biomedical text into machine-readable structured data, such as an interaction between compounds.

Information extraction systems applied to the biomedical domain have been developed and are available to the community [5]. However, their performance depends on the machine used by the user, usually requiring external libraries and specific installation instructions. A more practical solution is releasing the system as a web tool, with a front-end enabling any user to test and experiment with it.

We developed the IICE framework (Identifying Interactions between Chemical Entities), for automatic annotation of biomedical documents. IICE is based on

© Springer International Publishing Switzerland 2015
A. Bifet et al. (Eds.): ECML PKDD 2015, Part III, LNAI 9286, pp. 285–288, 2015.
DOI: 10.1007/978-3-319-23461-8_31

supervised machine learning algorithms and semantic similarity between ontology concepts. We have evaluated the framework with the CHEMDNER [7] dataset, for the recognition of chemical entities, and with the DDIExtraction dataset [8], for extraction of drug-drug interactions. The F-measure obtained for each dataset was of 78.26% and 72.52%, respectively, which can be considered nearly state-of-the-art.

The IICE framework can be accessed by a web tool[1], with several configuration options available to the user. These options enable the user to obtain different results by adjusting the methods and thresholds applied. As such, it is possible to set the options for higher recall or precision, depending on the specific needs of the user. The results may be given in the form of HTML tables, or a XML file.

2   Architecture Overview

The IICE framework is based on three components which take as input biomedical text, to accomplish distinct tasks. With this modular approach, it is possible to run the framework only using some of the modules, which may be useful if the text was already partially annotated, or if it is going to serve as input to another framework.

Entity recognition This module recognizes the chemical entities mentioned in each sentence. If the input consists of more than one sentence, we split the text in sentences, and process each one individually. The input text is classified using Conditional Random Fields classifiers [2], trained with data sets from community challenges [7,8]. We have trained classifiers for specific types of chemical entities, in order to obtain higher recall and also provide a type for each entity recognized.

Validation The chemical entities recognized in the text are normalized to ChEBI ontology [6] identifiers. Using the ChEBI ontology, it is possible to validate the entities recognized in the same sentence. Our assumption is that entities that were correctly recognized in a given sentence should share more similarity than recognition errors. Therefore, we implemented a filter to exclude entities with low semantic similarity to other recognized entities in the same fragment of text. This approach obtained high precision values.

Relation extraction The relation extraction module identifies pairs of entities in the text that are described as interactions. We trained a classifier with the DDIExtraction dataset [8], using kernel-based learning algorithms [9,10]. This type of algorithm has been successfully applied to other relation extraction tasks. The input text for this module should be already annotated with chemical entities, either by the previous module, with a different framework, or manually. Each interaction is also labeled with one of the types of interactions considered in the DDIExtraction dataset.

---

[1] http://www.lasige.di.fc.ul.pt/webtools/iice/

3   Web Tool

The IICE web tool can be used to automatically annotate the abstract of a scientific article with chemical entities and interactions. This can be useful for applications such as developing a network of interactions based on the literature, or finding articles relevant to a particular chemical compound.

Figure 1 shows the options that are available to the user. Using these options, it is possible to recognize only the chemical entities in the text (NER), or only the chemical interactions (RE) if the text is already annotated with chemical entities, or both. The input text can be annotated with the "<entity>" tag. We have trained classifiers for entity recognition with two datasets, annotated with different criteria: the CHEMDNER corpus considers various types of chemical entities, while the DrugNER corpus is focused only on drugs. The user may choose to use only the set of classifiers trained with one of the datasets, if annotations similar to that dataset are preferred. We also provide several options related to the validation module, in order to tune the framework for higher recall or precision. Finally, it is also possible to choose which types of machine learning algorithms to use for Relation Extraction. The user may select the classifier we have trained with the Shallow Language kernel [9] or with the Subset Tree kernel [10]. The ensemble classifier combines the results of the kernel classifiers with other domain-specific features to obtain better results. We have previously described in detail how these algorithms were applied and how they may influence the results [11].

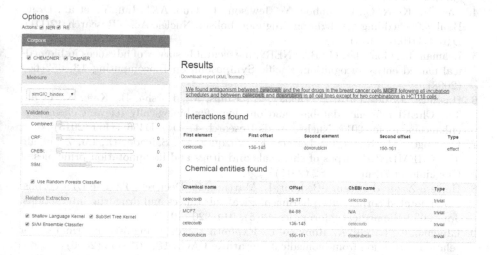

Fig. 1. Screenshot of the options panel and results obtained with the IICE web tool.

The results obtained with the web tool are shown on Figure 1. First we provide a link to the results in the XML format used by the DDI Extraction dataset. Then, the original text is shown, with the chemical entities highlighted. We organize the interaction and the chemical entities found in two distinct tables. The interactions table provides the two elements of the interactions, and the type of chemical interaction. The entities table provides the name, offset and type of chemical entity, as well as the ChEBI ontology identifier mapped to that entity.

Using only one set of NER classifiers, the system takes about 10 seconds to process one sentence. This value increases as more options are activated, taking as long as 60 seconds if all classifiers are used. We plan on improving the speed performance of the web tool by pre-loading the classifiers and ontologies and deploying the tool on the server as a background service.

Acknowledgments. This work was supported by the Fundação para a Ciência e a Tecnologia (https://www.fct.mctes.pt/) through the PhD grant PD/BD/106083/2015 and LaSIGE Unit Strategic Project, ref. PEst-OE/EEI/UI0408/2014 and by the European Commission (http://ec.europa.eu) through the BiobankCloud project under the Seventh Framework Programme (grant #317871).

References

1. Hunter, L., Cohen, K.B.: Biomedical language processing: what's beyond PubMed? Molecular Cell **21**(5), 589–594 (2006)
2. McCallum, A.K.: MALLET: A Machine Learning for Language Toolkit (2002)
3. Kerrien, S., Aranda, B., Breuza, L., Bridge, A., Broackes-Carter, F., Chen, C., et al.: The IntAct molecular interaction database in 2012. Nucleic Acids Research, gkr1088 (2011)
4. Law, V., Knox, C., Djoumbou, Y., Jewison, T., Guo, A.C., Liu, Y., et al.: DrugBank 4.0: shedding new light on drug metabolism. Nucleic Acids Research **42**(D1), D1091–D1097 (2014)
5. Leaman, R., Gonzalez, G.: BANNER: an executable survey of advances in biomedical named entity recognition. Pacific Symposium on Biocomputing **13**, 652–663 (2008)
6. Hastings, J., de Matos, P., Dekker, A., Ennis, M., Harsha, B., Kale, N., et al.: The ChEBI reference database and ontology for biologically relevant chemistry: enhancements for 2013. Nucleic Acids Research **41**(D1), D456–D463 (2013)
7. Krallinger, M., Rabal, O., Leitner, F., Vazquez, M., Salgado, D., Lu, Z., et al.: The CHEMDNER corpus of chemicals and drugs and its annotation principles. J Cheminform **7**(Suppl 1), S2 (2015)
8. Herrero-Zazo, M., Segura-Bedmar, I., Martnez, P., Declerck, T.: The DDI corpus: An annotated corpus with pharmacological substances and drugdrug interactions. Journal of Biomedical Informatics **46**(5), 914–920 (2013)
9. Giuliano, C., Lavelli, A., Romano, L.: Exploiting shallow linguistic information for relation extraction from biomedical literature. EACL **18**, 401–408 (2006)
10. Moschitti, A.: Making Tree Kernels Practical for Natural Language Learning. In: EACL, vol. 113, no. 120, p. 24 (2006)
11. Lamurias, A., Ferreira, J.D., Couto, F.M.: Identifying interactions between chemical entities in biomedical text. Journal of Integrative Bioinformatics **11**(3), 247 (2014)

Interactively Exploring Supply and Demand in the UK Independent Music Scene

Matt McVicar[✉], Cédric Mesnage, Jefrey Lijffijt, and Tijl De Bie

Intelligent Systems Lab, University of Bristol, Bristol, UK
{mattjamesmcvicar,tijl.debie}@gmail.com,
{cedric.mesnage,jefrey.lijffijt}@bristol.ac.uk

Abstract. We present an exploratory data mining tool useful for finding patterns in the geographic distribution of independent UK-based music artists. Our system is interactive, highly intuitive, and entirely browser-based, meaning it can be used without any additional software installations from any device. The target audiences are artists, other music professionals, and the general public. Potential uses of our software include highlighting discrepancies in supply and demand of specific music genres in different parts of the country, and identifying at a glance which areas have the highest densities of independent music artists.

1 Introduction and Motivation

Supply and demand form the basis of many products in microeconomics, affecting the price consumers are willing to pay for an item or service. With respect to music, the 'supply' of a particular music genre[1] can be thought of as the number of artists actively recording and performing music in this style. 'Demand' on the other hand refers to a population's desire to spend resources (time, money) purchasing and listening to a given music genre, as well as attending live concerts in this style.

Quantification of the supply and demand of music is not straightforward; until very recently only the demand for the *most popular* genres was known, through charts produced by companies such as Billboard[2] or The Official Charts Company[3]. Our aim is to provide estimates of supply and demand at much finer granularities by using simple statistical techniques combined with the massive amount of data available on the web. In particular, music artist platforms such as SoundCloud, MixCloud and YouTube[4] present us with a unique chance to study the supply of particular music genres outside the mainstream, whilst fan activity on (micro)blogging sites such as Twitter and Facebook give us a window into the demand for particular musical styles.

---

[1] 'style' and 'genre' are considered to be synonyms in this paper, examples of which include *Classical*, *Rock* etc.
[2] http://www.billboard.com/
[3] http://www.officialcharts.com/home/
[4] https://soundcloud.com/,https://www.mixcloud.com/,https://www.youtube.com

© Springer International Publishing Switzerland 2015
A. Bifet et al. (Eds.): ECML PKDD 2015, Part III, LNAI 9286, pp. 289–292, 2015.
DOI: 10.1007/978-3-319-23461-8_32

We are particularly interested in lesser-known artists as they may represent fresh talent or emerging music styles. Musicians such as these may never reach commercial breakthrough if they are not identified and linked with their fans, and our tool can help record labels identify growing genres and thus artists at an early stage in their career, leading to greater profit potential. In contrast to existing work on geo-located twitter music [1], [2], our system pairs artists active on the web with their fans, and is also concerned exclusively with non-mainstream music. Furthermore, we provide an intuitive system for browsing the patterns seen in the geographical distribution of these niche music artists.

The remainder of this demo paper is organised as follows. In Section 2 we outline our data collection methods. Section 3 gives an overview of our system, and we conclude in Section 4.

2   Data Collection

We collected data about 54,192 UK independent artists from Reverbnation.com[5]. In particular, we collected artist name, genre(s), location and links to their Twitter, Soundcloud and/or Youtube accounts. Using Twitter handles of the artists (21,616 artists volunteered this information) we periodically gathered tweets using this handle. Currently, our MongoDB database is storing 10,005,640 tweets, tweeted from 1,844,278 different accounts. 181,042 of these tweets we have found to be geolocalised and can therefore be used to assess the demand for a genre in a particular location.

To localise our data, we obtained the location from ReverbNation (free text) and Twitter (curated by Twitter) and semi-automatically binned locations into the 83 counties of England, 26 districts of Northern Ireland, 32 unitary authorities of Scotland and the 32 principal areas of Wales. We found that many artists listed their location simply as 'London', which is not itself a well-defined area. We therefore collected all London buroughs (Camden, Tower Hamlets etc.) into one region called simply *London*.

Supply (demand) for each genre and region in our maps was then estimated simply as number of artists (tweets about artists) divided by population of region. These were generally on different scales, making a direct comparison challenging. Instead, we employed the following statistical approach which aims to quantify how balanced the supply and demand is given a background balance given by the UK as a whole.

Let a_i and t_i be the number of artists and tweets for region i. Furthermore, suppose that a and t are the number of artists in this genre for the UK in its entirety. Considering a_i and t_i as samples from independent Poisson random variables, we then regard the respective rate parameters $\lambda_{a,i}$ and $\lambda_{t,i}$ as proxy measures for supply and demand for a given genre in region i. The degree to which region i's music market is out of equilibrium can then be quantified by comparing $\log(\lambda_{a,i}/\lambda_{t,i})$ with $\log(\lambda_a/\lambda_t)$; the latter is considered the equilibrium.

---

[5] http://www.reverbnation.com/

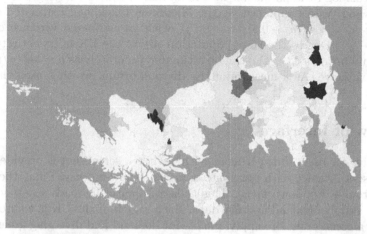
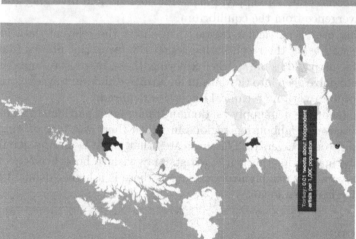
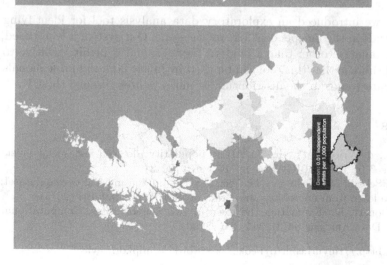

Fig. 1. Exploring supply, demand, and disequilibria of/for Dubstep music in the UK. Left map: supply of Dubstep per region, as measured by number of independent artists active on the web divided by population. Center map: demand for dubstep, measured by density of tweets about independent Dubstep artists. Right map: state of net supply vs. demand, showing that the two neighbouring regions of Devon and Torbay have a surplus of and demand for Dubstep respectively.

Estimating this log ratio parameter in a maximum likelihood setting can be conducted following Wald's approach [3], the Maximum Likelihood Estimate for $\log(\lambda_{a,i}/\lambda_{t,i})$ is $\mu_i = \log((a_i + 0.5)/(t_i + 0.5))$, with an estimated variance of $\sigma_i^2 = \frac{1}{a_i+0.5} + \frac{1}{t_i+0.5}$. Due to the large population size of the UK in its entirety, the 'prior' disequilibrium $\log(\lambda_a/\lambda_t)$ can be estimated accurately as $\mu = \log(a/t)$ with negligible variance. As a measure of the disequilibrium, we thus used the z-score of μ_i: $z_i = \frac{\mu_i - \mu}{\sigma_i}$.

3   Overview of Interface

Our system is live for users to experiment with[6], while a screenshot is provided in Figure 1. Users interact with our system in the following way. First, they select a genre from a dropdown menu. This then automatically loads a new set of maps for the supply, demand and net data. Each region of the UK is colored by intensity on a green scale for supply, purple for demand and diverging green-purple for net divergence from the equilibrium.

Mouse-hovering over a particular region then shows the relevent data in a clear and concise way using the recently-developed D3 Javascript library (data-driven documents, http://d3js.org/). By exploring these maps, we hope that users will be able to delve deep into the data in an intuitive and fun way, exploring many gigabytes of data with no technical expertise required.

The final map (showing net supply vs. demand) may be of particular interest, since users can discover neighbouring regions in which there is a lack of supply and excess demand (or vice-versa). This indicates that there is an opportunity for *arbitrage* in the market for this genre: artists may wish to book live concerts or promote their work in a neighbouring county if there is a higher demand for their style than in their own locality.

4   Conclusions

In this demo, we introduced an exploratory data analysis tool for identifying geographic trends in the independent UK music scene. Our system is web-based, requires no installation other than a modern browser, and is highly intuitive to use. We are optimistic that this tool can be used by music fans and professionals alike, to navigate trends in the distribution of music genres through the UK.

References

1. Bellogín, A., de Vries, A.P., He, J.: Artist popularity: do web and social music services agree? In: Proc. of ICWSM, pp. 673–676 (2013)
2. Hauger, D., Schedl, M., Košir, A., Tkalčič, M.: The million musical tweets dataset: what can we learn from microblogs. In: Proc. of ISMIR, pp. 189–194 (2013)
3. Price, R., Bonett, D.: Estimating the ratio of two poisson rates. Computational Statistics & Data Analysis **34**(3), 345–356 (2000)

[6] available at: http://tinyurl.com/qybeoke, password "ecmlpkdd_review"

Monitoring Production Equipment in Remote Locations

Ujval Kamath, Ana Costa e Silva[✉], and Michael O'Connell

Tibco Software Inc., Palo Alto, USA
ansilva@tibco.com

1 Objectives

Within the context of Remote Equipment Monitoring, the specific customer implementation of this project has been in the area of an upstream oil and gas process. The goal was to improve the efficiency of ESP (Electric Submersible Pump) oil & gas production, by predicting (rather than just reacting to) ESP shutdown and failure and thus avoiding downtime which results in a loss of production as well as repair costs. Please see Figure 1 for an illustration of an ESP.

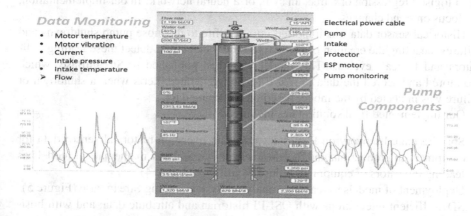

Fig. 1. Electric Submersible Pumps (ESPs)

"Replacing a single failed pump costs us more than $100K and days of production loss. We have hundreds of failures a year. Preventing even 10% of these failures represents a massive reduction in costs", well operation engineer of major Oil & Gas company.

2 Project Implementation

A methodology and solution for real-time monitoring of production equipment in remote locations is presented. The solution is developed on sensor data, transmitted from equipment to field information systems; and analyzed using visual/predictive applications connected to a central Historian data source. Resulting mathematical

© Springer International Publishing Switzerland 2015
A. Bifet et al. (Eds.): ECML PKDD 2015, Part III, LNAI 9286, pp. 293–297, 2015.
DOI: 10.1007/978-3-319-23461-8_33

models, developed and validated on historical data, are used to monitor new sensor data as they arrive in real-time.

Our solution is illustrated using data from Electric Submersible Pumps across multiple fields. The pumps are fitted with downhole monitoring units that transmit streams of data back to surface including: motor temperature, motor frequency, motor current, pump intake pressure and intake temperature. Data is aggregated from multiple sensors with a resulting data rate of several thousand readings per minute from the Historian data source.

This data was combined with subject matter expertise to improve detection and provide event classification. Examples of this include slipping conditions indicated by pressure and flow remaining constant while energy consumption is increasing; motor temperature decrease and pressure increase indicating gas buildup; motor temperature increases at a certain rate over time leading to a motor burnout

There are several possible strategies for monitoring such variables over time, specifically a trend analysis approach, i.e. monitoring for changes in location of distribution, or changes in variability, or slope, or a statistical approach, based on statistical (e.g. Shewhart control chart) or machine learning models (y (0/1) = f (X, b) + e;where f is a logistic regression or a tree, an svm, or a neural network. In our implementation, we focus on trend changes.

Historical sensor data was imported into Spotfire to diagnose pump shutdowns and failures, enabling the creation of predictive model that could detect these leading indicators and forecast events. This model was then published to StreamBase, which monitored and scored the data in real time and generated alerts when a shutdown or failure was indicated by the model based on incoming data.

Our implemented final solution has included:

- Spotfire data discovery on ESP data: Data discovery on historical sensor data for equipment in production, and development of hierarchical mathematical models as leading indicators of equipment failure conditions (Figure 4)
- Deployment of models to real-time analytics systems, using Streambase (Figure 5)
- SDK-efficient integrations with OSI-PI historian and attribute data; and with business process management systems for management of equipment maintenance
- Understanding failure events: alerting the engineers with an email that has a picture of the variables related to root-cause of failure. (Figure 6)
- Live monitoring of real-time sensor signals using StreamBase LiveView (Figure 7)
- Comprehensive geo and location analytics mapping
- Analysis of real time alerting data for classification and prioritization, and continuous model improvements leading to reduction of false positives

3   Results

The solution performs remarkably well, identifying a variety of anomalous equipment behaviour states, and preventing multiple shutdowns and pump failures, with false positive rates close to zero.

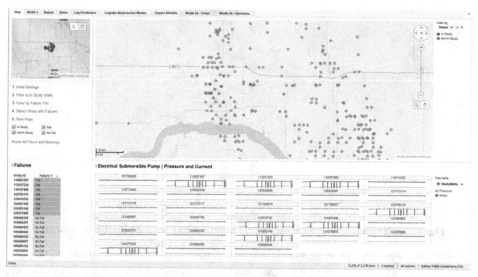

Fig. 2. Historical Data Analysis in Spotfire

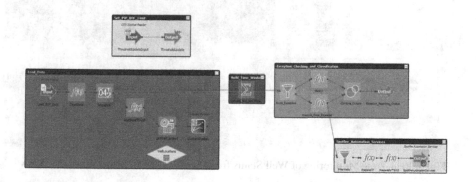

Fig. 3. Set-up for Real-Time Sensor Monitoring and Alerting in StreamBase

An ROI model for the equipment monitoring is developed as a companion to the methodology and solution. The ROI model indicates savings of up to 400 hours of production per day per thousand wells, which conservatively equates to $40M/yr regarding ESP lift alone.

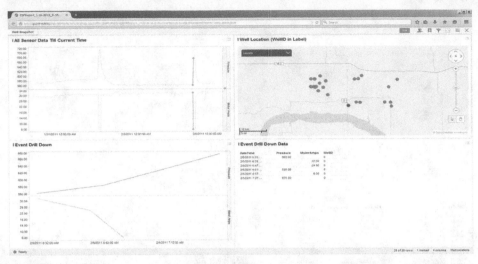

Fig. 4. Alerting View in Web Browser (via Spotfire) to Facilitate Root-Cause Analysis

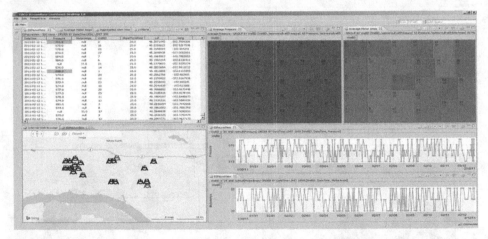

Fig. 5. Real Time Monitoring of Well Status from Sensor Data in StreamBase LiveView

4   Innovative Aspects of the Project

Data are integrated from thousands of wells across multiple plays.   Data from equipment sensors are analysed and visualized. The resulting mathematical models comprise combinations of sensor data summaries, e.g. slopes and excursions beyond empirically derived thresholds set up to detect anomalous equipment behavior. The models are developed as hierarchies and back-tested on recent equipment sensor data. The models are subsequently applied to new sensor data arriving in real time.

Seamless integration of historic and real-time data, connecting big data to fast data.

Often well products are highly custom-developed, making it difficult to reapply a model or solution from one piece of equipment to another. Conversely, the products involved in our solution are highly flexible, it being easy and fast to re-implement a model or workflow onto another type of equipment, which results in very high ROI.

Actionability of the solution - alerting becomes very actionable and efficient, with human involvement only when it is required.

5   Future Work

Right now the predictive model is based on data that is selected by subject matter experts but our future state is a large scale automated back-testing framework to determine the best parameters for models.

MultiNets: Web-Based Multilayer Network Visualization

Matija Piškorec[1]($\boxtimes$), Borut Sluban[2], and Tomislav Šmuc[1]

[1] Rudjer Bošković Institute, Zagreb, Croatia
matija.piskorec@irb.hr
[2] Jožef Stefan Institute, Ljubljana, Slovenia
borut.sluban@ijs.si

Abstract. This paper presents MULTINETS: a Javascript library for multilayer network visualization. MULTINETS provides reusable HTML components with functions for loading, manipulation and visualization of multilayered networks. These components can be easily incorporated into any web page, and they allow users to perform exploratory analysis of multilayer networks and prepare publication quality network visualizations. MULTINETS components are easily extendable to provide custom-based visualizations, such as embedding networks on geographical maps, and can be used for building complex web-based graphical user interfaces for data mining services that operate on multilayered networks and multirelational data in general.

Keywords: Network visualization · Multilayer networks · Graph mining · Network mining

1 Introduction

Network science is becoming an important multi-disciplinary research area including mathematics, physics, life sciences, social sciences, and computer science. Complex systems can often be represented as networks of interacting components, where each aspects of the system can be presented as an individual network, hence together forming a multi-aspect or multilayer network. The wide applicability of multilayer networks for solving and understanding different problem scenarios is gaining recognition in the scientific community [2,6]. In machine learning and data mining relevant research includes multilayer clustering [5], multi-network link prediction [11], mining heterogeneous networks [10] and meta-learning from network induced features [3].

Network visualizations offer an unique way to understand and analyze complex systems by enabling users to more easily inspect and comprehend relations between individual units and their properties [8]. The majority of programming languages offer libraries, modules and extensions for visualizing static networks. Many stand-alone programs for network visualization are available, for example

© Springer International Publishing Switzerland 2015
A. Bifet et al. (Eds.): ECML PKDD 2015, Part III, LNAI 9286, pp. 298–302, 2015.
DOI: 10.1007/978-3-319-23461-8_34

Gephi[1] [1], Pajek[2], LaNet-vi[3], GraphViz[4], and *ORA[5]. Some extensions for data mining tools also exist: ORANGE offers the Network add-on[6] for visualizing and analyzing networks, and KNIME has a network mining plug-in[7]. Recently, few stand-alone programs for multilayer network visualizations were developed, for example MuxViz [4] and Arena3D [9]. In comparison, there are far less tools for network visualization in web browsers, which is an oddity because web-based platforms are becoming increasingly common. Although some *Javascript* libraries offer basic support for network visualizations in web browsers, for example D3.js[8], vis.js[9], and sigma.js[10], they are not tailored to support out-of-the box multilayered network visualizations. An exception is the HiveGraph application[11] which renders multilayer networks in a special *hive plot* layout [7].

This is why we developed MULTINETS[12] - a Javascript library that allows easy integration of interactive multilayer network visualization into any web page. MULTINETS consists of reusable HTML components that provide functions for visualization, loading, manipulation, filtering, and exporting of multilayer network data. It presents an unique and easy to use tool for the exploration of multilayer networks and multi-relational datasets in general. Because it is built using well known web technologies like HTML, CSS and Javascript, it is easily extendable, making it a perfect building block for complex web-based interfaces that use visualization as an aid in data analysis. Furthermore, the ubiquity of web browsers ensures that visualizations built with MULTINETS will be accessible to the widest possible audience.

The source code of MULTINETS is publicly available on the Github repository https://github.com/matijapiskorec/multinets.

2 Implementation

The MULTINETS library is implemented as a set of Angular.js directives. Angular.js[13] is a Javascript framework that supports the Model-View-Controller pattern (MVC) for building highly complex web-based user interfaces. Directives act as reusable components which users can easily include into their own web pages by using appropriate HTML elements. The main layout for presenting the multilayer network is based on the *force layout* from the D3.js Javascript

---

[1] http://gephi.github.io/
[2] http://pajek.imfm.si/
[3] http://lanet-vi.soic.indiana.edu/
[4] http://www.graphviz.org/
[5] http://www.casos.cs.cmu.edu/projects/ora/
[6] http://orange.biolab.si/download
[7] https://tech.knime.org/network-mining
[8] http://d3js.org/
[9] http://visjs.org/
[10] http://sigmajs.org/
[11] http://wodaklab.org/hivegraph/graph
[12] http://matijapiskorec.github.io/multinets/
[13] https://angularjs.org/

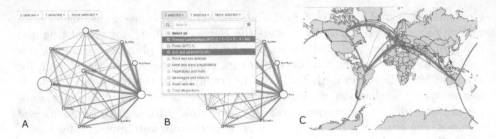

Fig. 1. Screenshots of MULTINETS visualization of trade relations between countries, which is one of the use cases for multilayer network visualization. Panel **A** shows the visualization of the multilayer network of trade relations between countries. Sizes of the nodes are proportional to the population of each countries, colors of the connections correspond to various products and their thickness to the total trading volume. Other mappings from data to visual elements can be easily defined by modifying the existing components. Panel **B** shows the menu for selecting various aspects of the multilayer network, in this case the products traded between the countries. Multiple layers can be selected at once. Panel **C** shows geographical embedding of the given multilayer network. Users can extend MULTINETS to provide similar embeddings for their specific domains.

library. In addition to visualization elements there are also elements responsible for selecting the layers of the network, loading network data from the server either statically or dynamically through REST API, loading network data from the user, and printing retrieved data. Users can easily extend the MULTINETS library by defining custom directives that can generate visualizations tailored to specific domains or manipulate them in any desired way, as well as load and export networks in different data formats. To demonstrate this we implemented a special geographic layout that embeds the network of trade relations between countries on the World map. This layout also uses the D3.js library as a backbone for visualization, along with open source geographical data. The MULTINETS library provides the following HTML elements:

- `<force-layout>`, `<geo-layout>` Generate an interactive visualization of the multilayer network either as a force layout or embedded on the World map.
- `<aspect-selector>` Gives multi-choice menus for selection of various layers of networks.
- `<server-load>`, `<rest-load>`, `<file-load>` Allows loading of network data from server either statically or dynamically through REST API, and loading data from user.
- `<network-dump>` Outputs the retrieved network in text format.

3   Summary

Our motivating use case was to provide an easy to use web-based tool for exploratory data analysis of multilayered networks. By simple composition of

HTML components users can build interfaces for comparison of different multilayered network datasets, exploratory data analysis, as well as prepare publication quality visualizations. Although MULTINETS can visualize networks from any domain, the components of the library are easily extendable to support development of highly custom network visualizations tailored to specific domains. For example, the data mining community can use it to visualize large collection of multi-relational rules as an aid in various data analysis tasks. Also, in the biological domain users can build custom visualizations of gene regulatory networks or networks of metabolic pathways, and expose the visualization along with the data. In addition to providing easy to use exploratory analysis of different datasets, the high quality visualizations can be directly hosted on websites describing the research, or published as figures in articles and reports. The MULTINETS library can also be extended with elements that expose meta information of the network dataset, including network based statistics like degree distribution of nodes, PageRank, betweenes, as well as many other statistics defined on the attributes of nodes or connections. Moreover, MULTINETS components can be used for building web-based graphical user interfaces for advanced data mining services that operate on multilayered networks and multi-relational data in general.

Acknowledgments. This work was supported in part by the European Commission under the FP7 projects MULTIPLEX (grant no. 317532) and MAESTRA (grant no. 612944), Croatian Science Foundation (grant no. I-1701-2014) and by the Slovenian Research Agency programme Knowledge Technologies (grant no. P2-103).

References

1. Bastian, M., Heymann, S., Jacomy, M.: Gephi: An open source software for exploring and manipulating networks (2009)
2. Boccaletti, S., Bianconi, G., Criado, R., del Genio, C., Gmez-Gardees, J., Romance, M., Sendia-Nadal, I., Wang, Z., Zanin, M.: The structure and dynamics of multilayer networks. Physics Reports **544**(1), 1–122 (2014)
3. Cheng, J., Adamic, L., Dow, P., Kleinberg, J., Leskovec, J.: Can cascades be predicted? In: Proceedings of the 23rd International Conference on World Wide Web, WWW 2014, pp. 925–935 (2014)
4. De Domenico, M., Porter, M.A., Arenas, A.: MuxViz: a tool for multilayer analysis and visualization of networks. Journal of Complex Networks (2014)
5. Gamberger, D., Mihelčić, M., Lavrač, N.: Multilayer clustering: a discovery experiment on country level trading data. In: Džeroski, S., Panov, P., Kocev, D., Todorovski, L. (eds.) DS 2014. LNCS, vol. 8777, pp. 87–98. Springer, Heidelberg (2014)
6. Kivelä, M., Arenas, A., Barthelemy, M., Gleeson, J.P., Moreno, Y., Porter, M.A.: Multilayer networks. Journal of Complex. Networks **2**(3), 203–271 (2014)
7. Krzywinski, M., Birol, I., Jones, S.J.M., Marra, M.A.: Hive plots-rational approach to visualizing networks. Briefings in Bioinformatics **13**(5), 627–644 (2012)
8. Rossi, L., Magnani, M.: Towards effective visual analytics on multiplex and multilayer networks. Chaos, Solitons & Fractals **72**, 68–76 (2015)
9. Secrier, M., Pavlopoulos, G.A., Aerts, J., Schneider, R.: Arena3D: visualizing time-driven phenotypic differences in biological systems. BMC Bioinform. **13**, 45 (2012)

10. Sun, Y., Han, J.: Mining Heterogeneous Information Networks: Principles and Methodologies. Morgan & Claypool Publishers (2012)
11. Zhang, J., Yu, P.S., Zhou, Z.H.: Meta-path based multi-network collective link prediction. In: ACM SIGKDD International Conference on Knowledge Discovery and Data Mining, KDD 2014, pp. 1286–1295. ACM (2014)

OMEGA: An Order-Preserving SubMatrix Mining, Indexing and Search Tool

Tao Jiang[✉], Zhanhuai Li, Qun Chen, Zhong Wang,
Kaiwen Li, and Wei Pan

School of Computer Science and Technology,
Northwestern Polytechnical University, Xi'an 710072, China
jiangtao@mail.nwpu.edu.cn

Abstract. Order-Preserving SubMatrix (OPSM) has been accepted as a significant tool in modelling biologically meaningful subspace cluster, to discover the general tendency of gene expressions across a subset of conditions. Existing OPSM processing tools focus on giving a or some batch mining techniques, and are time-consuming and do not consider to support OPSM queries. To address the problems, the paper presents and implements a prototype system for OPSM queries, which is called OMEGA (Order-preserving subMatrix mining, indExinG and seArch tool for biologists). It uses Butterfly Network based BSP model to mine OPSMs in parallel. Further, it builds index based on prefix-tree associated with two header tables for gene expression data or OPSM mining results. Then, it processes exact and fuzzy queries based on keywords. Meanwhile, the vital query results are saved for later use. It is demonstrated that OMEGA can improve the effectiveness of OPSM batch mining and queries.

Keywords: Order-Preserving SubMatrix · Indexing and search · Tool

1   Introduction

DNA microarray enables simultaneously monitoring of the expression level of tens of thousands of genes over hundreds of experiments. Gene expression data on DNA microarrays can be viewed as an $n \times m$ matrix with n genes (rows) and m experiments (columns), in which each entry denotes the expression level of a given gene under a given experiment. Existing clustering methods do not work well for gene expression data, due to that most genes are tightly coexpression only under a subset of experiments, and are not necessarily expression at the same or similar expression level. Thus, it makes *Order-Preserving SubMatrix* (OPSM)

This work was supported in part by the National Basic Research Program 973 of China (No. 2012CB316203), the Natural Science Foundation of China (Nos. 61033007, 61272121, 61332006, 61472321), the National High Technology Research and Development Program 863 of China (No. 2013AA01A215), the Graduate Starting Seed Fund of Northwestern Polytechnical University (No. Z2012128).

© Springer International Publishing Switzerland 2015
A. Bifet et al. (Eds.): ECML PKDD 2015, Part III, LNAI 9286, pp. 303–307, 2015.
DOI: 10.1007/978-3-319-23461-8_35

[1,2], a special model of pattern-based clustering, as the popular tool to find meaningful clusters. In essence, an OPSM is a subset of rows and columns in a data matrix where all the rows induce the same linear ordering of the columns, e.g., rows g_2, g_3 and g_6 have an increasing expression level on columns 2, 7, 5, and 1. And OPSM cluster model focuses on the relative order of columns rather than the actual values. As the high-rate increasing of the numbers and sizes of gene expression datasets, there is an increasing need for the fast mining techniques to handle the massive gene datasets. Further, OPSM mining results are accumulated and not efficiently utilized, thus it is urgent to build a tool for biologists to find supporting rows or columns based on keyword queries, which plays an important role in inferring gene coregulated networks.

Traditional OPSM processing tools such as *BicAT* [1] and *GPX* [2] are developed for single machine, and cannot work well on parallel distributed platform. For example, how to reduce the communication time, workload of bandwidth and percent of duplicate results. And they are mainly focusing on providing batch OPSM mining techniques, and have very limited consideration on supporting OPSM search, even if *GPX* uses a graphical interface to drill down or roll up.

To solve the problems, we present and implement a prototype system for OPSM queries called *OMEGA* (Order-preserving subMatrix mining, indExinG and seArch tool). The major features of *OMEGA* can be described as follows:

(1) *OMEGA* provides a Butterfly Network based parallel OPSM mining framework [4]. Existing batch OPSM mining methods only employ a machine, but *OMEGA* utilizes multi-machines to mine OPSMs.

(2) *OEMGA* supports OPSM indexing and search. It builds index based on prefix-tree with two header tables, then processes queries based on keywords [3], meanwhile saves the vital query results for later use. It is demonstrated that *OMEGA* can improve the effectiveness of OPSM batch mining and queries.

2   The Architecture and Key Technologies of OMEGA

The architecture of *OMEGA* can be divided into four major shown in Fig. 1(a).

(1) Permutation of Columns. This module permutates columns based on expression values, under which row sequences are in ascending/descending order.

OPSM model focuses on the relative order of columns rather than actual values. By sorting row vectors and replacing entries with their corresponding column labels, data matrix can be transformed into a sequence database, and OPSM mining is reduced to a special case of sequential pattern mining problem with distinctive properties. Actually, we can use any sort method to permutate the columns of each row, and quick sort algorithm is employed in the paper.

(2) Parallel Mining OPSMs. This module employs Butterfly Network based BSP model for OPSM mining.

The number of nodes of Butterfly Network is N ($N = 2^n$, where n is the maximum number of super-steps). For simplicity, the names of nodes are denoted by integers which are from 0 to $2^n - 1$. In the ith super-step ($i \geq 1$), each

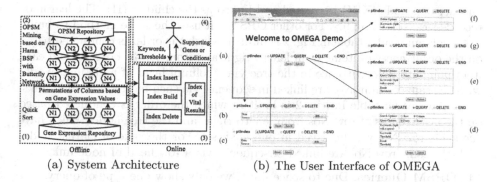

(a) System Architecture   (b) The User Interface of OMEGA

Fig. 1. System Architecture and User Interface

node firstly does local computation, then N nodes are divided into $(log_2 N)/2^{i-1}$ groups, where $1 \le i \le n$, i.e., each group has 2^i members which have continuous integers, further the members in each group are divided into 2 partitions, the members in first half partition communicate or transfer data with the nodes in the last half partition with 2^{i-1} steps, and vice versa, finally the nodes go into barrier synchronization. Once there are no data to transfer or the number of super-steps is equal to $log_2 N$, the computational work of the nodes on Hama will be stopped. For more details about the method, please refer to work [4].

(3) Indexing of Datasets. The module uses prefix-tree with two header tables to index the permutation of columns or OPSM mining results.

A compact index can be designed based on three observations described in work [3]. We use dataset in Table 1 to show how to build index in Fig. 2.

Table 1. Dataset

Row No.	Column No.
1,2,5	VI,III,I,VIII,XVI
3,6,9	VI,III,I,II,XIII
7,10,11	VI,II,III
4,8,12	III,II,XVI
4,6	VI,III,I,VIII,XVI

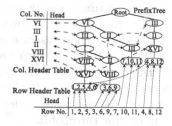

Fig. 2. pIndex

First, one may create the root of a tree, labelled with "null". Then, scan OPSM dataset. The scan of 1st OPSM leads to the construction of 1st branch of the tree: <VI, III, I, VIII, XVI>. Notice that we keep item order in an OPSM. And add a leaf node (1, 2, 5). For the 2nd OPSM, since its item list <VI, III, I, II, XIII> shares a common prefix <VI, III, I> with existing path <VI, III, I, VIII, XVI>. One new node (II) is created and linked as a child of (I), and another new node (XIII) is created and linked as the child of (II). The leaf node records row No. <3, 6, 9>. For other OPSMs, it uses the same method to build.

A column header table in Fig. 2 is built in which the order is conducted based on the occurrences of items from left to right and from top to bottom, and each item points to its occurrence in the tree via a column head of node-link. Nodes with the same column No. are linked in sequence via such bidirection node-links.

A row header table in Fig. 2 is built in which the order is conducted based on the occurrences of row Nos from left to right, and the tree nodes which have the same row No. will be saved in one hash set. For the sake of clarity, nodes with the same row No. are linked in sequence via a row head of node-link.

(4) OPSM Queries. Due to space limit, we only show one type of query.

For fuzzy queries on conditions FQ_c, first rotate the 1st element of keywords. Then, locate column keywords with column header table, from located node to tree root. If the number of keywords in the same branches is above length threshold τ, get gene names in the leaves. And test whether the number of gene names above size threshold δ, if true, add keyword set (gene name) as key (value) into final results. Otherwise, test until each keyword as first element one time.

When getting query result, and if query time is above a threshold, we save it in memory. If the memory is scarcity, we remove the longest unused result. When it does queries, we first test whether the memory saves the query results, if true, we return query result soon. For more details, please refer to work [3].

3   Demonstration

The gene expression repository used for demonstration consists of 6 documents (http://www.broadinstitute.org/cgi-bin/cancer/datasets.cgi). The offline part (permutation and mining) is conducted on Hama 0.4.0, and the online part (indexing and queries) is running on a PC with a 1.86GHz CPU, 2.9GB RAM, Ubuntu 14.04 system, and Firefox 28.0 web browser.

We use a web interface to demonstrate our system. Fig. 1(b) shows user interface of OMEGA. For more details, we provide a video to show the demonstration process (https://sites.google.com/site/jiangtaonwpu/). In the demonstration, we will show the permutation of columns, parallel OPSM mining, indexing of datasets, and searching examples under different configurations, such as different keyword and result thresholds.

References

1. Barkow, S., Bleuler, S., Prelić, A., Zimmermann, P., Zitzler, E.: BicAT: A biclustering analysis toolbox. Bioinformatics **22**(10), 1282–1283 (2006)
2. Jiang, D., Pei, J., Zhang, A.: GPX: interactive mining of gene expression data. In: Nascimento, M.A., Özsu, M.T., Kossmann, D., Miller, R.J., Blakeley, J.A., Schiefer, K.B. (eds.) VLDB, pp. 1249–1252. Morgan Kaufmann (2004)

3. Jiang, T., Li, Z., Chen, Q., Li, K., Wang, Z., Pan, W.: Towards order-preserving submatrix search and indexing. In: Renz, M., Shahabi, C., Zhou, X., Chemma, M.A. (eds.) DASFAA 2015. LNCS, vol. 9050, pp. 309–326. Springer, Heidelberg (2015)
4. Jiang, T., Li, Z., Chen, Q., Wang, Z., Pan, W., Wang, Z.: Parallel partitioning and mining gene expression data with butterfly network. In: Decker, H., Lhotská, L., Link, S., Basl, J., Tjoa, A.M. (eds.) DEXA 2013, Part I. LNCS, vol. 8055, pp. 129–144. Springer, Heidelberg (2013)

Probabilistic Programming in Anglican

David Tolpin(✉), Jan-Willem van de Meent, and Frank Wood

Department of Engineering Science, University of Oxford, Oxford, UK
{dtolpin,jwvdm,fwood}@robots.ox.ac.uk

Abstract. Anglican is a probabilistic programming system designed to interoperate with Clojure and other JVM languages. We describe the implementation of Anglican and illustrate how its design facilitates both explorative and industrial use of probabilistic programming.

Keyword: Probabilistic programming

1 Introduction

For data science practitioners, statistical inference is typically but one step in a more elaborate analysis workflow. The first stage of this work involves data acquisition, pre-processing and cleaning. This is often followed by several iterations of exploratory model design and testing of inference algorithms. Once a sufficiently robust statistical model and corresponding inference algorithm have been identified, analysis results must be post-processed, visualized, and in some cases integrated into a wider production system.

Probabilistic programming systems [1–3,9] represent generative models as programs written in a specialized language that provides syntax for the definition and conditioning of random variables. The code for such models is generally concise, modular, and easy to modify or extend. Typically inference can be performed for any probabilistic program using one or more generic inference techniques provided by the system backend, such as Metropolis-Hastings [3,8, 10], Hamiltonian Monte Carlo [7], expectation propagation [5], and extensions of Sequential Monte Carlo [4,6,9] methods. Although these generic techniques are not always as statistically efficient as techniques that take advantage of model-specific optimizations, probabilistic programming makes it easier to optimize models for a specific application in a manner that is efficient in terms of the dimensionality of its latent variables.

While probabilistic programming systems shorten the iteration cycle in exploratory model design, they typically lack basic functionality needed for data I/O, pre-processing, and analysis and visualization of inference results. In this demonstration, we describe the implementation of Anglican (http://bitbucket. org/dtolpin/anglican/), a probabilistic programming language that tightly integrates with Clojure (http://clojure.org/), a general-purpose programming language that runs on the Java Virtual Machine (JVM). Both languages share a common syntax, and can be invoked from each other. This allows Anglican programs to make use of a rich set of libraries written in both Clojure and Java. Conversely Anglican allows intuitive and compact specification of models for which inference may be performed as part of a larger Clojure project.

A. Bifet et al. (Eds.): ECML PKDD 2015, Part III, LNAI 9286, pp. 308–311, 2015.
DOI: 10.1007/978-3-319-23461-8_36

2   Design Outline

An Anglican program, or *query*, is compiled into a Clojure function. When inference is performed with a provided algorithm, this produces a sequence of return values, or *predicts*. Anglican shares a common syntax with Clojure; Clojure functions can be called from Anglican code and vice versa. A simple program in Anglican can look like the following code:

```
1  (defquery models
2    "chooses a distribution which describes the data"
3    (let [;; Model -- randomly choose a distribution and parameters
4          dist (sample (categorical [[uniform-discrete 1]
5                                     [uniform-continuous 1]
6                                     [normal 1]
7                                     [gamma 1]]))
8          a (sample (gamma 1 1)) b (sample (gamma 1 1))
9          d (dist a b)]
10     ;; Data  --- samples from the unknown distribution
11     (observe d 1) (observe d 2) (observe d 4) (observe d 7)
12     ;; Output --- predicted distribution type and parameters
13     (predict :d (type d))
14     (predict :a a) (predict :b b)))
```

Internally, an Anglican query is represented by a computation in *continuation passing style* (CPS), and inference algorithms exploit the CPS structure of the code to intercept probabilistic operations in an algorithm-specific way. Among the available inference algorithms there are Particle Cascade [6], Lightweight Metropolis-Hastings [8], Iterative Conditional Sequential Monte-Carlo (Particle Gibbs) [9], and others. Inference on Anglican queries generates a lazy sequence of samples, which can be processed asynchronously in Clojure code for analysis, integration, and decision making.

Clojure (and Anglican) run on the JVM and get access to a wide choice of Java libraries for data processing, networking, presentation, and imaging. Conversely, Anglican queries can be called from Java and other JVM languages. Programs involving Anglican queries can be deployed as JVM *jars*, and run without modification on any platform for which JVM is available.

3   Usage Patterns

Anglican is suited to rapid prototyping and exploration, on one hand, and inclusion as a library into larger systems for supporting inference-based decision making, on the other hand.

For exploration and research, Anglican can be run in Gorilla REPL (http:// gorilla-repl.org/); a modified version of Gorilla REPL better suited for Anglican is provided. Gorilla REPL is a notebook-style environment which runs in browser and serves well as a workbench for rapid prototyping and checking of ideas. Figure 1 shows a fragment of an Anglican worksheet in the browser:

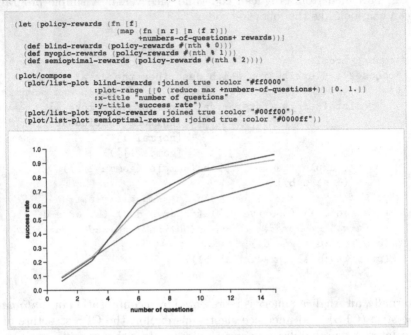

Let us visualize the results: red is blind, green is myopic, blue is semioptimal.

```
(let [policy-rewards (fn [f]
                        (map (fn [n r] [n (f r)])
                          +numbers-of-questions+ rewards))]
  (def blind-rewards (policy-rewards #(nth % 0)))
  (def myopic-rewards (policy-rewards #(nth % 1)))
  (def semioptimal-rewards (policy-rewards #(nth % 2))))

(plot/compose
  (plot/list-plot blind-rewards :joined true :color "#ff0000"
                  :plot-range [[0 (reduce max +numbers-of-questions+)] [0. 1.]]
                  :x-title "number of questions"
                  :y-title "success rate")
  (plot/list-plot myopic-rewards :joined true :color "#00ff00")
  (plot/list-plot semioptimal-rewards :joined true :color "#0000ff"))
```

Fig. 1. Anglican worksheet fragment. Post-processed inference results shown in a plot.

Library use is inherent to the Anglican's design for interoperability with Clojure. An Anglican query, along with supporting functions written in either Anglican or Clojure, can be encapsulated in a Clojure module and called from other modules just like Clojure function. Additionally, Anglican functions common for queries of a particular type or structure, such as state-space models or decision-making queries, can be wrapped as libraries and re-used.

4   Anglican Examples

Anglican benefits from a community-maintained collection of problem examples (https://bitbucket.org/fwood/anglican-examples), styled as Gorilla REPL worksheets. Each example is a case study of a problem involving probabilistic inference, includes problem statement, explanations for the solution, and a graphical presentation of inference results. Some of the included examples are:

- Indian GPA,
- Complexity Reduction,
- Bayes Net,
- Kalman Smoother,
- Gaussian Mixture Model,
- DP Mixture Model,
- Hierarchical Dirichlet Process,
- Probabilistic Deterministic Infinite Automata,
- Nested Number Guessing,
- Maximum Likelihood for Logistic Regression.

Anglican users are encouraged to contribute examples, both demonstrating advantages of probabilistic programming and presenting challenges to the current state-of-art of inference algorithms, to the repository.

Acknowledgments. This work is supported under DARPA PPAML through the U.S. AFRL under Cooperative Agreement number FA8750-14-2-0004.

References

1. Goodman, N.D., Stuhlmüller, A.: The Design and implementation of probabilistic programming languages (2015). http://dippl.org/ (electronic; retrieved March 11, 2015)
2. Goodman, N.D., Mansinghka, V.K., Roy, D.M., Bonawitz, K., Tenenbaum, J.B.: Church: a language for generative models. In: Proc. of Uncertainty in Artificial Intelligence (2008)
3. Mansinghka, V.K., Selsam, D., Perov, Y.N.: Venture: a higher-order probabilistic programming platform with programmable inference (2014). CoRR abs/1404.0099
4. van de Meent, J.W., Yang, H., Mansinghka, V., Wood, F.: Particle gibbs with ancestor sampling for probabilistic programs. In: Artificial Intelligence and Statistics (2015). http://arxiv.org/abs/1501.06769
5. Minka, T., Winn, J., Guiver, J., Knowles, D.: Infer.NET 2.4. Microsoft Research Cambridge (2010)
6. Paige, B., Wood, F., Doucet, A., Teh, Y.: Asynchronous anytime sequential Monte Carlo. In: Advances in Neural Information Processing Systems (2014)
7. Stan Development Team: Stan: A C++ Library for Probability and Sampling, Version 2.4 (2014)
8. Wingate, D., Stuhlmüller, A., Goodman, N.D.: Lightweight implementations of probabilistic programming languages via transformational compilation. In: Proc. of the 14th Artificial Intelligence and Statistics (2011)
9. Wood, F., van de Meent, J.W., Mansinghka, V.: A new approach to probabilistic programming inference. In: Artificial Intelligence and Statistics (2014)
10. Yang, L., Hanrahan, P., Goodman, N.D.: Generating efficient MCMC kernels from probabilistic programs. In: Proceedings of the Seventeenth International Conference on Artificial Intelligence and Statistics, pp. 1068–1076 (2014)

ProbLog2: Probabilistic Logic Programming

Anton Dries$^{(\boxtimes)}$, Angelika Kimmig, Wannes Meert, Joris Renkens,
Guy Van den Broeck, Jonas Vlasselaer, and Luc De Raedt

KU Leuven, Leuven, Belgium
{anton.dries,angelika.kimmig,wannes.meert,joris.renkens,
guy.vandenbroeck,jonas.vlasselaer,luc.deraedt}@cs.kuleuven.be
https://dtai.cs.kuleuven.be/problog

Abstract. We present ProbLog2, the state of the art implementation of
the probabilistic programming language ProbLog. The ProbLog language
allows the user to intuitively build programs that do not only encode
complex interactions between a large sets of heterogenous components
but also the inherent uncertainties that are present in real-life situations.
The system provides efficient algorithms for querying such models as well
as for learning their parameters from data. It is available as an online tool
on the web and for download. The offline version offers both command
line access to inference and learning and a Python library for building
statistical relational learning applications from the system's components.

Keywords: Probabilistic programming · Probabilistic inference ·
Parameter learning

1 Introduction

Probabilistic programming is an emerging subfield of artificial intelligence that
extends traditional programming languages with primitives to support probabilistic inference and learning. Probabilistic programming is closely related to
statistical relational learning (SRL) but focusses on a programming language
perspective rather than on a graphical model one. The common goal is to provide powerful tools for modeling of and reasoning about structured, uncertain
domains that naturally arise in applications such as natural language processing,
bioinformatics, and activity recognition.

This demo presents the ProbLog2 system, the state of the art implementation of the probabilistic logic programming language ProbLog [2–4]. Probabilistic logic programming languages and systems such as ProbLog2, PRISM and
CPLint, cf. [1] for an overview, combine ideas from both SRL and probabilistic
programming. They are thus related both to SRL systems such as Alchemy and
Primula, and to probabilistic programming languages rooted in other programming paradigms, such as Church, Venture, Infer.net, and Figaro.

ProbLog2 supports marginal inference, i.e., computing the conditional probabilities of queries given evidence, parameter learning from data in the form

© Springer International Publishing Switzerland 2015
A. Bifet et al. (Eds.): ECML PKDD 2015, Part III, LNAI 9286, pp. 312–315, 2015.
DOI: 10.1007/978-3-319-23461-8_37

of (partial) interpretations, sampling of possible worlds, and interaction with Python.

ProbLog2 is available on `https://dtai.cs.kuleuven.be/problog`.

2   Language

ProbLog is a probabilistic programming language that extends Prolog along the lines of Sato's distribution semantics. Its development focusses especially on machine learning techniques and implementation aspects.

From a modeling perspective, a ProbLog program has two parts: (1) a probabilistic part that defines a probability distribution over truth values of a subset of the program's atoms, and (2) a logical part that derives truth values of remaining atoms using a reasoning mechanism similar to Prolog. While the latter part simply contains Prolog clauses, the former is specified by *probabilistic facts* p :: `fact`, meaning that `fact` is true with probability p. All these are probabilistically independent; in case they contain variables, all ground instances are independent as well.

For ease of modeling, ProbLog also allows the use of *annotated disjunctions* p_1 :: h_1; ...; p_n :: h_n :− body with $\sum_{i=1}^{n} p_i \leq 1$, meaning that if body is true, one of the h_i will be true according to the specified probabilities p_i. If the probabilities do not sum to one, it is also possible that none of the h_i is true (with probability $1 - \sum_{i=1}^{n} p_i$).

The following ProbLog program models a small social network, where people's smoking behaviour is influenced by the behaviour of their friends. And indirectly, also by the behaviour of friends of friends.

```
0.4::asthma(X) :- smokes(X).
0.3::smokes(X).
0.2::smokes(X) :- friend(X,Y), smokes(Y).
friend(1,2). friend(2,1). friend(2,4). friend(3,2). friend(4,2).
```

The main inference task addressed in ProbLog is that of calculating the probability that a query succeeds, for instance ?- `asthma(2)` succeeds with a probability 0.15. If we add `asthma(3)` as evidence, the conditional probability of ?- `asthma(2)` is 0.19.

3   System Blocks

ProbLog2 is the successor of ProbLog1 [2], which was completely integrated in YAP Prolog and performed BDD-based probabilistic inference. ProbLog2 inference consists of a series of transformation steps as shown in Figure 1. The first step is to take the weighted logic program (ProbLog model) and ground it using a Prolog-based grounder. The ground program that is obtained can be represented as a logical formula which may contain cycles. The next steps convert the ground program to a formula in propositional logic, which involves handling

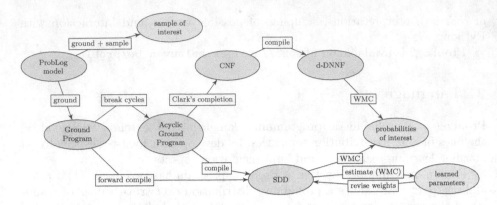

Fig. 1. Overview of the primary ProbLog pipelines.

cycles. Different options are available for this conversion, using different logical reasoning techniques. Forward compilation directly compiles to sentential decision diagrams (SDD). Alternatively, cycles can be removed first, and the resulting acyclic ground program can then be either transformed into conjunctive normal form and compiled into a d-DNNF, or compiled directly into a SDD. Both these normal forms support efficient weighted model counting to obtain the final probabilities of interest.

ProbLog also supports query-based sampling in which it assigns a truth value to each of the queries based on their joint probability in the model. This algorithm operates directly during the grounding phase and does not require knowledge compilation.

Parameter learning from interpretations takes a base model and a set of examples as sets of evidence and compiles these into an SDD. These SDDs are then evaluated repeatedly until the weights in the model converge.

4   System Usage

The ProbLog system can be used through three channels:

Online: The web version of ProbLog allows the user to enter and solve ProbLog problems without the need to install any additional software. It offers an interactive tutorial that illustrates the key modeling concepts in ProbLog through a range of examples, which can be edited on the fly, as well as a separate editor. Examples in the tutorial range from traditional probabilistic models such as Bayesian networks (with plates) and Hidden Markov models to relational probabilistic models that use the full flexibility of ProbLog.

Command line tool: The backend of the online interface is also available as a command line tool, which in contrast to the online version does not impose resource restrictions. It further offers a number of extra options such as, for example, flags to select the required pathway through Fig. 1. It is written in Python and is easy to install on multiple platforms. Additionally, the command line tool allows one to execute ProbLog programs that make use of external functionality written in Python. This makes it possible to, besides the declarative modeling capacities of Prolog, also harness the full power of the Python programming language and its extended ecosystem (e.g. scikit-learn, NLTK). For example, in a natural language processing application, the probability associated with a probabilistic fact P::similar(d1,d2) could be defined as the edit-distance between the two string arguments d1 and d2 as computed by a corresponding Python function.

Library: ProbLog can be used as a Python package for expressing and querying probabilistic concepts and models. The library provides data structures and algorithms that represent all the components shown in Figure 1. It allows the user to build Statistical Relational Learning (SRL) applications that reuse ProbLog's components.

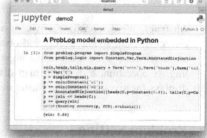

Fig. 2. The online interface (left) and as a Python library in a Jupyter Notebook (right)

References

1. De Raedt, L., Kimmig, A.: Probabilistic (logic) programming concepts. Machine Learning (2015)
2. De Raedt, L., Kimmig, A., Toivonen, H.: ProbLog: a probabilistic prolog and its application in link discovery. In: Proceedings of the 20th International Joint Conference on Artificial Intelligence (IJCAI) (2007)
3. Fierens, D., Van den Broeck, G., Renkens, J., Shterionov, D., Gutmann, B., Thon, I., Janssens, G., De Raedt, L.: Inference and learning in probabilistic logic programs using weighted Boolean formulas. Theory and Practice of Logic Programming 15(03), 358–401 (2015)
4. Vlasselaer, J., Van den Broeck, G., Kimmig, A., Meert, W., De Raedt, L.: Anytime inference in probabilistic logic programs with Tp-compilation. In: Proceedings of 24th International Joint Conference on Artificial Intelligence (IJCAI) (2015)

Real Time Detection and Tracking of Spatial Event Clusters

Natalia Andrienko[1,2]([⊠]), Gennady Andrienko[1,2], Georg Fuchs[1], Salvatore Rinzivillo[3], and Hans-Dieter Betz[4]

[1] Fraunhofer Institute IAIS, Sankt Augustin, Germany
{natalia.andrienko,gennady.andrienko,
georg.fuchs}@iais.fraunhofer.de
[2] City University London, London, UK
{natalia.andrienko,gennady.andrienko}@iais.fraunhofer.de
[3] CNR ISTI, Pisa, Italy
rinzivillo@isti.cnr.it
[4] nowcast GmbH, Munich, Germany
hdbetz@nowcast.de

Abstract. We demonstrate a system of tools for real-time detection of significant clusters of spatial events and observing their evolution. The tools include an incremental stream clustering algorithm, interactive techniques for controlling its operation, a dynamic map display showing the current situation, and displays for investigating the cluster evolution (time line and space-time cube).

1   Problem Setting

Spatial events are physical or abstract entities with limited existence times and particular locations in space, for example, lightning strikes or mobile phone calls. A spatial event is characterized by its start and end times (which may coincide), spatial coordinates, and, possibly, some thematic attributes. We assume that occurrences of spatial events are registered, e.g., by sensors, and corresponding data records are immediately sent to a server. The resulting data stream needs to be monitored.

We consider monitoring scenarios in which each individual event is not significant whereas spatio-temporal event clusters (i.e., occurrence of multiple events closely in space and time) may require observer's attention. For example, moving vehicles may emit low speed events when their speed drops below a certain threshold. It is neither feasible nor meaningful to attend to every such event, but a spatio-temporal cluster of low speed events sent by several cars may deserve observer's attention as a possible indication of a traffic jam. After detecting a cluster, the observer may need to trace its further evolution, i.e., changes in the number of events, number of vehicles involved, spatial location, shape, and extent. The task is to support the observer in detecting the emergence and tracking the evolution of spatial event clusters in real time.

A. Bifet et al. (Eds.): ECML PKDD 2015, Part III, LNAI 9286, pp. 316–319, 2015.
DOI: 10.1007/978-3-319-23461-8_38

2   Approach

We apply clustering techniques to separate spatio-temporal event concentrations (clusters) from scattered events (noise). In fact, the problem setting requires an analog of a density-based clustering method capable to process a data stream in real time. However, the existing stream clustering methods are oriented to somewhat different problem settings. The main problem they address is the memory limitation. Assuming that all data cannot fit in the memory, the methods summarize incoming data on the fly and keep only the summaries (*micro-clusters*) but not the original data items. Many streaming algorithms assume a two-phase approach: micro-clusters are created and maintained during an online phase and post-processed (e.g., merged into larger clusters) during an offline phase. This general framework is instantiated with different approaches to creating micro-clusters. The main representatives are CluStream [1] and DenStream [2] doing partition-based and density-based clustering, respectively.

CluStream partitions an initial portion of a stream into k micro-clusters. When a new data point d appears, it tries to fit d into one of the current micro-clusters, while satisfying the constraints on the maximum number of clusters k and maximum boundary R. DenStream identifies micro-clusters with a maximal radius *Eps* based on the concepts of core object and density adopted in density-based clustering. Unlike in CluStream, the number of micro-clusters is not bounded. Both approaches rely on a following offline phase, in which micro-clusters are merged into macro-clusters.

Our problem setting and requirements differ from those of the existing methods in several respects. First, emerging significant clusters need to be detected in real time and immediately shown to the observer, permitting no reliance on off-line post-processing. Second, clusters may emerge, evolve (grow, shrink, move, change shape, split, merge), and disappear, excluding the approaches assuming a constant number of clusters, like [1]. Third, the main memory limitations is not our primary focus. We assume that the available memory is sufficient for keeping all micro-clusters that may co-exist within a certain time interval ΔT. Fourth, old micro-clusters (where the latest event is older than ΔT) are not of interest anymore and may be discarded.

We propose a hybrid approach in which micro-clusters are built and updated similarly to [1], but without limiting the maximal number. They are merged online into larger clusters of arbitrary sizes and shapes by exploiting *k-connectivity*, as in [2].

In our approach, a *micro-cluster* consists of events fitting in a circle with a user-specified maximal radius R. Another user-specified constraint is the maximal temporal gap ΔT that may exist between the events within a micro-cluster. When a new event comes, the algorithm checks whether it fits in one of existing micro-clusters, i.e., whether the event's distance to the micro-cluster center does not exceed R. If so, the event is added to this micro-cluster, and the position of the center is updated. If not, a new micro-cluster consisting of only this event is created. For effective search of candidate micro-clusters for including new events, we use a spatial index [3].

The algorithm keeps in the memory only events that occurred within the time interval $[t_c\text{-}\Delta T, t_c]$, where t_c is the current moment. Older events are erased from the micro-clusters, and micro-clusters that become empty are removed from the memory. A *connecting event* of two or more micro-clusters is an event located sufficiently close to their centers. By default, it means that the distance does not exceed R, but the user may specify a different connection distance threshold R_c. When micro-clusters have at least k connecting events, where k is a user-specified parameter, these micro-clusters are merged into a *macro-cluster*. A similar idea is employed in algorithm AING [4], where analogs of micro-clusters are treated as graph nodes. When a connecting data point appears, an edge is created between the respective nodes.

A macro-cluster may include an arbitrary number of micro-clusters provided that each of them is k-connected to some other member micro-cluster. Hence, macro-clusters may have arbitrary shapes and spatial extents. As soon as the size (i.e., the number of the member events) of some micro- or macro-cluster reaches a user-chosen minimum N_{min}, this cluster is visually presented to the observer on a map display.

The algorithm can additionally account for thematic attributes of the events. Thus, low speed events from moving vehicles may have an attribute 'movement direction'. The algorithm can ensure that only events with similar directions are put together (the user needs to specify the maximal allowed difference in the directions). Another extension is accounting for the event sources (e.g., the vehicles sending low speed events). For each micro- and macro-cluster, the algorithm can maintain a list of distinct event sources. Clusters where all events come from a single source or from too few distinct sources may be disregarded as insignificant.

The map display used for the event stream monitoring shows the spatial positions, extents, and shapes of the significant event clusters that have been existing during the time interval $[t_c\text{-}TH, t_c]$, where t_c is the current time moment and TH is a chosen time horizon of the observation. The evolution of the clusters, i.e., changes in the number of events, spatial extent, density, and/or other characteristics of the clusters, can be explored using additional time line and space-time cube displays. The parameter settings of the clustering algorithm can be interactively modified while the algorithm works. For presenting the tools, an event stream is simulated using real data.

3   Example

Figure 1 illustrates an application of the tools to a stream of lightning strike events simulated based on real data collected by nowcast GmbH (www.nowcast.de). The following parameter settings have been used: $R = R_c = 3$ km; $\Delta T = 20$ minutes; $k = 1$ event; $N_{min} = 50$ events. In the upper part of Fig. 1, two screenshots of the map display show situations at two different time moments of the stream monitoring process with the time horizon TH = 20 minutes.

The clusters of lightning strike events are visually represented using several complementary techniques. The semi-transparent polygons represent the spatial convex hulls of the clusters. The violet-colored polygons show the cluster shapes and extents at the current time moment t. The light gray polygons enclose the cluster states that took place during the interval [t-TH, t]. The cluster sizes (i.e., the total event counts) are represented by proportional sizes of the red circles. The purple lines show the trajectories of the cluster centers. By hovering the mouse cursor over the trajectories, as shown on the top left of Fig. 3, the user can obtain information about the cluster states at different times.

The map screenshots clearly show that the event clusters move over time in the northeastern direction. The big cluster on the top right is a result of several neighboring clusters having merged.

In the lower part of Fig. 3, the cluster histories from a 2-hour time interval are presented on a map (left) and in a space-time cube (right). The clusters are differently colored, to enable easier distinguishability. On the map, the colored polygons show the latest states of the clusters and the light grey polygons enclose all their member events. In the space-time cube, the colored three-dimensional shapes show the spatio-temporal extents and evolutions of the clusters.

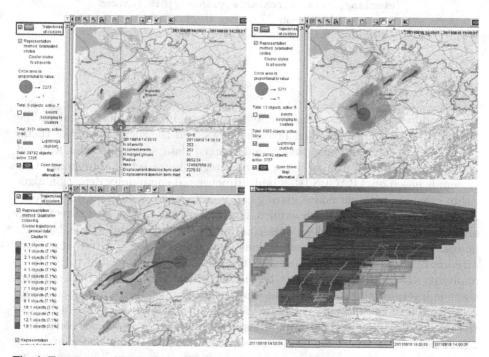

Fig. 1. Top: two screenshots of the map display showing the situations in different time intervals. Bottom: the final clusters are represented on a map and in a space-time cube.

References

1. Aggarwal, C.C., Han, J., Wang, J., Yu, P.S.: A framework for clustering evolving data streams. In: Proc. 29th Int. Conf. Very Large Data Bases, Berlin, Germany (2003)
2. Cao, F., Ester, M., Qian, W., Zhou, A.: Density-based clustering over an evolving data stream with noise. In: Proc. 6th SIAM Int. Conf. Data Mining, SIAM, Bethesda, Maryland, USA (2006)
3. Andrienko, N., Andrienko, G.: Spatial generalization and aggregation of massive movement data. IEEE Trans. Visualization and Computer Graphics **17**(2), 205–219 (2011)
4. Bouguelia M.-R., Belaïd Y., Belaïd, A.: An adaptive incremental clustering method based on the growing neural gas algorithm. In: Proc. 2nd Int. Conf. Pattern Recognition Applications and Methods - ICPRAM 2013, pp. 42–49 (2013)

S&P360: Multidimensional Perspective on Companies from Online Data Sources

Michele Berlingerio[1]([✉]), Stefano Braghin[1], Francesco Calabrese[1],
Cody Dunne[2], Yiannis Gkoufas[1], Mauro Martino[2],
Jamie Rasmussen[2], and Steven Ross[2]

[1] IBM Research Ireland, Dublin, Ireland
{mberling,stefanob,fcalabre,yiannisg}@ie.ibm.com
[2] IBM Watson, Yorktown Heights, USA
{cdunne,mmartino,jrasmus,steven_ross}@us.ibm.com

1 Introduction

We introduce S&P360, a system to analyse and explore multidimensional, online data related to companies, their financial news, and the social impact of them. Our system combines official and crowd-sourced data sources to offer a broad perspective on the impact of financial newsregarding a set of companies. Our system is based on ABACUS [1], a multidimensional community detection algorithm grouping together nodes sharing communities across different dimensions. ABACUS is able to find both explicit connections that appear as direct links among entities, but also hidden ones coming from indirect interactions between nodes. We enrich structural connections (co-occurrence in Twitter and news articles, hyperlinks in Wikipedia) with latent semantics associated to them by applying NLP techniques such as Latent Dirichlet Allocation (LDA), guiding users in interpreting results. We add a powerful visualization interface enabling users to query the data by company, time, and dimension. Users can browse the results and explore the communities along with their associated semantics. "Evidence" of the structural connections (i.e., the source documents supporting the explicit connections) are shown in the user interface, as well as community

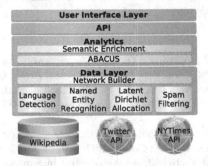

Fig. 1. S&P360 architecture

© Springer International Publishing Switzerland 2015
A. Bifet et al. (Eds.): ECML PKDD 2015, Part III, LNAI 9286, pp. 320–324, 2015.
DOI: 10.1007/978-3-319-23461-8_39

topics that summarize both explicit and latent semantic associated to the multidimensional relationships. The challenges faced in the development of S&P360's technology and architecture arise mainly from the noise contained in online data, particularly significant in sources like Twitter, and from the difficulty in displaying and interacting with multidimensional and semantically enriched data effectively. The demo uses publicly available data from Twitter, Wikipedia, and the New York Times (hereafter, NYTimes). We obtained multidimensional communities showing interaction between companies around important events like acquisition of companies or products, change of core business, stock market fluctuations, and the launch of new products. All these events are not hard-coded, but emerge directly from the data and interpretation of the semantics associated to the communities.

2   Architecture

S&P360 is modular, and the architecture of the main modules is depicted in Figure 1.

2.1   Data Layer

This layer implements five main modules: language detection, named entity recognition, latent dirichlet allocation, anti-spam filtering, and network builder.

Language Detection. We used a public library for language detection that is based on a Naive Bayes approach (code.google.com/p/language-detection), for its very good accuracy-scalability trade-off. We focused on English tweets, for the sake of a better manual validation of the next steps, but there are no particular constraints in adding other languages already available in the library.

Named Entity Recognition. After filtering by language, we reduced problematic ambiguous terms like "Apple" or "Coach" by running named entity recognition on the English tweets. We used Stanford's NER (nlp.stanford.edu/software/CRF-NER.shtml) library for its popularity and its specific training for organizations.

Latent Dirichlet Allocation. In order to select tweets relevant to finance, we ran Latent Dirichlet Allocation (LDA) to assign to each tweet a scored set of related topics, in a semi-supervised fashion. We used MALLET (mallet.cs.umass.edu/topics-devel.php) as it provides a well established set of tools and links topics back to source documents.

Anti-Spam Filtering. The next step was filtering out "bots", i.e. automatic Twitter accounts talking about companies. A simple and natural way to do this is to rely on their very limited dictionary, so we just kept the tweets maximizing the ratio $\frac{\#distinct\ words}{\#tweets}$.

Network Builder. We take the results of the four previous modules and build a multidimensional network between the companies found. Links are assumed to be bidirectional by default. For Wikipedia, we link companies if the wiki page of one had the other as an external link. In Twitter, we link companies mentioned together and weight by the number of such tweets. In NYTimes, we link companies if they were mentioned together in an article, or if one of the two was used to tag an article regarding the other one. The number of articles for which this happens is used to weigh the edges.

2.2 Analytics

This layer implements two main modules: ABACUS and semantic enrichment.

ABACUS. ABACUS [1] is an innovative algorithm for community detection in multidimensional networks that introduces a new definition of community, i.e. a group of nodes sharing memberships to the same community in the same dimensions. This property allows ABACUS to relax the requirement for dense multidimensional connections between nodes across dimensions, while finding nodes that are related together but not necessarily directly connected in any of the dimensions [1]. This means that we can find a group of companies that are related to each other regarding a set of topics, although they do not have to be directly connected in Wikipedia, or Twitter, or NYTimes.

Semantic Enrichment. Two steps are performed in this block: i) for each edge in a resulting community, we retrieve all the associated information including topics with associated score; ii) we compute aggregated topics and scores (min and max across edges) for the community. In this way, we have a semantic associated to the entire community rather than to a single edge, and we can sort or query the communities using the semantics rather than only time or dimension.

2.3 API

S&P360 analytics block exposes a set of API methods to the User Interface Layer. Such calls have been developed as a RESTful Web service, deployed in IBM BlueMix (console.ng.bluemix.net), an implementation of the Cloud Foundry PAAS specification.

2.4 User Interface Layer

S&P360 provides a visual interface to allow interactive exploration of the ABACUS algorithm's results. The user interface is built on HTML5 and JavaScript technologies. Though the visualizations are two-dimensional they are implemented as planar shapes within a 3D scene rendered on WebGL canvases using Three.js (threejs.org), allowing for high rendering performance and smooth zooming. The D3.js (d3js.org) library is used for network layout. Upon user selection of a company and time range, the interface queries the API for up to ten top scoring community matches, which are shown in a ranked list (see Figure 2). Each community result indicates the data source(s), number of companies, and representative topics as determined by semantic enrichment.

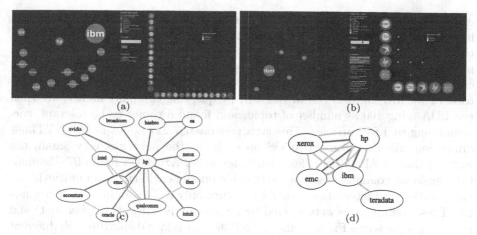

Fig. 2. Examples of resulting communities visualized by S&P360 (top) and with GraphViz (bottom)

Table 1. Characteristics of the two communities. First row is community in Fig 2a,c. Second is Fig 2b,d.

Time	Companies	Dimensions	Topics	Evidences (subset)
Q1 2011	Accenture, Broadcom, EA, EMC, Hasbro, Hp, IBM, Intel, Intuit, Nvidia, Oracle, Xerox	Wikipedia NYTimes	investment, technology	Wikipedia: IT / electronics companies NYTimes: "Recent Acquisitions by Major Technology Companies http://www.nytimes.com/aponline/2011/01/05/[...]" "[..] Four American giants - Microsoft, Oracle, Amgen and Pfizer - were among the fund's top five [..] The fifth is Accenture" "Computing has shifted to phones, but the leading maker of processors has not. Now Intel is trying to catch up to Qualcomm, Nvidia and Marvell."
Q2 2013	EMC, Hp, IBM, Teradata, Xerox	Wikipedia, NYTimes, Twitter	cloud computing, investment, deal, common interest	Wikipedia: IT / electronics companies NYTimes: "International Business Machines Corp and EMC Corp are among parties in talks to buy privately held database web hosting company SoftLayer Technologies Inc, in a deal that could fetch over $2 billion, three sources close to the matter said." Twitter: "Webinar replay: IBM & Teradata Compared: A Total Cost of Ownership Study http://t.co/JPMUpDHjhU" "RT @mcgoverntheory: Big #Amazon Customer Moves To HP Public #Cloud http://t.co/OaVLS7f6Wt #cio #entarch #ibm #emc #gartner #forrester #ensw.." "Teradata, IBM and EMC are at #hadoopsummit. Oracle and HP are missing."

3 Demo on Real World Data

We tested S&P360 on real data from Wikipedia, Twitter and NYTimes. We started from the list (en.wikipedia.org/wiki/List_of_S%26P_500_companies) of the 500 S&P companies available in Wikipedia, reporting stock ticker symbol, name, sector, address, website, and date entered. We then crawled the landing 500 links from this page and built the Wikipedia network dimension out of it. For sake of simplicity, we ignored the temporal dimension in Wikipedia. This step generated 3,637 edges in the Wikipedia dimension. We used the list of names and tickers to query the NYTimes and Twitter APIs. For Twitter, we had access to 10% of the world wide stream from 2011 to 2014. From this collection, we ran string matching on tickers and names, and ended up with 5,724,590 tweets. Out of these, the language detection took 3,276,367 English tweets. We ran LDA on the resulting dataset using 200 topics. We manually annotated all of them, ending up with 19 topics relevant to the financial space, with 231,075 corresponding tweets. By running the named entity recognition task we selected 79,428 tweets. From those, we filtered out the tweets coming from around 100 automated bots, ending up with a final collection of 55,678 tweets. This step generated 11,966 edges in the Twitter dimension. From NYTimes, we retrieved a total of 103,676 snippets (all in English) by querying company names. We then ran LDA using 100 as number of topics and found 6 topics to be relevant, corresponding to 1,010 articles. This step generated 7,227 edges in the NYTimes dimension. We then ran ABACUS on each (as the networks very small, the running time of ABACUS is less than 2 seconds). ABACUS found 97–158 multidimensional communities (we excluded monodimensional communities). For visualization and interaction, our APIs return only 6–34 communities per quarter. These communities were filtered by number of dimensions (higher first) and minimum topic score. Figure 2 shows S&P360's interface displaying two different results. On the top, we see the two communities in the S&P360 user interface. They were found by querying "IBM" and narrowing the search to Q1 2011 (left) and Q3 2013 (right). Since multidimensional properties, topics, and evidences are available upon user interaction, we report their structure on the bottom of Figure 2 and their characteristics in Table 1.

4   Video and Requirements

A video of S&P360 in action can be seen at vimeo.com/117626196, showing the visual querying interface; the result browsing capability; and the interaction with multiple dimension, topics, and evidences. For best performance, the demo would need a widescreen external monitor and Internet access.

References

1. Berlingerio, M., Pinelli, F., Calabrese, F.: ABACUS: frequent pattern mining-based community discovery in multidimensional networks. Data Min. Know. Dis. **27**(3), 294–320 (2013)

Scavenger – A Framework for Efficient Evaluation of Dynamic and Modular Algorithms

Andrey Tyukin , Stefan Kramer, and Jörg Wicker[✉]

Johannes Gutenberg-Universität Mainz, Staudingerweg 9, 55128 Mainz, Germany
tyukiand@students.uni-mainz.de, {kramer,wicker}@informatik.uni-mainz.de

Abstract. Machine Learning methods and algorithms are often highly modular in the sense that they rely on a large number of subalgorithms that are in principle interchangeable. For example, it is often possible to use various kinds of pre- and post-processing and various base classifiers or regressors as components of the same modular approach. We propose a framework, called Scavenger, that allows evaluating whole families of conceptually similar algorithms efficiently. The algorithms are represented as compositions, couplings and products of atomic subalgorithms. This allows partial results to be cached and shared between different instances of a modular algorithm, so that potentially expensive partial results need not be recomputed multiple times. Furthermore, our framework deals with issues of the parallel execution, load balancing, and with the backup of partial results for the case of implementation or runtime errors. Scavenger is licensed under the GPLv3 and can be downloaded freely at https://github.com/jorro/scavenger.

1 Introduction

Consider the following example: Suppose we want to compare different instances of a modular algorithm $A_{f,g}$ that consists of two major parts: a preprocessing part f, and a core algorithm g: $A_{f,g}(x) = g(f(x))$, with $f \in \{f_1, \ldots, f_n\}$ and $g \in \{g_1, \ldots, g_m\}$. We could evaluate each combination separately, and then choose the best performing combination. However, this results in a repeated computation of each $f_i(x)$, one time for each possible core-algorithm g_j. Hence, it is beneficial to cache and to keep intermediate results, and to share these results between the members of the algorithm family $\{A_{f,g}\}_{f,g}$ to avoid unnecessary recomputation. Yet, we want to keep parallel execution, although the evaluation of the family of algorithms with shared subalgorithms is no longer *embarrassingly parallel*.

In the demo, we present Scavenger, a framework that simplifies the representation and evaluation of families of algorithms that share common subalgorithms. The user of Scavenger formulates a family of algorithms in a declarative style as a directed acyclic graph (DAG) of subalgorithms which can in turn spawn arbitrarily complex DAGs to carry out its computation. The whole job, represented as a DAG is then evaluated in a way that is suitable for single or

© Springer International Publishing Switzerland 2015
A. Bifet et al. (Eds.): ECML PKDD 2015, Part III, LNAI 9286, pp. 325–328, 2015.
DOI: 10.1007/978-3-319-23461-8_40

multiple computers (including clusters). In contrast to cluster engines or map-reduce approaches that process large amounts of data with a single algorithm, our framework is focused on the use case with a moderate amount of data, but a large number of algorithms.

Workflow systems like Taverna[1], Knime[2], WEKA[3], Pipeline Pilot[4], or ADAMS[5] provide an easy-to-use interface to carry out experiments with repetitive steps. This makes them closely related to SCAVENGER. Yet, caching and reusing of intermediate results is not a central aspect of these systems, which is the core functionality and benefit of SCAVENGER. Additionally, SCAVENGER was designed with an easy-to-use mechanism to run on clusters in contrast to workflow systems, which tend to orient more on the single desktop use case (note that we talk about the emphasis of the system, exceptions do exist).

2   The SCAVENGER Framework

The back-end of the SCAVENGER is built on top of the Akka framework[6]. Akka provides an implementation of an actor system that can be distributed across multiple physical compute nodes connected by a network. Actor systems are hierarchical collections of actors. Actors are lightweight entities that are characterized by their internal state and their reactions on incoming messages. Every actor has its own mailbox, and communicates with the outside world exclusively by sending and receiving messages. Conceptually one can think that each actor is executed on its own thread, however, in reality actors are much more lightweight than threads of the operating system. The messages are usually simple, immutable, serializable JVM objects which can be either passed by reference within a single JVM or serialized and sent over network via the TCP/IP protocol.

The Akka framework ensures that all actors get enough CPU time for the whole system to stay responsive, delivers the messages sent by actors, handles error propagation within the actor system, and also takes care about the communication between multiple actor systems that are running on different physical machines. Users have only to specify various types of actors with their behaviors as well as the kinds of messages that are exchanged between the actors.

The SCAVENGER back-end uses multiple types of virtual nodes in order to do its job. Each node is controlled by an actor that is responsible for communication with other nodes. For simplicity, one can think of each virtual node running on a separate physical machine, but this is actually not required: multiple SCAVENGER nodes can coexist on the same physical machine, or even be executed within a single JVM. The back-end implements the classical Master-Worker pattern. There are currently three types of nodes: the seed node, the master node, and

---

[1] see http://www.taverna.org.uk/

[2] see http://www.knime.org/

[3] see http://www.cs.waikato.ac.nz/ml/weka/

[4] see http://accelrys.com/products/pipeline-pilot/

[5] see https://adams.cms.waikato.ac.nz/elgg/

[6] see http://akka.io

the worker node. When a SCAVENGER service is running, a single seed node, a master node, and multiple worker nodes must be active.

The seed node is responsible for establishing connections between all other nodes. The main purpose of the seed node is to wait for a handshake message from the master and then to tell each worker node where the master node is.

The master node communicates directly with the client application. It translates the incoming computation-valued requests from the client into a form that is suitable for parallel computation by subdividing it into smaller tasks, and then schedules these tasks for execution. It coordinates the work of the worker nodes, and is responsible for load balancing and dealing with failure of single worker nodes. It is also responsible for caching and backing up of intermediate results.

The worker nodes provide the raw computing power. They receive internal jobs, compute the result, and send it back to the master node.

3 Case Study: Autoencoders

Autoencoders are neural networks that can be trained in an unsupervised manner [1]. The input and the output clamped to both ends of the autoencoder are the same. The information is pressed through the central bottleneck layer. The activations of neurons in the central layer can be considered a compressed representation of the data. We train autoencoders as follows. We start with a single layer of neurons. Then we keep unfolding the innermost layer, eventually train every new layer separately, and then tune the whole network using backpropagation. Furthermore, we have the possibility to train the final inner layer as an Restricted Boltzmann Machine (RBM) or to tune it as a separate autoencoder with one single layer. We can also tune the whole network using backpropagation. The trained network can have different depths and different dimensions of layers. Obviously, if one autoencoder arises as an unfolded version of another autoencoder, models will share some common intermediate results when executed. We want to briefly sketch how one would approach the training of the family of autoencoders using the SCAVENGER framework. First, we have to define the atomic algorithms unfold, trainInnerAsRbm tuneInnerAsAnn, tuneWholeNetwork:

```
1   case class TuneInner extends Algorithm[(Data, Autoencoder), Autoencoder] {
2       def identifier = formalccc.Atom("tuneInner")
3       def difficulty = Parallel
4       def cachingPolicy = CacheGlobally
5       def apply(dataAndEnc: (Data, Autoencoder), ctx: Context):
6       Future[Autoencoder] = {...}
7   }
```

The implementation of iterative optimization methods requires some more work. The reason is that we have to transform synchronous while loops into asynchronous code. For this, the SCAVENGER framework provides asynchronous control structures like async_while. In an iterative optimization method like backpropagation, we have to replace all while that have a body that requires substantial computation time by async_while. Furthermore, we have to chain asynchronous calls by Futures flatMap instead of usual semicolons:

```
1  case class Backpropagation extends Algorithm[(Data, Autoencoder), Autoencoder] {
2      ...
3      def apply(da: (Data, Autoencoder), ctx: Context):
4      Future[Autoencoder] = {
5          async_while(predicate) {/* submitting ad-hoc jobs to `ctx`, waiting for
                  results*/}.flatMap{...}
```

When all the basic atomic algorithms are defined, we might want to partially apply some of them to the data, so that we have a uniform collection of algorithms that map **Autoencoders** to new **Autoencoders**, and do not require any extra input:

```
1  val data = Computation("theData") { /* load data from file */ }
2  val tuneWholeWithData = tuneWholeNetwork.partialFst(data)
```

Now suppose that we have an implementation that trains an autoencoder with a specified training strategy:

```
1  def autoencoder(trainingStrategy: Algorithm[Autoencoder, Autoencoder],numLayers):
      Computation[Autoencoder] = {...}
```

Since we do not know which combination performs best according to some error measure, we generate every possible combination of **trainInnerAsRbm**, **tuneInnerAsAnn** and a identity operation **Id**:

```
1  val strategies = for {
2      f <- List(trainInnerAsRbmWithData, Id)
3      g <- List(tuneInnerAsAnnWithData, Id)
4      h <- List(tuneWholeWithData, Id)
5  } yield (h o g o f)
```

Now we submit multiple autoencoder-jobs to the SCAVENGER framework. We vary the depth and the training strategy.

```
1  val futures = for {
2      s <- strategies
3      n <- List(2,3,4,5)
4  } scavengerContext.submit(autoencoder(s, n))
5  val allResults = Future.sequence(futures)
6  val results = Await.result(allResults)
```

After this, depending on how we are executing the application, we could issue some cleanup instructions and shut down the system.

4   Conclusion

Our framework combines the aspects of caching, persistence of intermediate results, and parallelism. It provides a simple and minimalist user API and can be used on single computers or clusters. It can simplify and accelerate the development and execution of experiments that involve a large number of algorithms that share common subalgorithms. In the demo, we will show the simplicity and power of the framework by means of simple examples that will be developed step by step.

References

1. Hinton, G.E., Salakhutdinov, R.R.: Reducing the dimensionality of data with neural networks. Science **313**(5786), 504–507 (2006)

UrbanHubble: Location Prediction and Geo-Social Analytics in LBSN

Roland Assam[✉], Simon Feiden, and Thomas Seidl

RWTH Aachen University, Aachen, Germany
{assam,seidl}@cs.rwth-aachen.de, simon.feiden@rwth-aachen.de

Abstract. Massive amounts of geo-social data is generated daily. In this paper, we propose UrbanHubble, a location-based predictive analytics tool that entails a broad range of state-of-the-art location prediction and recommendation algorithms. Besides, UrbanHubble consists of a visualization component that depicts the real-time complex interactions of users on a map, the evolution of friendships over time, and how friendship triggers mobility.

1 Introduction

The volume of data generated from human social interactions in Location-Based Social Networks (LBSN) is breathtaking. Such data encapsulates all visited locations and mimics the identity, behaviors, and affiliations of an individual or group. This has fueled enormous research interests to study location-based social interactions or group dynamics. One profound user behavior that has emerged during mobile social networking is the generation of *check-in*. Check-in is a phenomena whereby a person deliberately broadcasts her current location to a group of friends in an LBSN.

Numerous location prediction techniques have been proposed. To the best of our knowledge, there is no platform that consists of a broad array of innovative state-of-the-art location prediction techniques such as [1,3,5,6,8,9]. The availability of such a framework would assist researchers to quickly compare and evaluate state-of-the-art prediction techniques. Thus, saving their time and allowing them to focus more on the new techniques they aspire to develop.

Towards this end, we were motivated to create UrbanHubble[1], an innovative LBSN predictive analytics tool, which entails a broad spectrum of state-of-the-art LBSN prediction algorithms. Specifically, the algorithms include [1,3,5,6,8,9]. While Spot [4] also provides a platform to analyze LBSN, it consists of three algorithms. In contrast, we provide more algorithms than [4] and most importantly, our framework contains the most recent or relevant location prediction techniques. In addition to the aforementioned algorithms, UrbanHubble consists of a visualization component that shows the real-time complex social interactions of users on a map, the evolution of friendships over time, and how friendship triggers mobility or vice-versa.

---

[1] http://dmc.rwth-aachen.de/en/urbanhubble

© Springer International Publishing Switzerland 2015
A. Bifet et al. (Eds.): ECML PKDD 2015, Part III, LNAI 9286, pp. 329–332, 2015.
DOI: 10.1007/978-3-319-23461-8_41

While UrbanHubble is primarily intended for researchers, its visualization interface can also be used across an array of industries such as in location-based advertisement, where customers behaviors' strongly depend on the location context, and advertisers are interested to efficiently identify patterns to hyper target such customers, or in urban planning or for traffic monitoring. Given the importance and enormous potentials of LBSN research, we believe these use-cases connote the demand for UrbanHubble.

2   UrbanHubble Tool

In this section, we provide a detailed description of the UrbanHubble predictive analytics Java platform. We provide four real datasets that are publicly available. They include, two versions of the LBSN Gowalla[2] check-in datasets (where one Gowalla dataset has category information while the other does not), the Brightkite LBSN dataset[3] and the non-LBSN San Francisco taxi cab[4] dataset.

2.1   Architecture

The architecture of UrbanHubble is depicted in Figure 1. It entails 4 layers, namely, the Persistence, Scalability, Predictive Engine and User Interface layer.

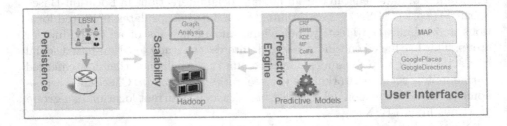

Fig. 1. Urban Hubble Architecture.

Persistence Layer: is home to the previously alluded datasets. To determine important social-relationship properties of users, we represent the collection of users as a simple graph G.

Scalability Layer: The number of check-ins from the Gowalla and Brightkite datasets exceeds a million. To rapidly analyze the check-ins correlations and the complex interconnections between the users in the huge graph G, a scalability layer is required. The scalability layer entails several Hadoop clusters to process G during Matrix Factorization and other exhaustive computations.

Predictive Engine Layer: consists of a wide variety of state-of-the-art location prediction algorithms that use Hidden Markov Model (HMM), Conditional

---

[2] https://snap.stanford.edu/data/loc-gowalla.html
[3] https://snap.stanford.edu/data/loc-brightkite.html
[4] http://cabspotting.org/

Random Fields (CRF), Matrix Factorization (MF), Collaborative Filtering and Kernel Density Estimation (KDE). Specifically, the UrbanHubble framework consists of six algorithms. They include [1,3,5,6,8,9]. Due to space constraints, we briefly describe a few of the algorithms packaged in UrbanHubble. [1] is our predictive model that uses Lipschitz exponent and CRF to determine the top-k future locations. [5] utilizes matrix factorization and KDE to predict and recommend the top-k locations.

User Interface: The user interface consists of two tabs. They include, the Geo-Social Analysis tab and the Location Prediction tab as illustrated in Figure 2a. The former shows the real-time dynamics between mobility and friendship at a city resolution, while the later illustrates the detailed trajectory paths taken by users. The user interface of the *Geo-Social Analysis* tab provides a researcher the possibility to select an algorithm and the option to analyze geo-social mobility on weekdays or weekends. Furthermore, there is an option to chose either to analyze only trajectories (i.e., Show Route) or the social interactions between users. After selecting the desired configurations, the Play button can be clicked to run the selected algorithm. As the algorithm runs, the movements of users and their social connections are displayed on the map. On this tab, the small green balls in Figure 2a correspond to the end destinations of users, the dark cyan lines are the routes, while the red edges represent the friendships between users. In addition, the large circles denote nodes or users.

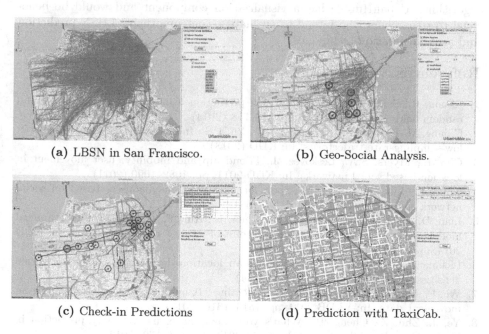

(a) LBSN in San Francisco.   (b) Geo-Social Analysis.

(c) Check-in Predictions   (d) Prediction with TaxiCab.

Fig. 2. Urban Hubble User Interface.

On the *Location Prediction* tab, each hexagon (e.g., in Figure 2c and Figure 2d) is a candidate next location. If the category information is absent from the dataset, to run the WhereNext [8] algorithm, UrbanHubble queries Google Places API to determine the category of each cell as shown in Figure 1. As mentioned earlier, we provide a default check-in dataset with category. After an algorithm runs to completion, the results of the algorithm are displayed as shown in Figure 2c using the precision, recall and accuracy evaluation measures. Figure 2d shows a scenario where the San Francisco TaxiCab dataset is used.

2.2 Related Works

[2] presented a reachability-based predictive model to predict check-in locations for distant-time queries. Our work differs from [2,4] since the algorithms packaged in our framework are different from theirs. Besides, our framework has more visualization functionalities that is not only limited to prediction analysis but also depicts geo-social interaction and group dynamics. MoveMine 2.0 [7] is a framework that focuses on trajectory clustering.

3   Conclusions

We propose an innovative LBSN predictive analytics java framework called UrbanHubble, which consists of a wide variety of state-of-the-art prediction algorithms. UrbanHubble has a visualization component and would be beneficial to researchers who intend to evaluate recent innovative LBSN prediction and recommendation techniques.

References

1. Assam, R., Seidl, T.: Check-in location prediction using wavelets and conditional random fields. In: ICDM 2014, pp. 713–718 (2014)
2. Chiang, M.-F., Lin, Y.-H., Peng, W.-C., Yu, P.S.: Inferring distant-time location in low-sampling-rate trajectories. In: KDD (2013)
3. Cho, E., Myers, S.A., Leskovec, J.: Friendship and mobility: user movement in location-based social networks. In: KDD 2011, pp. 1082–1090 (2011)
4. Kong, L., Liu, Z., Huang, Y.: Spot: locating social media users based on social network context. In: VLDB 2014, pp. 1681–1684 (2014)
5. Lian, D., Zhao, C., Xie, X., Sun, G., Chen, E., Rui, Y.: Geomf: joint geographical modeling and matrix factorization for point-of-interest recommendation. In: KDD 2014, pp. 831–840 (2014)
6. Lichman, M., Smyth, P.: Modeling human location data with mixtures of kernel densities. In: KDD 2014, pp. 35–44 (2014)
7. Wu, F., Lei, T.K.H., Li, Z., Han, J.: Movemine 2.0: mining object relationships from movement data. In: VLDB 2014, pp. 1613–1616 (2014)
8. Ye, J., Zhu, Z., Cheng, H.: What's your next move: user activity prediction in location-based social networks. In: SDM 2014, pp. 171–179 (2013)
9. Ye, M., Yin, P., Lee, W.-C., Lee, D.-L.: Exploiting geographical influence for collaborative point-of-interest recommendation. In: SIGIR (2011)

VIPER – Visual Pattern Explorer

Matthijs van Leeuwen[✉] and Lara Cardinaels

Department of Computer Science, KU Leuven, Leuven, Belgium
matthijs.vanleeuwen@cs.kuleuven.be

Abstract. We present VIPER, for *Visual Pattern Explorer*, an innovative, browser-based application for interactive pattern exploration, assisted by visualisation, recommendation, and algorithmic search. The target audience consists of domain experts who have access to data but not to –potentially expensive– data mining experts. The goal of the system is to enable the target audience to perform true *exploratory* data mining. That is, to discover interesting patterns from data, with a focus on subgroup discovery but also facilitating frequent itemset mining.

1 Introduction

Pattern mining concerns the discovery of patterns in data, where a pattern is a succinct description of some structure that occurs locally in the data. Two typical pattern mining tasks are frequent itemset mining (FIM) and subgroup discovery (SD). The task of FIM is to find combinations of items, such as products in retail data, that frequently occur together. The task of SD is to find subsets of the data for which a certain target of interest has a deviating distribution. Given a dataset containing movie ratings, for example, one might find that women tend to give higher ratings to movies with genre 'drama' than men.

Many interestingness measures and even more algorithms have been proposed for both tasks. See [1] and [2] for recent overviews. One of the key challenges of pattern mining is the infamous *pattern explosion*, i.e., the huge numbers of patterns that are commonly found by pattern mining algorithms. A recent trend to alleviate this issue is to mine pattern sets [5] rather than individual patterns.

Unfortunately, existing systems are hard to use due to the many parameters and other choices that have to be made. As a result, pattern mining as a technology is only accessible to data mining experts, strongly limiting its potential impact. The average person who has access to data and wants to analyse it, whether it is in academia or industry, is a domain expert, *not* a data mining expert. She knows what patterns she finds interesting and doesn't need someone to determine this, but she does need help in finding these interesting patterns.

For exactly that purpose we present VIPER, for *Visual Pattern Explorer*, an innovative, browser-based application for exploring the space of all possible patterns, assisted by visualisation, recommendation, and algorithmic search. The target audience are domain experts who have access to data but not to –potentially expensive– data mining experts. The goal of the system is to enable the target audience to discover interesting patterns from data, with a focus on subgroup discovery but also facilitating frequent itemset mining.

© Springer International Publishing Switzerland 2015
A. Bifet et al. (Eds.): ECML PKDD 2015, Part III, LNAI 9286, pp. 333–336, 2015.
DOI: 10.1007/978-3-319-23461-8_42

2   Related Work

Throughout the years several pattern mining systems with a graphical user interface have been developed, such as MIME [8] for FIM and Cortana[1] for SD. These systems, however, first mine a (large) number of patterns and then give the user the opportunity to browse this collection; search and interaction are completely decoupled. These tools are generally inaccessible to domain experts, our target audience, due to the large number of algorithms, measures, and parameters.

More recently, interestingness measures have been investigated that can adapt to the background knowledge and/or feedback of a user. Bhuiyan et al. [3] proposed to use user feedback to adapt the sampling distribution of itemsets. Dzyuba et al. [7] proposed to learn pattern rankings using techniques from preference learning. Orthogonally, De Bie [6] focused on a theoretic framework for iterative data mining and a formalisation of subjective interestingness. None of these works presented a working system though. Boley et al. [4] did present a system for 'one-click-mining', in which the preferences of the user for certain algorithms and patterns are learned. Still, objective interestingness measures are used to mine patterns, which are then presented to the user.

3   VIPER – Visual Pattern Explorer

We present VIPER, a browser-based application that allows the user to explore the pattern space, assisted by visualisation, recommendation, and search. Given that the target audience are domain experts rather than data mining experts, one of the primary design goals is to keep it as simple as possible.

Implementation. The web application is publicly available[2] and has been implemented in JavaScript and runs locally in the browser. It has been developed and tested using Chrome, making it available on many platforms.

Features. The features of our initial implementation are focused on the application's core functionality, i.e., visually assisted pattern exploration.

Measures To avoid overwhelming the user with a plethora of interestingness measures, VIPER uses just two: 1) coverage (a.k.a. support or frequency) for itemsets and 2) weighted relative accuracy (WRAcc) for subgroups.

Data VIPER can be used for the analysis of categorical datasets (in a simple text format). This allows to use bitvector computations for many operations, making the application respond nearly instantly for datasets of moderate size.

Pattern exploration Once a dataset has been loaded, the user can start exploring the pattern space; see Figure 1. A key principle is that the user is always in control. At any time, the current pattern can be specialised (extended) by clicking an attribute-value combination in one of the charts, or be generalised by removing an attribute-value combination by clicking on it.

The charts indicate all possible specialisations: each chart represents an attribute, and each bar in each chart a possible value for that attribute. The bar

---

[1] datamining.liacs.nl/cortana.html
[2] Demo and source code at www.patternsthatmatter.org/viper

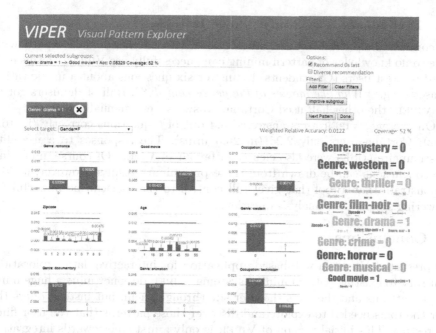

Fig. 1. Screenshot of VIPER. Example using movie rating data.

heights represent the interestingness that is obtained by the respective specialisations. For reference the horizontal red lines show the current interestingness.
For SD, a target can be chosen (and changed!) at any time; WRAcc is used
as interestingness measure when a target is chosen, otherwise coverage is used.
To assist the user to assess patterns, a 'word cloud' shows those attribute-values
that occur more frequently within the pattern than in the remainder of the data.

Specialisation Recommendation. To enable the user to efficiently explore the
pattern space, by default all specialisations are ordered by attainable interestingness descending. Moreover, several filtering and specialisation recommendation
options are offered. For example, text-based filters can be applied to exclude
specialisations from the view. Finally, datasets tend to contain strongly correlated attributes, which may result in very similar specialisations ranked first. To
remedy this, VIPER offers a diverse recommendation option, which is based on
the cover-based beam selection procedure from the DSSD algorithm [9].

Greedy Search. A user may be interested in quickly finding the best possible
specialisation of the current pattern. To facilitate this, we implemented a greedy
hillclimbing search strategy ('Improve subgroup') that iteratively specialises a
given pattern until weighted relative accuracy cannot be improved.

Iterative Mining. Once the user has found a pattern she likes, she can move to
the next pattern or end the session. In the former case, the pattern is stored and
the user can opt to apply sequential covering, meaning that the data selected by
the pattern is removed and exploration continues on the remainder of the data.
After finishing, the final set of patterns can be inspected and exported.

3.1   Preliminary User Study: Movie Ratings

We conducted a preliminary user study to evaluate whether domain experts with little to no knowledge of pattern mining can successfully use VIPER. To this end, asked ten test users (CS students) to answer six questions about a movie rating dataset[3], e.g., *"Who like movies of the genre comedy?"*. Half of the users got to use VIPER, the other half used Cortana. Answers were manually checked.

On average, VIPER users answered 5.4 out of 6 questions correctly (in 16.4 mins), Cortana users only 2.25 (in 26.5 mins). The responses to a usability questionnaire (USE) were also clearly in favour of VIPER. Of course, no definitive conclusions can be drawn from these preliminary results. However, it does demonstrate that a tool that gives full control to the user can be powerful for answering simple data analysis questions.

4   Conclusions

We presented VIPER, a web-based application for interactive data exploration. Contrary to existing pattern mining systems, VIPER does not mine a huge number of patterns and then lets the user sift through them, but instead offers the user the tools needed to explore and discover those patterns that the user finds interesting. This initial version of VIPER is only a first step towards integrating visually assisted data exploration with pattern mining and we hope to extend it in the future. But before that, we hope to get feedback from the community.

Acknowledgments. Matthijs van Leeuwen is supported by a Postdoctoral Fellowship of the Research Foundation Flanders (FWO).

References

1. Aggarwal, C., Han, J. (eds.): Frequent Pattern Mining. Springer (2014)
2. Atzmueller, M.: Subgroup discovery. Wiley Interdisc. Rew.: Data Mining and Knowledge Discovery **5**(1), 35–49 (2015)
3. Bhuiyan, M., Mukhopadhyay, S., Hasan, M.A.: Interactive pattern mining on hidden data: a sampling-based solution. In: Proc. of CIKM 2012, pp. 95–104 (2012)
4. Boley, M., Mampaey, M., Kang, B., Tokmakov, P., Wrobel, S.: One click mining: Interactive local pattern discovery through implicit preference and performance learning. In: Proceedings of IDEA 2013, pp. 27–35. ACM, New York (2013)
5. Bringmann, B., Nijssen, S., Tatti, N., Vreeken, J., Zimmermann, A.: Mining sets of patterns: next generation pattern mining. In: Tutorial at ICDM 2011 (2011)
6. De Bie, T.: An information theoretic framework for data mining. In: Proceedings of KDD 2011, pp. 564–572 (2011)
7. Dzyuba, V., van Leeuwen, M., Nijssen, S., Raedt, L.D.: Interactive learning of pattern rankings. International Journal on Artificial Intelligence Tools **23**(6) (2014)
8. Goethals, B., Moens, S., Vreeken, J.: MIME: a framework for interactive visual pattern mining. In: Gunopulos, D., Hofmann, T., Malerba, D., Vazirgiannis, M. (eds.) ECML PKDD 2011, Part III. LNCS, vol. 6913, pp. 634–637. Springer, Heidelberg (2011)
9. van Leeuwen, M., Knobbe, A.: Diverse subgroup set discovery. Data Mining and Knowledge Discovery **25**, 208–242 (2012)

---

[3] grouplens.org/datasets/movielens/

Visualization Support to Interactive Cluster Analysis

Gennady Andrienko[1,2(✉)] and Natalia Andrienko[1,2]

[1] Fraunhofer Institute IAIS, Sankt Augustin, Germany
[2] City University London, London, UK
{gennady.andrienko,natalia.andrienko}@iais.fraunhofer.de

Abstract. We demonstrate interactive visual embedding of partition-based clustering of multidimensional data using methods from the open-source machine learning library Weka. According to the visual analytics paradigm, knowledge is gradually built and refined by a human analyst through iterative application of clustering with different parameter settings and to different data subsets. To show clustering results to the analyst, cluster membership is typically represented by color coding. Our tools support the color consistency between different steps of the process. We shall demonstrate two-way clustering of spatial time series, in which clustering will be applied to places and to time steps.

1 Introduction

Our system V-Analytics [1] enables analytical workflows involving partition-based clustering by methods from an open-source library Weka [2] combined with interactive visualizations for effective human-computer data analysis and knowledge building. According to the visual analytics paradigm, knowledge is built and refined gradually by iterative application of analytical techniques, such as clustering, with different parameter settings and to different data subsets. A typical approach to visualizing clustering results is representing cluster membership on various data displays by color-coding [3-5]. To properly support a process involving iterative clustering, the colors assigned to the clusters need to be consistent between different steps. We have designed special color assignment techniques that keep the color consistency.

When data are stored in a table, clustering can be applied to the table rows or to the columns [6]. Spatial time series, i.e., attribute values referring to different spatial locations and time steps, can be represented in a table with the rows corresponding to the locations and columns to the time steps. Two-way clustering groups the locations based on the similarity of the local temporal variations of the attribute values and the time steps based on the similarity of the spatial situations, i.e., the distributions of the attribute values over the set of locations [5].

2 Interactive Two-Way Cluster Analysis of Spatial Time Series

To support iterative data analysis and knowledge building with the use of clustering, V-Analytics provides the following functionality:

© Springer International Publishing Switzerland 2015
A. Bifet et al. (Eds.): ECML PKDD 2015, Part III, LNAI 9286, pp. 337–340, 2015.
DOI: 10.1007/978-3-319-23461-8_43

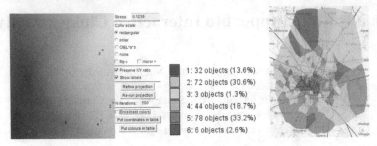

Fig. 1. Left: a projection of cluster centers onto a color plane; right: the spatial distribution of the cluster membership; center: a legend showing cluster colors and sizes.

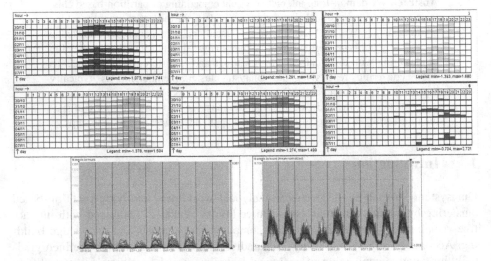

Fig. 2. Top: the 2D time histograms (9 days x 24 hours) correspond to different clusters; the bars in the cells represent the cluster means. Bottom: the time graphs show the variations of the absolute (left) and transformed (right) values for cluster 1.

- Colors are assigned to clusters based on Sammon's projection [7] of cluster centers (provided by the clustering algorithm or computed) onto a plane with continuous background coloring. This ensures that close clusters receive similar colors.
- After data re-clustering with different parameter settings, the projection of the new cluster centers is aligned with the previous projection, aiming at minimizing the distances between the positions of the corresponding clusters in the two projections. In this way, corresponding clusters receive similar colors, which makes the clusters traceable throughout the analysis process.
- Interactive techniques support progressive clustering [8], in which further clustering steps are applied to selected clusters obtained in previous steps. This allows controlled refinement of the clustering results and, on this basis, refinement of the analyst's knowledge.

This functionality is generic, i.e., applicable to various types of data and diverse partition-based clustering methods. We shall demonstrate it on an example of spatial time series obtained by spatio-temporal aggregation of records about mobile phone calls made in Milan (Italy) during 9 days by 235 spatial regions and 216 hourly time steps. To focus on the temporal variation patterns rather than purely quantitative differences between the call counts in different regions, we transform the original absolute values to the region-based z-scores. As a clustering method, we shall apply k-means.

Figures 1 and 2 show an example of partitioning the set of regions into 6 clusters based on the similarity of the region-associated temporal variations of the transformed call counts. Figure 1 shows the projection of the cluster centers onto a color plane, the colors assigned to the clusters on the basis of the projection, and a map with the regions colored according to their cluster membership. To see and compare the temporal variation patterns corresponding to the different clusters, we use temporal displays shown in Fig. 2. In the upper part, there are 2D time histograms with the rows corresponding to the days, columns to the hours of a day, and colored bars in the cells representing the average values for the clusters. Each histogram corresponds to one cluster, the bars being painted in the color of this cluster. The histograms demonstrate prominent daily and weekly periodic patterns and clearly show the pattern differences between the clusters. In the lower part of Fig. 2, the original and transformed time series from one of the clusters are shown on time graphs.

We iteratively increase the number of clusters and observe the impacts. In this example, the major clusters mostly keep unchanged, but several singletons emerge. They consist of regions with unusual time series, for example, having peaks of call counts at times of some public events. Hence, in this example, we have found four major common patterns of the temporal variation of the calling activities (clusters 1, 2, 4, and 5) and several more specific patterns.

Next, we consider the same data set from a different perspective by clustering the table columns (i.e., the hourly time intervals) according to the similarity of the distributions of the attribute values over the set of regions. We start with five clusters, the centers of which are projected on a color plane in Fig. 3 (upper left). On the upper right, there is a calendar display with the rows corresponding to the days, columns to the hours of a day, and cells colored according to the cluster membership of the corresponding time intervals. A prominent periodic pattern can be seen. The upper two rows correspond to Thursday and Friday, rows 3 and 4 to the weekend, and the following rows to five week days from Monday to Friday. In the center of Fig. 3, the spatial distributions of the calling activities corresponding to the five time clusters are shown on maps. The shades of blue and red represent, respectively, values below and above the means.

The observed temporal periodicity corresponds to our background knowledge about human activities. The pattern is preserved with increasing the number of clusters. Thus, the lower part of Fig. 3 shows the result for ten clusters. The temporal pattern has been refined while preserving the same main features as for five clusters.

To conclude, V-Analytics supports cluster analysis by providing immediate visual feedback allowing the analyst to interpret clusters, assess their similarity, identify major patterns and separate outliers, and understand the impact of clustering method parameters.

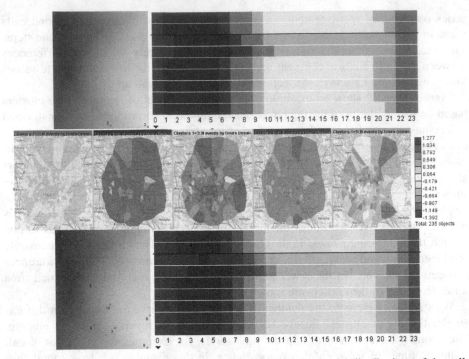

Fig. 3. Clustering of the hourly time intervals according to the spatial distributions of the calling activities. Top: projection and calendar displays for 5 clusters; center: the spatial distributions corresponding to the clusters; bottom: projection and calendar displays for 10 clusters.

References

1. Andrienko, G., Andrienko, N., Bak, P., Keim, D., Wrobel, S.: Visual Analytics of Movement. Springer (2013)
2. Witten, I.H., Frank, E., Hall, M.A.: Data Mining: Practical machine learning tools and techniques. Morgan Kaufmann (2011)
3. Andrienko, G., Andrienko, N.: Blending aggregation and selection: Adapting parallel coordinates for the visualization of large datasets. The Cartographic Journal 42(1), 49–60 (2005)
4. van Wijk, J.J., van Selow, E.R.: Cluster and calendar based visualization of time series data. In: Proc. Information Visualization, pp. 4–9 (1999)
5. Andrienko, G., Andrienko, N., Bremm, S., Schreck, T., von Landesberger, T., Bak, P., Keim, D.: Space-in-Time and Time-in-Space Self-Organizing Maps for Exploring Spatiotemporal Patterns. Computer Graphics Forum 29(3), 913–922 (2010)
6. Seo, J., Shneiderman, B.: Interactively exploring hierarchical clustering results. Computer 35(7), 80–86 (2002)
7. Sammon, J.W.: A nonlinear mapping for data structure analysis. IEEE Transactions on Computers 18, 401–409 (1969)
8. Rinzivillo, S., Pedreschi, D., Nanni, M., Giannotti, F., Andrienko, N., Andrienko, G.: Visually driven analysis of movement data by progressive clustering. Information Visualization 7(3–4), 225–239 (2008)

Author Index

Printed in the United States
By Bookmasters